COMPETITION FOR TECHNOLOGICAL LEADERSHIP

EU POLICY FOR HIGH TECHNOLOGY

JOHAN LEMBKE

Institute for European, Russian and Eurasian Studies
Elliott School of International Affairs
George Washington University
Washington, DC, USA

Edward Elgar
Cheltenham, UK · Northampton, MA, USA

Published by
Edward Elgar Publishing Limited
Glensanda House
Montpellier Parade
Cheltenham
Glos GL50 1UA
UK

Edward Elgar Publishing, Inc.
136 West Street
Suite 202
Northampton
Massachusetts 01060
USA

A catalogue record for this book
is available from the British Library

Library of Congress Cataloguing in Publication Data

Lembke, Johan, 1969-
Competition for technological leadership : EU policy for high technology / Johan Lembke. p.cm.
Includes bibliographical references and index.
ISBN 1-84064-792-2
1. Technological innovations--Economic aspects-- European Union countries--Congresses. 2. Transportation--Technological innovations--Government policy--Europe. 3. Telecommunication--Technological innovations--Government policy--Europe. 4. Information technology--Government policy--Europe--Congresses. 5. Artificial satellites in telecommunication--Government Policy--Europe. 6. Infrastructure (Economics)--Government Policy---Europe--Congresses. I. Title: EU policy for high technology. II. Title.

HC240.9.T4 L45 2002
338'.064'094--dc21

2002020237

ISBN 1 84064 792 2

Printed and bound in Great Britain by MPG Books Ltd, Bodmin, Cornwall

Table of Contents

List of Figures

LIST OF TABLES

Preface

In 2000, the European Union (EU) governments announced a long-term strategic commitment to transform the European economy into the most competitive, dynamic knowledge-based economy in the world. Technology is an integral part of the new economy and of the EU's strategy for economic development. This book sets out to specify conditions under which the EU, and in particular the EU Commission (EC), makes policy choices to support large-scale, technology infrastructures, and to provide an explanation for why EU political intervention in seemingly similar infrastructure projects varies. It is concerned with the development of large-scale technology infrastructures in the telecommunications, transport (including electronics and aerospace) and audiovisual broadcasting sectors (wireless Internet, UMTS; satellite navigation and positioning, Galileo; and terrestrial digital audio broadcasting, DAB-T). The analytical framework examines the mechanisms under which the EU undertakes (or does not undertake) high technology policy to support the market adjustment of large-scale industrial projects. The variation in EU policy demands explanation in that EU high technology policy regarding three seemingly similar technology-intensive infrastructure projects has differed, although those projects are all claimed to provide significant benefits for the European economy.

The exploration of EU high technology policy through a comparative analysis will provide insights into the political economy and the purposive-strategic behavior of major public and private actors. The analysis of EU policy for wireless Internet and satellite navigation, in particular, provides an assessment of what may be viewed as a new form of critical industry support in the 'information age', which contrasts with the subsidy-based industry support promulgated by the first wave of strategic trade theorists in the 1980s and 1990s. The EU strategy for wireless Internet and satellite navigation has been driven by an internationally oriented approach. It will also show the differential role of the EU as a more organized supranational political actor by offering a differentiated account of EU policy. In addition, it will illustrate the role of the private sector in the 'new economy' and the need for clearly expressed strategies when interacting with the EU institutions. Moreover, it includes an examination of European security policy and transatlantic relations. In sum, this study provides an analysis that should appeal to scholars and students in political economy, technology policy, foreign economic and security policy, and corporate strategy. It should also provide interesting insights for the business and policymaking communities.

ACKNOWLEDGEMENTS

I wish to thank a number of people who have supported me in the writing process and contributed to this book. Jan Hallenberg has always been prepared to read my manuscript closely and has unstintingly offered comments on the entire project. I also acknowledge, with gratitude, the valuable advice offered by Andrew Davies, Mark Hallenberg, Richard Hawkins, Giandomenico Majone, Ulrika Mörth, Simon Reich, Razeen Sally, Scott Siegel, Martin Staniland, Martin Weiss and Sandra Williamson. In particular, I benefited from discussions with Guy Peters, Jon Pierre, Thomas Plümper and Robert Walters.

I am also grateful for generous administrative and financial support from the European Union Center (EUC) at the Center for West European Studies (CWES), the Department of Political Science, University of Pittsburgh, as well as the Institute for European, Russian and Eurasian Studies (IERES) at the George Washington University's Elliott School of International Affairs. From 1999 to 2002, these institutes provided me with both office space and access to various activities. The assistance of Tom Allen, Michelle Amato, Marina Barnett, Chad Damro, Amy Eckhardt, James Goldgeier, Lena Helldner, Claes Linde and Alberta Sbragia is highly appreciated.

I am thankful for the financial support I received from the Institute for European, Russian and Eurasian Studies (IERES). In particular, I am indebted to Dr. Marcus Wallenberg's Foundation for research and studies in international affairs and international industrial business.

Furthermore, this project would have been much more difficult without the support of a large number of people in the EU institutions. A special thanks is owed to people in the EU institutions. The assistance from representatives and members of the European Telecommunications Standards Institute (ETSI), the European Space Agency (ESA), the WorldDAB Forum and the European Broadcasting Union (EBU) has been very important. I have also benefited from material collected from national and international governmental organizations and private sector organizations throughout Europe and the US. I thank all the officials in the public and private sector who graciously supported my efforts and gave me permission to use some of the material that appears in this book.

Finally, this book would have been much harder to complete without the support of Tracy who has put up with a great deal.

Glossary

ACTS	Advanced Communications Technologies and Services
AEA	Association of European Airlines
AER	Association of European Radio
ANSI	American National Standards Institute
ARIB	Association of Radio Industries and Businesses
ARTES	Advanced Research and Telecommunication Systems
CDMA	Code Division Multiple Access
CEPT	Conference Européenne des administrations des Postes et Télécommunications/European (Conference of Postal and Telecommunications Administrations)
CNES	Centre National d'Etudes Spatiales
CNS/ATM	Communication, Navigation, Surveillance/Air Traffic Management
DAB	Digital Audio Broadcasting
DECT	Digital Enhanced Cordless Telephone
DG EAC	Directorate-General for Education and Culture
DG INFSO	Directorate-General for Information Society
DG TREN	Directorate-General for Transport and Energy
DVB	Digital Video Broadcasting
EACEM	European Association for Consumer Equipment Manufacturers
EBU	European Broadcasting Union
EC	Commission of the European Communities
ECAC	European Civil Aviation Conference
EGNOS	European Geostationary Navigation Overlay System
EIB	European Investment Bank
EMAC	European Parliament Committee on Economic and Monetary Affairs and Industrial Policy
EMRF	European Maritime Radionavigation Forum
EOIG	Egnos Operators Interest Group
EP	European Parliament
ESA	European Space Agency
ESDP	European Security and Defense Policy
ETG	European Tripartite Group

ETSI	European Telecommunications Standards Institute
EUROCONTROL	European Association for Air Traffic Control
EUROSPACE	European Space Industry Association
FAMOUS	Future Advanced Mobile Universal Systems
FPLMTS	Future Public Land Mobile Telecommunication System
GISS	Galileo Interim Support Structure
GLONASS	Global Navigation System Satellite
GMDSS	Global Maritime Distress and Safety System
GNSS-1	First Generation Global Navigation Satellite Systems
GNSS-2	Second Generation Global Navigation Satellite Systems
GSM	Global System for Mobile Communications
GPS	Global Positioning System
ICAO	International Civil Aviation Organization
ICT	Information and Communication Technologies
IMO	International Maritime Organization
IMT-2000	International Mobile Telecommunications 2000
INMARSAT	International Maritime Satellite Organization
IPR	Intellectual Property Rights
ISDN	Integrated Services Digital Network
ITU	International Telecommunication Union
JRC	Joint Research Center
MoU	Memorandum of Understanding
NATO	North Atlantic Treaty Organization
NMT	Nordic Mobile Telephone System
PMB	Program Management Board
RACE	Research and Technology Development in Advanced Communications in Europe
R&D	Research and Development
SAG	Space Advisory Group
SMG	Special Mobile Group
TDMA	Time Division Multiple Access
UGG	UMTS Globalization Group
UMTS	Universal Mobile Telecommunication System
USTR	US Trade Representative
WARC/WRC	World (Administrative) Radio Conference
WTO	World Trade Organization
3GPP	Third Generation Partnership Project

1 Introduction

This book sets out to specify conditions under which the European Union (EU), and particularly the European Commission (EC), makes policy choices to support large-scale high technology infrastructures, and to explain why EU support for seemingly similar high technology projects varies.

Defining the New Economy in Europe

The EU Member States have set up an ambitious goal to transform the EU into the world's most competitive, knowledge-based economy by 2010. The development of advanced technology is an integral part of this 'new economy'. Many industrial activities and programs directed to strengthen European industry and the economy, however, do not attract ordinary public attention in the initial phases. In early April 2001, for example, the Swedish Prime Minister Göran Persson reported to the European Parliament on the recently held EU summit in Stockholm (March 2001). Persson instructed the EU transport ministers to continue the work toward the implementation of a European satellite navigation system, referred to as Galileo. This industrial program represents one of the major public private undertakings in Europe, but how many people had heard about it by April 2001? In fact, the debate on the new economy in Europe is strikingly apolitical.

A strong infrastructure, and the institutional capacity to provide it, is the foundation upon which continuous economic growth and the development of society depends. Infrastructural development, and a strong industrial base of globally competitive industries that strive to improve productivity and quality, are important for employment and the tax base. The pursuit of technological leadership, through investments and trade in high technology goods and services, is key to economic, and perhaps political, strength. The objective of gradually channeling the industrial base toward high technology, high-value-added products, in support of the public good and/or international competitiveness, is central in the race for competitiveness in global markets, the outcome of which influences opportunities for corporate profits and government revenue.

Throughout history, large-scale infrastructure projects have been publicly supported all around the globe. Roman roads and solid bridges constructed of arches were 'built to last a century' in order to facilitate military movements, commerce and communications over vast territories. A more recent example is the 'pragmatic neo-mercantilist' cooperation in the US between government agencies, firms, experts and research institutes to transform the originally military-oriented Internet technology into a commercially oriented infrastructure (Hart, 2000). History provides abundant evidence that public authorities have offered support for technology close to the market, such as airframe manufacturing, digital television and semiconductor technology. In addition, the symbolic-political value of large-scale infrastructures in enhancing societal cohesiveness should not be underestimated.

High technology commonly refers to knowledge-intensive industries that have a significant share of investment in R&D, relatively fast technological innovation, lengthy development phases and a potential for fast-growing production volume (Muller, 1990). Scholars have studied the conditions under which support for advanced technology may occur and have cast important insights into justifications, opportunities and problems with public intervention in the market (for example, Krugman, 1986; Flamm, 1988; Ostry, 1990; Tyson, 1992; Sandholtz, 1992b; Peterson, 1993; Thornton, 1995; Thurow, 1996). Public intervention in markets is purposive action and confers differential benefits and costs across the economy and society (tangible or intangible). In this book, 'high technology policy' refers to active and purposive market adjustment intervention by public institutions to shape technology-intensive infrastructure projects in order to maximize economic, political and other gains. It will examine the purposive-strategic behavior of political and economic actors and show the specific mechanisms through which public authorities undertake significant intervention in the market.

EU High Technology Policy

The question of the role of government and the EU as providers of goods and services is at the heart of political economy and the relationship between the market (economic actors) and public authority. This book will steer clear of the normative aspects of that overall question, that is, whether public authorities should intervene in the market. Rather, it will examine the conditions under which the EU, and in particular the EC, intervenes in the market and why high technology policy varies across seemingly similar technology-intensive, European-wide infrastructure projects. The question, thus, is why intervention varies and what variables determine the variance. The structure of the book is based on a comparison of three infrastructure projects: satellite navigation (Galileo), wireless Internet (UMTS) and terrestrial digital audio broadcasting (DAB-T).

The variation in EU high technology policy (with regard to seemingly similar technology-intensive infrastructures) demands explanation since all those infrastructures are claimed by their respective supporters to provide significant benefits for the European economy. Explaining this variation will offer insights into political economy and the purposive-strategic behavior underpinning the EU's ambitions to promote the world's most competitive and knowledge-based economy. The overall research question can be divided into three more specific questions. Why have the EC and the EU undertaken significant intervention to promote satellite navigation through direct political support and significant public investment and resources when there is already a well-functioning US satellite navigation system? Why have the EC and the EU intervened to support wireless communications and wireless Internet when this is a highly commercial sector? Finally, why have the EC and the EU not intervened politically to promote digital audio broadcasting when they have provided public support to, for example, satellite navigation and wireless communications?

The rationale of the comparative design has thus been to compare three large-scale technology projects that share a number of similar characteristics, while allowing for variance in EU high technology policy to support market implementation. All the selected sectors have been subjected to major adjustment pressure and are associated with electronics, communications and the implementation of advanced technology. They represent technology-intensive infrastructures and therefore carry the potential of attracting public intervention due to macroeconomic and other benefits. In addition, as large-scale infrastructures, they require substantial upfront investments and long development cycles, which may trigger requests for political and financial support. They face similar constraints, such as the need for coordination, regulatory and legal certainty, technical standardization, allocation of radio spectrum, and so on, that impact on market implementation. In addition, supporters of all projects claimed that there was a 'window of opportunity' for Europe to become a world leader and that a precondition for success was political and regulatory support. All the infrastructure projects have been aimed at world export markets and have evolved within a similar time frame.

The large-scale infrastructure projects have enjoyed different levels of EU political and financial support and various degrees of EU management and coordination. Third-generation (3G) mobile technology and services, or IMT-2000, referred to in Europe as UMTS, will offer higher-speed, always-on Internet access, data, full-motion video applications and CD-quality music on mobile phones, and represent the convergence between two fast-growing markets: the Internet and mobile telephony. In the case of UMTS, the EC had no formal management role, but it promoted politically (together with the EU Member States) the creation of a regulatory framework dedicated to wireless communications and wireless Internet. When it comes to satellite navigation, the EU Member States granted a central role to the EC in organizing the Galileo project. The EC and the EU supported the attempts to introduce analog high definition television in the past (which failed) and have provided, for a number of years, political support for the so-called digital video broadcasting (DVB) consortium. Interestingly, the transition to digital audio broadcasting in Europe, however, did not enjoy any similar political, institutional or public financial support.

Objectives, Analytical Framework and General Propositions

By examining the role of the EU/EC, and its relationship to economic actors, we can gain insights into the larger issue of administrative and political capacity and willingness to steer and maintain some degree of control over society and the economy. Intervention by the EU/EC in the market raises issues about how the public and private sectors negotiate transitional strategies for sectors subjected to major market adjustments. A main purpose of this book is to develop an analytical framework from which we can logically derive some general propositions, to generate insights into the conditions under which the EU supports large-scale infrastructures, and to approach what needs to be explained: why technology infrastructure projects experience variation in EU high technology policy. This framework is intended to specify some central mechanisms under which effective public policy comes about. It provides theoretical modules and functions as a common denominator that helps us to capture the dynamics of interaction between the public and private sector within a more differentiated picture.

The analytical framework is briefly outlined below and a number of general propositions derived from it. The basic model of high technology policy is built around three cornerstones: incentives, lobbying and acceptance of lobbying. It suggests that there must be an economic incentive for industry to demand and for a public bureaucracy (the EU/EC) to opt for high technology policy. This incentive can be long-term economic growth, technological leadership, employment, profitability, and so on. The cornerstone of the framework is strategic trade theory and related literature concerned with public intervention in the market. In general, this literature discusses the circumstances under which market interventions by political institutions are welfare maximizing or not, when industry (and/or public actors) would see intervention to promote technology as 'rational' or 'irrational', and when industry would petition and probably get resonance for high technology policy. Strategic trade theory assumes that the more visible the welfare-enhancing implications for industry and political institutions the more likely it is that public intervention occurs.

Proposition 1 *The more the sector and technology is defined and perceived to have welfare-enhancing implications, the greater is the likelihood that public intervention occurs.*

Since all the cases could potentially be targets for, but have enjoyed variation in, EU high technology policy, there is a need to elaborate a more refined model that goes beyond a basic incentive-demand-supply logic. In the more specific model, high technology policy is shaped by the interaction of critical intervening variables at two levels, industry characteristics and political entrepreneurship. The notion of industrial characteristics refers to features that are assumed to impact on the probability of effective governance and public policy: the organizability of interests and the voice of industry (Hall, 1986; Grant, 1989; Cawson et al., 1990; Eden and Molot, 1993; Cowles, 1995; Grande, 1996; Mazey and Richardson, 1996). This perspective posits that the political cohesion of economic actors and the position of economic actors in the structure of international competition influence the provision of high technology policy (Gourevitch, 1986, pp.54–61; Milner, 1988).

Proposition 2 *If economic actors are not able to organize politically, and there is no coherent demand for high technology policy, it is less likely that it will be provided.*

Proposition 3 *If economic actors fear retaliatory action due to their international position, they will be less inclined to demand protectionist high technology policy.*

Regarding conditions related to the second analytical level (political entrepreneurship) the focus here is on the role of the EC in defining, pursuing and executing public policy. For example, the bureaucracy may 'steer' policy by (a) creating policy networks, (b) facilitating institutionalized arenas of policymaking, and (c) attracting a variety of socio-economic actors. This level includes two conditions for high technology policy: bureaucratic cohesion and external support. The bureaucracy is more likely to support high technology policy proactively when it offers a window of opportunity to pursue its basic preferences.

What are the conditions and general propositions we can derive from the level of political entrepreneurship? Firstly, bureaucratic cohesion is necessary for effective high technology policy to occur (Peters, 1992; Cram, 1994). Secondly, the external support condition for the provision of public policy suggests that, in the face of expected political opposition from EU Member States, the bureaucracy will be less willing to risk confrontation and promote high technology policy. Public intervention in the market to assist the implementation of technology ultimately rests with the EU Member States.

Proposition 4 *If the relevant parts of the bureaucracy do not agree on high technology policy provision, it is less likely that it will be provided.*

Proposition 5 *If there is an expectation of considerable external opposition, the public bureaucracy will be reluctant to promote high technology policy.*

The framework is based on an 'additive model', which implies that the two basic conditions – industry characteristics and political entrepreneurship – depend on one another; in short, they need to be positive for high technology policy to occur (see Chapter 2, Figure 2.2). In the overview of the framework (Table 1.1), the categorization into policy sectors is based on the parts of the EC that have been 'responsible' for the projects in the 1990s. Moreover, the table indicates that leading economic actors in all sectors have made strong cases for the importance of EU high technology policy (scoring 'high' in terms of arguments for public support in strategic trade literature), but that the outcome has varied (thus 'enjoying' different qualitative scores in terms of high technology policy, ranging from 'low' to 'high').

Table 1.1 Overview of Framework

Policy Sectors	Telecom	Transport	Audiovisual
		European Projects	
Theoretical Modules	UMTS wireless Internet	Galileo satellite navigation	DAB digital radio
Strategic Trade Theory	High	High	High
Industry Characteristics			
Political Entrepreneurship			
EU High Technology Policy	High	High	Low

In addition, this study departs from a rational choice approach based on the principles of 'strategic rationality' and 'strategic interdependence' (in order to avoid the more extreme 'economistic' versions in the rational choice, social choice and economics literature, which depart from 'instrumental' or 'parametric' rationality). The approach in this study does not assume that actors have complete and perfect information, although it adopts the motivational assumption (admittedly simplistic) that behavior is largely goal-oriented, beliefs are relatively consistent, or preferences are plausible (actors act with certain expectations of benefits/returns and costs/risks). The assumption of strategic rationality assumes that actors' behavior impinges on the behavior of other actors; actors thus have to take the expectations of others into account before making decisions in settings characterized by strategic interaction.

Furthermore, the central public actors in this study are not assumed to be self-interested in a narrow sense. It is plausible, for example, to assume that the EC bureaucracy has broad preferences, that is, that the various parts of the EC are interested in serving society and in maximizing domestic or 'European' economic and social welfare (Caporaso and Levine 1992, p.192; Gourevitch, 1986, p.230). At the same time, many scholars pursue the methodological assumption that the EC attempts to maximize institutional self-interests or prerogatives in order to strengthen its own influence (see, for example, Kenis and Schneider, 1991; Peters, 1992; Sandholtz, 1993; Christiansen, 1998; Harcourt and Radaelli, 1999). It is important to spell out the assumptions guiding the examination of public policymaking. This study is based on a motivational assumption of public actors that incorporates the two aforementioned variants, thus allowing for altruistic preferences. This assumption states that the EC prefers to defend or expand its prerogatives within the EU separation of powers regime and that it tries to assert itself in this strategically interdependent setting in order to maintain or strengthen its institutional and regulatory self-interests and powers. The EC is assumed to value a strong administrative and political position in order to build up and enjoy the capabilities that are necessary to pursue broader preferences. An actor-centered framework is developed, which is not concerned with major personalities but with the political articulation of policymakers and private sector actors who are embedded in a complex, strategically interdependent setting (Dyson, 2000, p.66; Knight, 1992).

Main Argument and Results

In general, when the sector or technology is associated (correctly or not) with welfare-enhancing implications, we can expect that the likelihood for high technology policy is greater. The main argument, however, is that effective public policy is shaped by the interaction of critical intervening variables and associated conditions: industry characteristics and political entrepreneurship. According to this scheme, the organizability of economic interests and the position of economic actors in the structure of international competition, together with bureaucratic cohesion and external support, will determine policy outcomes. This section provides an overview of some of the results of this study, including specific case information.

Firstly, European ambitions to promote satellite navigation were justified on a number of industrial, macroeconomic, social and political grounds, which helped actors to define it as a rational undertaking to pursue through public intervention. Central to the push for the European Galileo program was the argument that 'Europe' could not be allowed to remain under the yoke of the global US monopoly in satellite navigation and that an independent capability was imperative to ensure European control of a critical infrastructure. The dominant satellite hardware producers in Europe strongly supported the European undertaking, as did the European Space Agency (ESA), air traffic control organizations, experts and government officials. The background situation in satellite navigation was very different from that in mobile communications, since Europe had no collective presence in this field in the early 1990s. Throughout the 1990s, the EC bureaucracy was proactive in promoting satellite navigation and was granted a central management role by the EU Member States. However, it became clear that there were many problematic issues that had to be resolved before all the major actors would unconditionally commit to support satellite navigation. The European Galileo project turned into a burden-sharing issue where economic and political actors tried to convince one another about what would be better for Europe, for example, whether it would be a public service or a purely commercial enterprise. The political controversies escalated when it was time for the EU Member States to commit substantial public funds to the project in 2000–2002.

In November 2001, ministers responsible for space affairs in the ESA Member States approved the funds to cover ESA's contribution of Euro 550 million for the initial development of the Galileo constellation. Before a meeting of the EU Transport Council on 6–7 December 2001, a consortium headed by PricewaterhouseCoopers presented a report on the economic perspectives offered by Galileo. It showed projections that Galileo could achieve positive operating cash flow from 2011 onwards. The Galileo program was thus economically justified, the report argued. It would have a positive cost-benefit ratio, which would be significantly higher than for comparable large publicly funded transport infrastructure projects (for example, airports or high-speed railways). The EU Council of transport ministers, however, decided to again defer a decision on a Euro 450 million payment (~US$394) for the development and validation phase of the Galileo program (2002–2005). Just before the European Council Summit in Laeken, a letter signed by a large number of European Parliament (EP) Members called on the European Council to overcome the differences in order to launch the Galileo infrastructure in a timely manner. The letter stressed that the EU had to seize the opportunity to demonstrate political will (by supporting the Galileo program that could be as successful as Airbus and

Ariane) in order to strengthen its independence in aeronautics and space policy and to foster technology development and the creation of highly skilled jobs in light of global competition. In addition, DG Research issued a report entitled 'Towards a European Space Policy' that strongly promoted the Galileo program as a vital element of European aerospace initiatives.

In late February 2002, the German government approved the release of EU funds for the development of Galileo, matching the financial commitment of the ESA, thus ending a deadlock in the German government between the finance ministry (led by Hans Eichel) and the transport ministry (led by Kurt Bodewig). An agreement was reached that extra expenditure would come out of the transport ministry's budget. This decision strengthened the political coalition behind the Galileo program. As part of one of the 'priority actions' identified by the European Council meeting in Barcelona in mid-March 2002, the European Council stressed the importance of the Galileo program. It asked the EU Transport Council to take the necessary decisions in spring 2002 with regard to both the funding and launching of the program and the setting-up of a Joint Undertaking in cooperation with the European Space Agency. Following the European Council meeting in Barcelona, those countries that had stalled the project did no longer constitute a blocking majority under the EU's voting procedures. They decided to commit funding to the Galileo program, while stressing the need to promote Galileo as a civilian-oriented project and to exclude private companies from the formal management of Galileo until the tendering process was put in place. In late March 2002, the EU Transport Council formally released significant public funding for the development phase of the European Galileo program.

Secondly, while there was political competition between manufacturers in the development of UMTS, and some tensions between manufacturers and network operators, this situation did not jeopardize the overall preference among industry actors to push for regulatory certainty and public support. At a general level, industry cohesion was helped by the perception of a common threat from Asian, and in particular Japanese industry, by the regulatory push by the Japanese government, and by export opportunities in Asia. European firms stressed the industrial and economic importance of wireless Internet for the European economy and employment. The EC had an interest, as had many EU governments, in trying to repeat the success story of the second generation Global System for Mobile Communications (GSM). Those large manufacturers that were stronger in terms of international competitiveness would come to dictate the direction and timing of strategic adaptation and public intervention to promote the market implementation of mobile cellular communications and wireless Internet.

The result is that industry, by insisting on a specific European technology infrastructure, had to invest substantial resources into the creation of new equipment demanded by the adoption of new technology and in the cost of licenses for new radio frequencies. In addition, the tight schedule risked delaying the provision of new equipment on time as planned in the implementation schedule. The much-cited high costs of UMTS license auctions in Europe came to haunt the wireless industry and to challenge investments in the light of global economic slowdown and debt-ridden companies. It will take years before we know whether this international high-risk game will help to maintain the leadership of European industry in mobile communications, or whether Asian and US industry will overtake European industry. Industry cohesion facilitated the feasibility of EU high technology policy and the pursuit of the interest in repeating a past policy success and creating a new export engine.

In the battle over 'globalization' within and outside ETSI in the late 1990s, it was clear that there was a conflict between more European-oriented and internationally oriented players. Entrenched positions, invested time and money, status and power were at stake. The more international players and the mobile community gained the upper hand. This story indicates how the contest between rival technologies in the context of European UMTS deliberations became an international issue that went far beyond Europe. In addition, it shows the strategic use of formal international standardization bodies by multinational corporations in the context of global competition in the information and communications technology (ICT) industry.

The leading European approach was eventually characterized by an internationally oriented strategy aimed at ensuring the support of Asian industry and standards bodies. Similar European attempts such as the support of a European standard for high-definition television (HDTV) failed because the latter was a defensive standard intended to protect Europe against Japanese competition, whereas the introduction and implementation of GSM was promoted as a means to integrate European markets. To globalize from the start was a good insurance policy for internationally oriented wireless manufacturers and operators in order to gain first-mover advantages and access to Asian and other markets, to seek to have other regions adopt their preferred infrastructure and technology, and to support economies of scale in equipment markets and network economies in services. The prospect of vast lucrative Asian markets certainly prompted European manufacturers to seek a deal with Asian players and to prevent independent technology developments in Japan, China and Korea. GSM was the first viable digital wireless standard and is well entrenched worldwide and the European wireless industry, and their customers, wanted to reap the benefits of their past investments.

European manufacturers did not significantly fear that rival Asian 3G technologies could take over a large chunk of the European 3G market. If anything, European operators and the EC may have been concerned about market fragmentation due to incompatible systems. The idea that European activities and ambitions in the wireless sector was a defensive approach would be consistent with the old idea of using standards as non-tariff barriers for protectionist purposes. Such an interpretation, however, does not pay sufficient tribute to the aggressive and internationally oriented approach of private sector actors in the wireless domain or the specific nature of the support they received from the EU institutions. The strategy for UMTS and W-CDMA (wideband code division multiple access) technology in Europe should thus not be regarded as a story of defensive standardization to keep Asian competitors out of the European market. The policy toward UMTS represents primarily an offensive story of co-opting Asian developments in light of lucrative Asian markets.

Finally, major broadcasting industry actors pushed strongly for digital audio broadcasting and stressed the importance of EU regulatory intervention and support, which would spur investment and market implementation. Industry pointed out that Europe enjoyed a technology leadership and that the European DAB system had the potential to become the world standard, as GSM had become in mobile communications. Moreover, industry argued that radio was an important democratic, well-trusted and locally oriented medium that through the provision of information and communication would support diversity throughout Europe. In addition, there were substantial industrial, commercial and employment benefits. If Europe failed to implement digital radio in a timely manner, the argument went, these benefits would not materialize and radio could be replaced by other technologies (for example, mobile communications or digital video broadcasting). The European market could be dominated by non-European technology and industry. Other DAB systems had been developed in other parts of the world, notably the so-called ISDB-T in Japan and IBOC DAB in the US. The EuroDAB consortium, which consisted of broadcasters and producers that were relatively weak internationally, called for strong political support. Industry supporters pointed to the imbalance between EU support for television and radio and argued that radio was discriminated against. All the investments that had gone into digital radio and DAB would be lost. The arguments for supporting digital radio had many similarities with those for Galileo and UMTS.

A main industrial-political and industrial-economic argument, expressed by the proponents of DAB, was that digital radio represented the last chance for Europe to enhance its competitiveness in the consumer electronics sector, which had a fairly weak international record in terms of competitiveness. If the EU, national governments and national and European regulators would provide a stable regulatory and investment climate (and avoid piecemeal and uncoordinated implementation) the DAB story could well be as successful as the European GSM story, which was viewed as the European flagship in high technology industries. This was certainly quite an optimistic claim. The question was thus whether European public authorities would listen. However, the EuroDAB consortium was neither able to persuade the EC that there was any imminent Asian or US import-competition threat, nor that DAB and its supporters could be as successful a story as GSM.

In the late 1990s, the political activities of industry eventually resulted in the preparation of an action plan by the EC's Directorate General for Education and Culture (DG EAC) for the transition to digital radio in Europe. The action plan repeated many of the arguments why digital radio should enjoy political and regulatory support. However, the report was resisted within the EC bureaucracy. European equipment producers had a relatively weak competitive position internationally. In addition, manufacturers did not appear to respond to the calls by trade associations and broadcasters to produce higher volumes. The lack of regulatory and market certainty and a clear business model in the radio industry at large seemed to result in low political activity among producers. In combination with paralysis within the EC services, the lack of strong political support among producers impeded political entrepreneurship. When major manufacturers signed a common position in spring 2001, this seemed to provide an important push to activities within the EC to start supporting digital radio. The EC stressed, however, that it would not directly support the European DAB system.

There was a common interest at higher EC level as well as among staff in DG INFSO (one of the most powerful DGs in the EC) to steer clear of specific preferential treatment of the DAB industry since such support was perceived as less 'welfare enhancing'. In short, there were few identified domestic returns on the part of the EU Member States. There could have been other factors involved that further strengthened this passive posture, such as (1) the investment in resources by DG INFSO in related technology infrastructures (such as DVB and UMTS), (2) the past policy failure in high-definition television, and (3) the interests in DG EAC to support satellite digital broadcasting services such as the Luxembourg-based Global Radio (or WorldSpace that was supported by Alcatel of France). For example, the Luxembourg government explicitly supported the Global Radio initiative and this may certainly have played a major role in influencing the political treatment of DAB in Western Europe in terms of EU policymaking. The satellite digital broadcasting industry lobbied the EC and received support from the higher level in DG EAC and from its Commissioner (Reding). (For a more detailed examination of this issue, see Lembke, 2002d.)

Contributions

Public support for market implementation raises questions about the role of political coordination above the level of individual EU Member States. The further consolidation of the internal European market requires integrated transport, communications, information, energy and other types of large-scale infrastructures. We need to examine different types of processes and bargains by which the EU attempts to achieve coordination and congruence based on, as well as beyond, its original assignment. Our understanding of the interaction between the private and public sector, and economic and political factors, and how they come together or diverge in different sectors, will be strengthened by such an analysis.

The large-scale technology-intensive projects that are covered in this book can be viewed as both infrastructures that can be contributors to macro-economic and industrial benefits, and as 'competitors' in and of themselves (Strange, 1988/1994; Lawton, 2000). The former dimension focuses on the creation of value-added infrastructure and highly skilled jobs with potential knock-on effects for the rest of the economy. The latter examines the jockeying for competitive positions among industries as well as public institutions in a broader geo-economic wealth and power structure. It is here more about building infrastructures that will attract foreign direct investment and help European industry to build a strong domestic base as a launch pad for world markets. Corporate behavior and bureaucratic intervention are central elements in the competition for technological leadership.

Europe has lagged behind Japan and the US in a number of electronics and information technology sectors. Technological and infrastructure capabilities raise issues of the role of the EU as an international power in an evolving, 'globalizing' techno-economic environment. This book will contribute to literature that focuses on that topic by examining public intervention across a number of sectors subjected to major adjustment pressures (for example, Williams, 1984; Gilpin, 1987, 2001; Sandholtz and Zysman, 1989; Tyson and Chin, 1992; Peterson, 1992; Kozmetzky and Yue, 1997; Lawton, 1997). In the 1950s, the US emerged as the leading political, military and industrial world power (for example, aeronautics, computers, communications and nuclear energy), while in Europe people were struggling with the postwar reconstruction process. This posed a number of dilemmas and associated challenges specific to attempts to deepen European integration and to ensure a secure and successful place for European technologies and associated firms and industries (Servan-Schreiber, 1969; Fochs, 1970; Williams, 1973; Roobeek, 1990; Muller, 1990; Gold, 1995; Thornton, 1995; Flamm, 1996).

The bulk of the literature on public intervention and high technology policy has been concerned with national industrial policy. The 'second-generation' research that emerged in the 1980s and 1990s paid attention to the new European R&D programs (Sharp and Shearman, 1987; Sandholtz 1992a, 1992b; Peterson, 1996). In addition, a literature on European policymaking and EU-industry relations started to grow that focused on European projects such as high-definition television (Hart and Thomas, 1995; Kaitatzi-Whitlock, 1998) and semiconductor technology (Lawton, 1997). Much of the existing research consists of single case studies rather than comparative analyses of EU high technology policy. This literature contributes to our understanding of the interaction and dynamics of the market-public authority dimension at the EU level. This study intends to take things a step further to focus on the specific mechanisms under which EU TIP occurs by comparing three large-scale infrastructure projects. By comparing a number of technology infrastructure projects, we can deepen our understanding of the mechanisms by which high technology policy and intervention to support major market adjustments is likely to occur.

The Structure of the Book

This book is structured into three main parts. Part I consists of two chapters: Chapter 1 presents the overall scope and direction of the book and the second chapter is concerned with the analytical framework. Part II consists of the empirical chapters. Chapter 3 provides an examination of the first-generation satellite navigation capability in Europe, while Chapter 4 deals with the development of the EU strategy for a civilian-controlled and European-led global satellite navigation and positioning infrastructure, dubbed Galileo. Chapter 5 examines the EU strategy for mobile cellular communications and wireless Internet. Chapter 6 further deepens the analysis into technology collaboration and standardization in the wireless communications domain, including institutional-strategic relationships with other regions. The last empirical chapter, Chapter 7, covers the planned implementation of digital radio infrastructure in Europe. Finally, Part III (Chapter 8) concludes this study with a comparative political-economic analysis of the technology-intensive infrastructure projects informed by the analytical perspectives.

2. Analytical framework

What are the mechanisms that are at work in the creation of high technology policy? This chapter will start by laying out a basic template, which is built around three cornerstones: incentives, lobbying and acceptance of lobbying. This framework states that there must be an economic incentive for industry to demand and for political institutions to opt for high technology policy. The incentive can be long-term economic growth, technological leadership, employment, profitability, and so on.

The technology-intensive infrastructure projects (satellite navigation, wireless Internet and digital radio) that are examined in this study all carry economic potential and could in principle fit into the broader EU agenda of creating the world's most competitive economy. However, variance in EU public intervention has been observed empirically. This observation triggered a need to take a further step beyond the basic framework and search for intervening variables that cast further light on the mechanisms under which EU high technology policy occurs. The extended framework focuses on a number of elements related to industrial characteristics and political entrepreneurship: the organizability of economic interests, the voice of industry, bureaucratic cohesion and external support. This analytical template will allow a more specific model from which a number of arguments can be derived. The aim is to evaluate what factors were salient in explaining observable differences in high technology policy (Gourevitch, 1986, p.21).[1] The first step, though, will be to present the basic analytical framework.

---

1 It opens up opportunity for an active interplay between the variables in each of the instances of EU policy that are examined. '[W]ith careful pairings or triangulations of cases it is possible to explore some of the strengths and difficulties of the various analytical approaches. One can find limiting examples, comparisons that bring out a linkage, others that point another way. And so the tension between parsimony and elegance on the one side, and accuracy on the other, remains powerful' (Gourevitch, 1986, p.34). The research design compares cases within a similar time frame (1995–2001) rather than over different time periods and across countries, such as those of Gourevitch (1986), Hall (1986) and Milner (1988), but share the latter's interplay of factors assumed to be closer to the real world.

2.1 HIGH TECHNOLOGY AND STRATEGIC TRADE THEORY

High technology policy should be seen as distinct from 'industrial' and 'regulatory' policy. Industrial policy measures can be classified into two types of intervention: sectoral and horizontal (Peterson, 1996; Lawton, 1999). Sectoral measures are designed to influence the performance of industries or sectors within the economy, while horizontal measures aim to influence the overall performance of the economy and the competitive framework in which firms operate. In other words, sectoral or vertical policy targets specific industries or firms, while a horizontal policy is a generic instrument aimed at supporting and strengthening structural determinants of an economy such as R&D and infrastructure development.[2] Moreover, 'regulatory policy' is often linked to 'industrial policy'. It can refer to 'a process or activity in which government requires or proscribes certain activities or behavior on the part of individuals and institutions, mostly private but sometimes public, and does so through a continuing administrative process, generally through specially designated regulatory agencies' (Reagan, 1987, p.17).[3]

However, such definitions of industrial and regulatory policy are too broad for the purposes of this study. The concept of advanced or 'high technology' is often defined as applying to knowledge-intensive industries that have a

2 Traditional definitions of industrial policy focus on the objective of generating structural change in the organization of industry and in the manufacturing industry. Blais defines industrial policy as 'the set of selective measures adopted by the state to alter industrial organization' (1986, p.4). Moreover, traditional definitions are associated with objectives of economic growth, improvement of productivity, strengthening of competitiveness and higher living standards. Intervention may be demanded to correct real, perceived or claimed market failures (Geroski, 1989). Industrial policy can also be viewed as a microeconomic form of intervention, a policy instrument intended to resolve or attenuate problems in specific industrial sectors (Zysman and Tyson, 1983). Zysman has argued that industrial intervention is purposive and requires that public institutions justify intervention on a particular definition of public purpose (Zysman and Tyson, 1983, pp.85–86).

3 Regarding EU policymaking and policy effectiveness, it has been stressed that there is no clear trend toward centralization through the 'creeping of competences'. Some scholars, for example, have stressed that the EC's monopoly of policy initiation and legislative powers has been diluted and may result in a loss of 'political property rights' over particular policy areas. Autonomous initiatives (where the EC has exclusive competence) represent less than 10 per cent of all the EC's legislative proposals. The potential *de facto* powers to initiate legislation of the European Parliament (EP) by its right to request the EC to submit any appropriate proposal on matters on which it considers that a Community act is required can have potentially far-reaching implications for EU policymaking and governance. There are indeed coordination problems due to divergences in national objectives, philosophies, practices and institutional capabilities, which have an impact on the capacity to respond and adjust to external and internal pressures.

relatively significant share of investment in R&D, a relatively rapid rate of technological innovation and a relatively fast-growing production volume. It is difficult to provide a precise definition of what constitutes high technology. Even when it comes to sectors and technologies that the EU defines as high technology, such as computers, semiconductors, aerospace, and so on, there are high-end and low-end markets in all these sectors that make it difficult to label broader categories as high technology. Advanced technologies result in a large number of products with high technological content, production process and applications and they are applicable across a wide range of industrial sectors (Van Tulder and Junne, 1988, p.253). They often have broad downstream market impacts, create large upstream demands, require lengthy and costly development phases, and are subject to high market and technological risks (Muller, 1990). 'Where conventional infrastructure is aimed at meeting well-defined needs through known methods within existing markets, technological infrastructure often involves the articulation of new needs that can only be met through the generation of new capabilities within markets that have yet to be created' (Justman and Teubal, 1996, p.21).

In this book, 'high technology policy' refers to active and purposive market adjustment intervention by public institutions to shape large-scale technology-intensive infrastructure projects in order to maximize economic, political and other gains. It is related to the 'technology-trajectory theory', which assumes that public institutions may choose to support the development of advanced technology infrastructures where learning curves are steep and where the set of core technologies will not move easily across borders (Hart and Prakash, 1999). The location of value-added activities takes on strategic importance since industrial investments and public support may generate positive spillovers and specific knowledge that become embodied in the domestic industrial and technological base (for example, Borrus 1988; Tyson 1992; Weber and Zysman 1992).

High technology policy and public intervention will be used largely interchangeably. In general, they refer to the actions the EU uses to initiate, promote, guide, administer and regulate the development of technology-intensive infrastructure projects. Public intervention is the promotion of certain sectors, groups, activities or objectives at the expense of others since resources are scarce. Intervention and the choice of governing instruments, that is, the devices that governments have at their disposal to intervene, are therefore not neutral activities (Blais, 1986). High technology policy is here conceptualized as an interrelated series of events and multiple decisions by a set of actors over time and of goal-oriented behavior on the part of governments (Jenkins, 1978).[4] The focus is on public intervention in support of market implementation of leading-edge technology, the results of which can have enduring influences on the final decision and outcome (Cobb and Elder, 1972; Cobb, Ross and Ross, 1976; Howlett and Ramesh, 1995, p.116).

Significant cross-border cooperation and competition in R&D and production also characterize high technology sectors. Moreover, the creation of infrastructures requires coordination, since advanced and costly capabilities must be created (Nelson, 1984, 1986). 'Whenever technological infrastructure policy necessitates such targeting, a clearly stated "vision" or at least credible announcement may be critical. For example, the government may declare its determination to create a significant level of infrastructure no matter what level of industry financing and participation is forthcoming. Indeed, in some extreme cases user coordination may be unthinkable without some confidence that a critical mass of infrastructure will materialize' (Justman and Teubal, 1996, p.47).[5] It has been argued that the European (and Japanese) use of the concept 'strategic' industries differs somewhat from that used in the US, where the former has a more civilian-economic connotation and the latter is commonly associated with defense and military matters (Lawton, 1997, p.20). 'The word strategic is often used loosely, but the general meaning in the context of the competitiveness debate is leading-edge industries for which a domestic base is considered essential for economic or security reasons' (Ostry, 1990, p.11). The idea here is that certain technologies and sectors, assumed to have increasing returns and spillover effects across the value chain and across sectors, can have both commercial and military applications. Governments may act on the basis of different non-economic and security considerations.

Dynamic economies of scale, dynamic economies of scope and learning economies (although hard to measure) are considered to lead to positive externalities in high technology sectors, or 'high-externality sectors'

---

4 Jenkins (1978) defines public policy as a set of interrelated decisions taken by a political actor or group of actors concerning the selection of goals and the means of achieving them within a specified situation where those decisions should, in principle, be within the power of those actors to achieve.

5 In the context of European policymaking for information and communications technology (ICT), some scholars have argued that policymakers should facilitate coordination and innovation by an active adjustment policy (Dalum et al., 1999). The rationale behind high technology policy is that ICT is a central high-growth industry that generates a number of macroeconomic benefits. Other scholars take a broader approach to government support for high technology. According to Hilpert (1991), for example, the role of the state is to work toward capable national industrial structures and innovative academic research, in order to facilitate technology transfer between science and industry and to stimulate a competitive market large enough for the amortization of investments.

(Krugman, 1992; Soete, 1997). The learning curve model of strategic trade policy stresses the need to allow firms to move down their learning curves. It has similarities with infant-industry models and rationales for political intervention, which are concerned with short-term and temporary protection of a new industry or technology.[6] According to this logic, public actors may support wholly new technologies. 'Catch-up by the follower [...] tends to hurt the leader by eliminating the gains from trade. Thus there is a sense in which the leader's real income depends on preserving its lead, and there is an economic case for technological protectionism' (Krugman, 1986, p.164). It has been suggested that we can expect a continuing role for public policy in the development of large-scale, advanced technologies through micro-level intervention in order to create and support competitive industries or to foster a level playing field in relation to foreign industries and their associated firms (Johnson, 1982; Ostry, 1990, Porter, 1990; Hart, 1992).

Export subsidies, for example, are aimed at increasing production to capture the external benefits associated with higher levels of output (production externality) or at alleviating the problem of high upfront R&D costs. Public intervention (such as import restrictions and subsidies) may allow producers to accumulate production experience, to reduce production costs, and thereby facilitate relative competitiveness and opportunities to capture export markets with the prospect of economic profits. '[T]he critical mass associated with technological infrastructure derives from the need to provide a spectrum of linked but specialized and distinct capabilities on which firms in the industry draw in a variety of patterns', which makes advanced technology different from conventional infrastructures (Justman and Teubal, 1996, p.26). In addition, high-tech products are strategic in that they are 'pervasive' and are intermediate technological inputs of pivotal importance in a wide range of capital and final consumer products.[7] Furthermore, technology-intensive sectors are often associated with so-called 'locational' agglomeration effects, that is, there are strong increasing local returns linked to the actual production of large and complex internationally traded products (Soete, 1997, pp.10–12). It could have positive dynamic externalities or spillover for regional and national growth, which would justify intervention and support through industrial and technology policies.

---

6 Public intervention would here have net welfare gains for the economy if the domestic firms sold a large portion of the output in foreign markets to shift the cost of the resulting higher prices paid by foreign consumers.

7 Such arguments have often been used concerning large-scale collaborative programs, such as in the field of semiconductors (for example, JESSI in Europe, SEMATECH in the US, and VLSI in Japan).

> This means that high-technology industries provide social benefits not reflected simply by the magnitude of their value added. The contribution to high-technology industries, because of this hidden contribution to national technological capability and economic welfare, clearly goes beyond their direct contribution to a country's production and trade balance. Thus the study of the evolution of these industries assumes particular importance. Even more so since their role has been continuously growing over the past decade. In effect the shares of technology-intensive production and trade in total manufactured output and export have increased steadily during the past two decades (Guerrieri and Milana, 1998, pp.189–190).

An additional feature of large-scale infrastructure, which applies to technology-intensive sectors as well as more conventional industries, is sunk costs and capital entrenchment. Sunk costs and associated risks, current institutional arrangements, and the installed base (the amount of users of the technology and infrastructure) constrain the freedom and strategy of actors (Davies, 1996; Pierson, 1996, 1997).[8] Background conditions are important for understanding why particular technologies come to dominate, why a set of competing technologies come to share the market, and what determines actors' choices between competing technical designs and solutions for adoption. A number of conservative forces can constrain the development of new technology infrastructure projects and associated policies. Among these factors are backward compatibility requirements with existing systems, the expansion of installed networks, 'communities' of supporters who protect their interests, allocation of taxpayer money and substantial investments in infrastructure and following sunk costs (capital entrenchment). Actors that have invested heavily in infrastructure want to capitalize on and reap benefits from past investments. There are usually several technological options available to a decision maker and it is often difficult and costly to reverse

8 The role of increasing returns/path dependence has been more central in economic and technology history than in mainstream economics or political science (for example, Arthur, 1994; Pierson, 1996). In terms of technology choices, it is not necessarily the best solutions that are selected in a competitive situation. There is a considerable time lag between political action and policymaking, on the one hand, and outcome, on the other, which makes it very difficult to predict the effects of policy processes. Early choices can favor a certain solution and technology over another and thus create lock-in and change-resistant effects where initial choices help to determine what will be perceived as superior outcomes. Pre-existing arrangements can be highly difficult to alter under such circumstances. So-called 'frequency dependency' or more commonly 'path dependence' are recurrent sources of technological inertia and occur when a new technology cannot be widely deployed and may not catch on until it is adopted by a sufficiently large number of users (Liebowitz and Margolis, 1990; Arthur, 1994).

technology decisions once they have been made. This is particularly true for technology-intensive infrastructures where future investments must remain compatible with the currently chosen system, as capital is usually long-lived (Rosenberg, 1992, p.205).

The logic of the sunk cost and capital entrenchment argument, thus, is that the behavior and interests of economic actors are to different degrees influenced by the legacy of investment in, and deployment of, infrastructures. In a context of increasing returns, pure market mechanisms cannot guarantee that a superior technology wins at the same time as a central authority may face a problem of selecting between technological alternatives (Arthur, 1989, p.127, 1994). If a technology project or a large technology system gains momentum, it will be difficult to reverse in the short term. The outcome is not necessarily the adoption of superior technology in a context of multiple candidates.[9] The analysis of large-scale infrastructure projects should be sensitive to the power of countervailing forces striving for alternative choices and paths. 'If technical and economic forces set limits to possible trajectories of system development, the final selection of one path rather than another is dependent on politics; that is, the relative power of institutional interests and their commitments to particular systems of technology' (Davies, 1996, p.1145). Background conditions can impact on the future competitiveness of an entire industry (Botelho, 1995).

Access to knowledge and the core circle can be a real political asset (Hamlett, 1992). 'When there are major technical choices to be made that involve public expenditures, complementary or substitute relations with other technologies, or other types of spillover effects, they will end up being judged by non-market criteria [...] In those cases, political lobbying about the new technology is natural. The usual rules of political economy and collective decision-making by interests groups apply, with the additional

9 Examples of sophisticated technologies that never reached a large worldwide audience are IBM's OS/2 operating system and DAT tape players. Other cases of technological battles are that between the VMS, Betamax and V2000 formats in the 1980s for consumer videocassette recorders (VCRs), or the triumph of the analog PAL TV system over the arguably technically more advanced MAC system. Yet another example is the DVORAK keyboard that many thought was superior to the standard QWERTY system (David, 1986). Though 'policy resistance' could account for some aspects of the EU policy, a historical institutionalist approach that takes a static approach would have a more difficult time in accounting for variation in EU high technology policy across the three projects and sectors examined in this study. A dynamic perspective could move the theoretical discussion to a focus on the conditions under which 'lock-in effects' may help to define public policy, how political institutions respond to technological path dependence and trajectories, and how stability is reproduced. In such a framework, the reproduction and political articulation of path dependence would assume center stage.

complication that the introduction of a new technology is by definition a highly uncertain event' (Mokyr, 1998, p.49, 1990). In short, the large-scale system proposal may acquire the status of a 'political imperative'.

Throughout history, governments have played a major role in preparing the ground for new industries, infrastructures and technologies. In the past, industrial policy was intended to support and alleviate the problems of declining and distressed industries with a large number of employees. Governments targeted sunset industries, which received beneficial treatment and enjoyed a special status at the expense of other manufacturing or service-based industries (Krauss, 1978).[10] Moreover, reasons such as national defense or prestige might influence decisions to single out certain industries for preferential treatment. National security concerns spurred many of these government efforts and military technology spilled over to commercial technologies, and vice versa.[11] Muller (1990) has argued that the strongest incentive for European collaboration involving public actors has been funding objectives, economic growth and the supply of security, for example through reduction of dependence on foreign technology and ownership.[12]

Intervention and assistance to provide public and private goods, contracts or ideologies represent allocations of scarce resources and acts of purposeful and strategic action.[13] Political institutions may grant preferential treatment to

---

10 Some of the central features of such industries are that they have been the most spared from liberalization, have a large number of employees, face harsh competition from abroad, and are geographically concentrated (Finger et al., 1982). Some scholars have argued that the costs of competition and recession are more visible than the benefits from liberalization and that protectionist forces have been more efficient in terms of mobilization (Bauer et al., 1972; Warnecke, 1978; Lavergne, 1983). Among the sectors that have enjoyed high levels of protection are public service, transport and financial sectors. In the face of international competition, public institutions can play a role providing funds to support industry, either to help the industry to maintain or increase market shares or to protect industries that are not competitive in order to support a more gradual shift in terms of retraining, and so on.

11 The postwar dominance of the US in commercial aviation, for example, was based on military R&D and procurement contracts. For example, the Boeing 707 was designed in parallel with an aircraft procured by the Air Force and partially represents military R&D spillover into commercial aviation.

12 For example, the French policy for semiconductors and computers was spurred by the dependence upon US companies for integrated circuits for military equipment (Zysman, 1977a, 1977b; Dosi, 1981; Malerba, 1983; Nelson, 1986; Nelson and Wright, 1992).

13 Governments have often supported new industries and infrastructure through regulatory, R&D and procurement-oriented policies. There are many examples of political intervention in markets in the US, such as in the cases of semiconductors (once the transistor was invented through private capital) and computers, the Internet, interstate highways, public airports, atomic energy, and the exploration of space (Schnee, 1978; Thurow, 1996). National governments and European public

sectors that have potential to reach into the world markets to increase industrial effectiveness. They may decide that a proactive response is important to influence the distribution of gains and reduce the risk of falling behind in the international race for competitiveness. Cerny (1990) and Peterson (1993) have argued that there has been a shift from the 'welfare state' to the 'competition state', as industrial policy became a higher priority for governments. Since the 1980s, much of the accent has been on the development of high technology capabilities around the developed world. The European economy has become more internationalized. 'The European economy is undergoing major changes associated with the globalization of industrial and technological systems. This internationalization process will have major implications not only within the European Community itself but also in terms of Europe's long-term relationships and ties with other major trading blocs and countries' (Howells and Wood, 1993, p.4). Severe threats to a domestic sector can trigger public intervention. In the late 1970s and 1980s, government funding for advanced technology in Europe was steered by a debate on a 'technology gap' in the information technology industry, in particular *vis-à-vis* Japan (Patel and Pavitt, 1987). European policymakers responded to increasing foreign competition with government support in several areas to support the creation of technological capabilities.[14]

Proposition 1 *The more the sector and technology is defined and perceived to have welfare-enhancing implications, the greater is the likelihood that public intervention occurs.*

This framework is based on the assumption that actors – be they private or public – are purposeful and engaged in strategic behavior and interactions. Motivational assumptions can be operationalized in different ways (for example, re-election, budget maximization, or prestige). Political and economic actors calculate the additional benefit from spending financial, intellectual or other resources under the constraints of various parameters: the

---

bodies have many times intervened through education policy, tariffs, patent laws, R&D and the setting up of administrative structures to promote European and industry competitiveness in major future generations of technology (Williams, 1973; Sharp and Shearman, 1987; Muller, 1990).

14 The position of EU countries in the high technology trade structure has been lagging behind Japan and the US in some areas, notably in electronics and software, while chemicals and pharmaceuticals, nuclear energy, automobiles, and scientific and professional instruments have fared better. More recently, European industry has become stronger in aerospace (Thornton, 1995; Guerrieri and Milana, 1998, p.204), information technology and telecommunications, and in particular wireless communications (Dalum et al., 1999). The performance and relative competitiveness of European, Japanese and US industries differ across sectors, representing variance in trade specialization (Patel and Pavitt, 1987).

environment in which they operate, resource scarcity, uncertainty, and transaction costs in economic and political activities, decision making procedures, and the gathering of information. It does not mean that people always know what is in their best interests, but that they would be more able to know them better than other people.[15]

A smaller number of advanced economies have fostered capabilities that can affect the direction of technological change, leading some observers to speculate about an oligopolistic world economic structure (Hawkins, 1994; Guerrieri and Milana, 1998, p.192). In oligopolistic industries, a relatively small number of companies compete through trade and investments in the presence of economic profit. Factors such as domestic competition, country characteristics and natural comparative advantage may not be the main determinants in internationally oriented industries (Gilpin, 1987, 2001; Yoffie, 1993; Kozmetsky and Yue, 1997). This approach focuses on strategic bargaining, oligopolistic competition and those collective agreements and controversies that underpin the development of large-scale technology (see, for example, Spencer and Brander, 1983; Brander, 1986; Stegeman, 1989; Richardson, 1990; Scherer, 1992).[16]

---

15 When assuming that actors are purposively oriented and engaged in rational decision making, do we have to assume that their actions are based on self-interest? In public choice models, human beings are viewed as self-interested, while some rational choice models allow for altruistic conceptions, although most often the self-interest comes logically prior to social and welfare interests. Moreover, self-interested actions and proposals can be couched in terms of the national interest or other abstract interest categories. The political game is about the application of actors' strategies and the way they are applied in relationships with other actors (who are often assumed to be homogeneous for all actors within each category of actors) in a context where behavior is guided by incentives, opportunities and costs. Hindess (1989, p.48) has argued that the equation of rationality with the invariable pursuit of self-interest is not a prerequisite for rational choice analysis. 'I concur and emphasize that even if we often lack a convincing test that would make it possible to distinguish behaviors that are truly altruistic from those that are essentially self-interested' (Breton, 1996, p.7). The Weberian assumption states that a public bureaucracy is concerned to serve the public interest, is efficiently providing goods and services, and embodies 'a public interest logic', which is a universal assumption (Elster, 1989). This does not imply, however, that they do not also have institutional self-interests.

16 The creation and support for high technology infrastructures evolve at the intersection of two general trends in the world economy. On the one hand, there is an internationalization of business as a result of the operations of multinational firms/conglomerates (trade, foreign direct investment, mergers and acquisitions, and so on). On the other hand, there is a resurgence of a 'new regionalism' that develops under different conditions and is generally slower to adapt to new environments (Hine, 1992). The development of large-scale infrastructures does not necessarily lead to global welfare-maximizing outcomes. Hawkins has argued, with regard to ICT sectors, that 'the institutional mechanisms for negotiating the

Together, these factors may create high barriers to the entry of new competitors. The differences in specialization and trade patterns across countries are not only the result of factor endowments, but also differential technological knowledge and capabilities (Ostry, 1990; Tyson, 1992; Pavitt and Sharp, 1993). 'Given the special concern national governments place on competition in high-technology industries, trade friction and new forms of public intervention in attempts to correct these imbalances are bound to proliferate in the industrialized area in the near future' (Guerrieri and Milana, 1998, p.205). Governments may try and help their domestic industries to capture available or expected profits, to alter the strategic interaction between economic actors, and to accomplish profit shifting, hoping that foreign governments will not retaliate.

Public actors may have an interest in supporting sectors with expected increasing returns and thus become engaged in attempts to assist the creation of comparative advantages. Tyson and Chin (1992, p.160) argue that public intervention can be potentially effective in particular industries by creating high-wage job opportunities and substantial shifts in the international division of labor, accelerating innovation and increasing global competition. In particular, they argue, the government should encourage competition and innovation in times when an industry faces major market adjustments and is at a critical juncture.[17] Public intervention has been particularly common in advanced large-scale technology systems, due to long-term uncertainty or specific public considerations (Davies, 1997).[18] 'The development of cutting-edge technological infrastructure is subject to an enormous amount of target uncertainty, much more so than for regular product innovation. Initially,

---

technical and administrative agreements necessary to provide and sustain a global information and communication infrastructure appear to be fragmenting at the international level rather than consolidating' (Hawkins, 1997, p.178). According to Hawkins, the 'global' dimension of communication infrastructures is strengthening regional power centers. 'There appears to be no general pattern of globalization in terms of the distribution of control over the production and use of infrastructural technologies for information and communication systems. In spite of their cultivation of a global image, the information technology and telecommunication industries look set to remain tied to a select number of highly developed markets and political structures' (Hawkins, 1997, p.191).

17 In the commercial aircraft industry, for example, European decision makers stressed the concepts of 'managed competitive balance' and 'competitive inequity' (Tyson and Chin, 1992). European decision makers charged that the US commercial aircraft manufacturers benefited from political support, state aid through R&D programs organized by NASA, and funding from the Department of Defense (DoD).

18 Hart has demonstrated the 'pragmatic neo-mercantilist' interactions between the US government and non-state actors, which fomented the US achievements in Internet technology (Hart, 2000).

nobody knows the critical parameters of the new technology or how they are related to the desired performance dimensions of future products, and potential users cannot specify their needs in terms of the new technology' (Justman and Teubal, 1996, p.42).

The support for communications and transport infrastructures that link value chains and particular functions (for example logistics or R&D) should be placed in a broader context of the creation of comparative advantage to spur economic competitiveness (Lawton, 2000). 'Targeting for future development is another form of industrial aid. The main assistance here has been in the form of government coordination, and even subsidization, of research and development, capital formation, and industrial development strategies. These programs – such as Japanese development of large scale integrated circuits or French development of telecommunications or the [US] Defense Department's funding of large scale computer development in the US – often do not have an immediate impact on trade, but may over a longer period. The government role may be negligible by the time the industry becomes internationally competitive and begins to export' (Price, 1992).

In the literature on high technology policy, public intervention is often justified in the presence of market failure.[19] This means that the market by itself may not offer sufficient incentives when social returns are assumed to exceed private returns and long-term developments imply a high degree of uncertainty (Baumol, 1952/1965; Bator, 1958/1988; Coombs et al., 1987, pp.207–8). In particular, government intervention is deemed justified with regard to the provision of 'public goods' (Head, 1974). Governments may allocate resources more efficiently, distribute wealth more equally, reduce negative externalities, and thus increase total social welfare under market failures.[20] Moreover, a proactive government can be supported to stimulate

19 In both the private and public sector, the scale of capital and R&D investment required for industries based on new advanced technology could be so large as to prevent individual economic actors in raising the necessary funds or accepting the high risk involved. Governments may intervene when the rate of private return is lower than the perceived rate of social return to investment (for example, energy and transportation). In basic research, the length of time to the potential economic benefit can lead to a shift of resources away from basic research, although such research can be important in the long run. However, it appears less fruitful to analyze public policy in terms of 'market failures' and the associated debate on optimal outcomes along neo-classical economic reasoning since there are no definite and easily identifiable 'optimal outcomes' in the first instance.

20 Public intervention, however, can result in regulatory capture, government failures, negative consequences for other producers and consumers, and retaliation. Industrial policy is carried out under constraints of imperfect information and invites questions about who will pay and benefit (Eads and Nelson, 1971, 1977). Those who are critical of industrial policy argue that it will reduce the incentives of economic actors to adjust to world competitive conditions and that it will result

consumption, investment, and capital expansion. There are a number of other factors that invite temporary public intervention: externalities, indivisibilities, uncertainty, economies of scale, network economies in services, asymmetric information, oligopolistic power, and other imperfections. They suggest conditions under which intervention may be considered rational or irrational from a public policy point of view.

Demand for High Technology Policy: Economic Actors

The next step in the outline of the basic model of high technology policy creation is the role of economic actors (producers, operators and interest groups) and the EC bureaucracy, and their interaction at EU level. Basically, the existence of an economic incentive may trigger political activity by economic actors and demands for high technology policy if they perceive benefits from it, and if they have the required resources to engage in political activity. The higher the stakes involved for industry, the higher is the interest in engaging in political activities to influence outcomes. If there is an economic incentive, economic actors may assume that the EC is more likely to provide high technology policy (Figure 2.1).

Access to the political market is often restricted to firms, business groups and industrial alliances that can provide political support to policymakers and have relevant information that they require for making informed choices. 'The collective action logic of large firms in Europe is driven by a desire to improve their European credentials, and an attempt to improve their private access to the new EC policy forums rather than to facilitate the emergence of

---

in products priced at above market-competitive levels, which may have serious long-term implications for restructuring, employment and efficiency. The risks of political intervention include the difficulties associated with selecting candidates for support, the limitations of identifying industries in advance that have potential advantage, the high potential costs and inefficiency, and the risk of capture of and/or prolonged assistance by the government ('lock-in effects'). Intervention to support industry may divert resources away from more promising uses and a 'hands off' approach to industrial change may ensure a more optimal allocation of resources (Barfield and Schambra, 1986; Grant, 1989). The Chicago theory of regulatory capture views regulation not as a neutral activity but as a redistribution of wealth, which confers differential benefits on different groups (Stigler, 1975). The Virginia school (and the theory of rent seeking/rent protection) argues that government intervention creates rent-seeking behavior and lobbying for special privileges that gives rise to direct (for example, public contracts) and indirect (for example, subsidies) transfers of income. Another argument is that convergence of advanced economies, spurred on by the internationalization of production, trade and finance and technological sophistication and diffusion, renders public policy and intervention to support technology, large domestic firms and specific sectors inefficient (Carliner, 1987; Eliasson, 1991; Nelson and Wright, 1992).

European corporatism' (Coen, 1997, p.98).[21] Those economic actors who benefit from current regulatory arrangements and public assistance in some form may be assumed to engage in political activity. European administrators and policymakers need knowledge not only of European conditions but also of developments around the world to facilitate as many rational decisions as the constraints allow. Politicians or public bureaucracies (such as the EC services) can provide benefits to those actors who have the capacity to provide information and resources that they need to elaborate and execute policies (Marcus, 1991; Grossman and Helpman, 1994).

Many companies and industry associations strive to gain policy benefits and preferential treatment from public institutions, for example to permit economies of scale and network economies in services, or to maintain or raise entry barriers for domestic or foreign competitors (Mazey and Richardson, 1996). Some scholars have argued that the activity of transnational economic actors has been the predominant influence at the EU level (Sandholtz and Stone Sweet, 1997; Grant, 2000, p.81).[22] The lobbying activities of industry at EU level can be attributed to the increase in the scope and impact of the EC's and the EU's regulatory powers that have made the EU an important arena of negotiation and lobbying (Cowles, 1995; Grande, 1996; Mazey and Richardson, 1996).

The willingness or acceptance of public intervention on the part of economic actors, the capacity and willingness of public actors to intervene, and the ways in which such an intervention occurs, vary across different

---

21 The institutional opportunities to apply resources in political activities may vary (Dobbin, 1994). The political system affects the patterns of behavior and strategies of economic actors by functioning as mediating structures through which proposals are channeled (Hall, 1986, p.277). The policy outcome depends to a large extent on the influence of powerful groups and their ability to promote or stall developments (Commoner, 1979, pp.32–48; Reece, 1979). 'The sophistication and complexity of contemporary technology preclude direct involvement by all but those immediately concerned with them' (Volti, 1988, p.13). Support groups (notably major producers and customers, industry pressure groups, governments, regulators) have considerable power to shape specific developments.

22 Sandholtz and Stone Sweet (1997) argue that supranational governance serves (1) those individuals, groups, and firms who transact across borders, and (2) those who are advantaged by European rules, and disadvantaged by national rules, in specific policy domains (p.299). The higher the level of transnational exchanges, the more likely is the feasibility of integration. This view has been criticized: 'However, this fails to account for attempts by the European institutions and some member states to promote market-correcting integration precisely to compensate for the *very absence* of transnational exchange' (Branch and Ohrgaard, 1999, p.129). Branch and Ohrgaard have argued that in the models by Moravcsik and Stone Sweet and Sandholtz, the demand for integration 'emanates from the same source – societal groups – and is essentially functionally (in the sense of economic necessity), rather than politically driven' (Branch and Ohrgaard, 1999, p.138).

sectors and policy issues. People and organizations have vested interests in the survival, expansion and durability of a certain technology, policy or specific institutional arrangement.[23] Certain organizations and groups control technology projects and accordingly shape society by deciding which technologies (if their attempts are successful) will be implemented and selected (Cawson et al., 1990, p.19).[24] 'Technologies and their supportive institutions are strongly intertwined; a technology is supported by its sponsor, and its sponsor's existence is supported by the continuance of the technology. When these institutions change, technologies may also change, and to change the technologies, we may have to change the institutions' (Westrum, 1991, p.209). Therefore, the development of large-scale industrial projects should be placed in a political-economic context within which major protagonists strive to inscribe their agendas and objectives into the design and direction of the systems (Hughes, 1983; Fries, 1988; MacKenzie, 1990).[25]

Technology development, thus, is often the result of economic groups that defend or call certain technologies into question and that attempt to maintain or gain control over resources (Albury and Schwartz, 1982). 'Experts are in a position to influence the course of technological change, but at the same time, they face many difficulties in converting their knowledge to power. Having the correct technical answers is not enough; wielding power requires the mobilization of a constituency and the ability to create a broad base of support [...] if they are to be successfully implemented. Even outside government, technological decisions have the same political character, with

---

23 Vested interests (government bureaucrats, corporate managers or employees with knowledge and skills associated with the established system) seek to protect or enhance institutional arrangements and gained positions. The assessment of a certain technology infrastructure along technical, economic and environmental dimensions is a very complex matter and each system is represented by its own 'communities' that define and validate the value of a system from their own perspective of system persistence and by other groups that challenge those already established communities from a perspective of system transformation.

24 Thompson (1989) has argued that we need to combine an examination of the actions of governments and regulatory agencies, the role of other key actors, and the particular character of the technology involved. Technical considerations are not necessarily the overriding concern or determining factors (Street, 1992).

25 Gourevitch has argued that the 'power' of economic groups depends on the importance and function of economic actors and 'the resources that control of the function provides' (Gourevitch, 1986, p.59). The access of large firms to EU policymaking provides them with opportunities to influence public policy in both 'open' market-conforming (for example, breaking up monopolies, cartels and other forms of protection in telecommunications, air transport, or financial services) and 'closed', rent-seeking market-non-conforming manners (Sally, 1997, pp.74–75). Economic actors can also try to elicit support for their preferences by referring to broader macroeconomic or non-economic implications (for example, national security) to persuade the public authority that intervention create benefits.

the same need for gathering support for projects and policies. This means that except for fairly trivial matters, issues are not settled on the basis of purely technical considerations' (Volti, 1988, p.231). Political and economic considerations are often embedded in and influenced by non-technical factors (see, for example, Mackenzie, 1990; Sandholtz, 1992a, 1992b; Hart and Thomas, 1995; Lawton, 1997).[26] Historically 'embedded' economic groups in Europe may engage in attempts to accentuate their 'insider character' in order to gain preferential treatment and may be granted such treatment (Cawson et al., 1990, p.26; Eden and Molot, 1993; Hawkins, 1994).[27]

Figure 2.1 A basic model of high technology policy creation

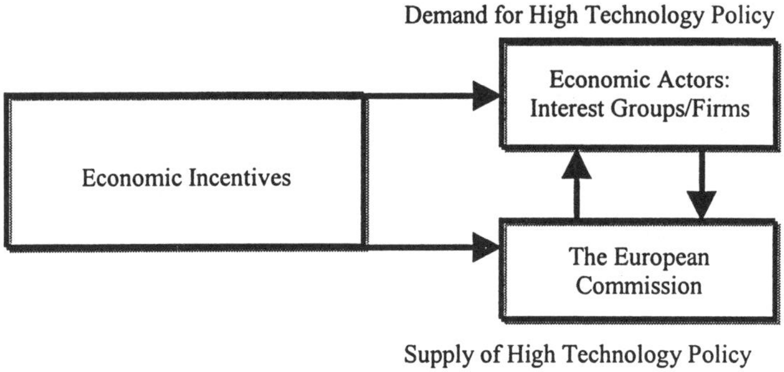

26 Certain policy choices are pursued and implemented that keep with the proposed views of powerful economic interests and that are in accordance with particular institutional arrangements (Mokyr, 1990, 1998). Public and economic actors strike sectoral bargains to implement technology and policy (Strange, 1988/1994). Technological development occurs as the result of negotiations among networks of sponsors surrounding certain technology proposals (MacKenzie and Wajcman, 1985). The adoption of the automobile, for example, was not inevitable but required extensive efforts and political lobbying by the automobile, rubber and petroleum firms that stood to gain from it. Another example is the relatively slow pace of solar power implementation that was influenced by political factors (Westrum, 1991, pp.77–78). Yet another example is the reluctance on the part of railroad interests in the US to support passenger trains since such interests made more money on freight transport, which has resulted in dependence on cars and airplanes (Fellmeth, 1970). It has been shown that the development of the Concorde supersonic airliner and ambitious atomic energy programs has evolved under such circumstances (Collingridge, 1980, pp.135–138; Volti, 1988, p.230).

27 Industry may be cautious, though. 'Companies are uneasy with the fluid, often highly personalized aspects of the Commission's approach to policymaking. In this sense, the Brussels-based aerospace lobby often appears to be more concerned to avoid "policy-ambush" or misrepresentation of their interests than a confident partnership with Commission officials' (Hayward, 1994, p.363).

Supply of High Technology Policy: The European Commission

The public bureaucracy (the EC) plays a central role in the provision of high technology policy in the broader political system and economy. Public intervention drives the equilibrium from one position to another, that is, from no EU intervention in high technology industries to intervention in 'carefully' selected industries, or through the provision of an overall regulatory and investment framework. Public actors are more than simply intermediaries of industry demand and public institutions are more than an arena of competing interests. The public bureaucracy aims to promote its goals, expand its influence, mandates and functions, and to strengthen its administrative capabilities (for example, Kenis and Schneider, 1991; Majone, 1992; Sandholtz, 1992b; Dang-Nguyen et al., 1993; Christiansen, 1998; Harcourt and Radaelli, 1999; Pierre, 2000). 'Like most bureaucracies, the ultimate aim of the Commission is to expand its powers' (Cram, 1994, p.199). The EC bureaucracy may be viewed as a purposive entrepreneur or opportunist, bringing together diverse actors and interests, aiding them to agree and providing common regulatory frameworks that support its organizational interests (Hayward, 1994).

The bureaucracy is interested in expanding or at least preserving the scope of its regulatory reach and carry out policy as cost-effectively as possible in the light of limited financial resources (Majone, 1994; Héritier, 1996; Hills and Michalis, 2000). It acts to preserve, modify or impose policies and to retain an active role. While it may support policies in the pursuit of broader welfare, it is likely to weigh the costs and benefits obtainable from supplying or not supplying intervention. The larger the industry and the more important it is in terms of benefits from executing the policies of the bureaucracy, the more likely it is that the latter will respond favorably. Public actors may prefer to provide divisible goods to special economic actors rather than indivisible goods to the wider society since the former choice could generate more support. The ability to mobilize private sector actors is a central element for governance, willingness to intervene and policy effectiveness. The EC can likewise be assumed to work toward the provision of policies and other activities aimed at meeting the demand from the general public, or to act in the belief that its policies will generate benefits to society.[28]

---

28 The bureaucracy has substantial control over policymaking through its control over specialized information vital to the debate on economic and social problems, which it can use to persuade other actors or trade for influence (Thompson, 1961; Lewis, 1980, p.6). It has the necessary managerial skills and 'a central role in the development and dissemination of policy-relevant information and thereby retains a powerful position in policy making' (Peters, 1989, p.190). In addition, it can be claimed to be relatively impartial. 'This partisan impartiality goes hand in hand

The EC has demonstrated a kind of 'institutional agency' and strategic behavior by organizing sectoral interests and by initiating industrial advisory groups (Grande, 1996, p.323). Hayward refers to the central role that the EC bureaucracy has played in the European aerospace sector: 'The Commission has begun to build up a competence in aerospace issues. In this respect, the pattern of events reflects the experience of other policy sectors – trading specialized knowledge held by companies and other actors such as research establishments and contracted personnel from the industry, for participation in Union-level policy formulation. Commission officials have learnt how to build effective bridges with industrialists – at several levels. The emerging policy network involves both technical specialists and senior industrial leaders, and has made impressive progress towards creating at least the structure needed for a more effective Union-level aerospace policy community' (Hayward, 1994, p.362).

The public bureaucracy (and the state) has a number of mechanisms at its disposal through which it can potentially influence other actors (Krasner, 1978; Nordlinger, 1981).[29] Information gathering, analysis and timing, and the promotion of policy networks are among the most important available instruments that the EC can use to identify political windows of opportunities in order to influence behavior (Kenis and Schneider, 1987; Peterson, 1992; Sandholtz, 1993; Cram, 1997; Sbragia, 2000). The opportunity for the EC to influence policymaking is its ability to respond rapidly to 'windows of opportunity' or to stimulate the creation of such opportunities. 'As rational actors, firms might have been expected to seek out the strategic and tactical alliances that best serve their technological, market and R&D requirements,

---

with the expertise of the bureaucracy to make a strong argument that their decisions will be superior on technical grounds to those that could be taken by political institutions' (Peters, 1989, p.197).

29 Cram (1994) has argued that the multi-actor participation in these networks has prepared the ground for emerging common business identities. While the importance of the EC as policy entrepreneur and purposeful opportunist in EU policymaking is widely acknowledged, the task as a policy manager has been less debated and researched. Some of the observations are that the organizational culture of the EC favors policy innovation and new policy ideas at the expense of policy management and that its segmented organization and lack of resources makes coordination and the fulfillment of responsibilities difficult (Laffran, 1997, p.424–5). A transnational policy network approach has been applied, for example, to analyze and explain EU policymaking for telecommunications (Dang-Nguyen et al., 1993; Sandholtz, 1993). The EC is interested in expanding regulatory policies in order to enhance its own power position in the light of the limited financial resources of the EU (Majone, 1994; Héritier, 1996). The use of steering and administrative guidance instruments, however, can fail as well as succeed. The relative balance of public and private sector actors is not least an empirical question (Wyatt-Walter, 1995, p.439).

irrespective of national or regional boundaries. These alliances would, of course, have included trans-European collaboration in any case, but would more likely have been predominantly global in orientation. DG XIII has, however, managed to "massage" the direction which this co-operation has taken and has encouraged increases in the extent of intra-EU co-operation between IT firms' (Cram, 1994, p.203).

2.2 INDUSTRY CHARACTERISTICS

The analytical framework examines the interaction of industry and the EC and the mechanisms under which high technology policy occurs to support the market implementation of large-scale industrial projects. The condition of industrial characteristics consists of two components: the ability of economic actors to organize (organizability of economic interests) and their relative position in the structure of international competititon (voice of industry).

The Organizability of Economic Interests

The ability of industry to engage in political action (the organizability of interests) is related to the extent to which the political interests within the sector are homogeneous and can overcome collective action problems, and how they evaluate the costs of engaging in political activity. Actors' interests and behavior are influenced by political-economic characteristics and their position in the regulated sector (Hall, 1986).

Proposition 2 *If economic actors are not able to organize politically, and there is no coherent demand for high technology policy, it is less likely that it will be provided.*

What literature exists to support this proposition? Both lobbying activities of organized industrial actors and structural characteristics (for example, the cohesion of organizations, resources and their access to EU institutions) play a role. Gourevitch (1986) and Hall (1986) stress the importance of the organization of the private sector. '[T]he organization of both economic and political arenas plays a critical role in determining which interests will be effectively articulated and what sort of response they will elicit from the state' (Hall, 1996, p.233). The analysis can be divided into two dimensions: the differential resources actors have and their access to policymaking. The difference in the magnitude of resources actors can generate, organize and apply is fundamental in political analysis.

Compliance or support from industrialists and consensus on the need for government intervention may be indispensable for the capacity of public institutions to intervene (Gourevitch, 1986, p.238; Samuels, 1987). Where interests are more organized we can expect more concerted actions. The smaller the number of firms, and the higher the prospective benefits and private good gains, the more likely it is that corporate actors will embark on collective action. Consultation, collaboration and a certain degree of mutual understanding are important contributors to the political feasibility of dealing with adjustment that involves substantial costs, although the process can involve conflicts and failures (Curzon Price, 1981; Hills, 1981). A coherent industrial strategy depends in part on a 'broad acceptance among political élites of the need for government to play an active role in helping industry adapt to change' (McKay and Grant, 1983, p.9). Fragmentation of economic actors, and tensions between economic actors and governments (or regulatory agencies), can impede the provision of high technology policy. Those who stand to win or lose more from regulation will be more inclined to respond in political activity (Stigler, 1975).

There may be problems of organization at the sectoral level due to divergence of interests of economic actors and expected differential impacts of public intervention (Noll and Owen, 1983). Producers may use political influence to encourage government policymakers to support their particular product line or development plans.[30] High technology policy can involve fundamental disagreements about the relative merits of the market mechanism and state intervention. Disunity and political posturing results from differential competitive positions, business strategies and technological orientation, and the expected implications of regulation on the competitive position of economic actors, be it deregulation, regulation or re-regulation (Posner, 1974; Spiller, 1983). Actors may thus perceive public intervention in different ways due to expected utility in terms of strategic advantage, which can outweigh the costs of political lobbying. 'The interest of individual firms may be at odds with the interest of an industry as a whole. Some firms are likely to break ranks with the industry position to gain competitive or other advantage' (Marcus, 1991, p.342).

The more heterogeneous the positions of economic actors involved in a specific policy issue within or across sectors, the harder it can get to organize. Common general goals among economic actors are thus important in presenting a coherent demand for high technology policy. Economic actors can have collective action problems in that they cannot agree on appropriate

---

30 The creation, for example, of standards for technology infrastructures implies substantial revenues for manufacturers backing the chosen technology. Operators, on the other hand, may be cautious and are likely to seek first to amortize their investment in existing infrastructure and build up new capacity step by step.

political actions and communication (not necessarily because of transaction costs). Moreover, industry disunity makes it more difficult to provide appropriate information and services to the public bureaucracy. Part of the collective action problems is that actors can calculate that they are better off if other actors provide collective goods and lobbying despite the existence of common goals (Olson, 1965/1971, 1982).

In addition, the more cohesive the support for a particular position, the more coalitional support a grouping enjoys, and the less resistance they meet from oppositional forces, the more likely it is that policymakers will take the arguments of organized industry into account (Potters and Sloof, 1996). Finally, organized political activity can be impeded since in principle the expected benefits need to be higher than the opportunity costs. Lobbying implies that capital and labor could have been invested in production but is instead invested in political activity. The more resources the industry has, and the more that is at stake for a particular industry or specific economic actors, the more this can impact on decisions to engage in political activities. This will proceed until the resources spent on political actions reach the so-called 'lobbying equilibrium'; that is, until the marginal gains and marginal costs equal one another, at least in principle.

The Voice of Industry

This condition focuses on the 'structure of international competition' and its relation to the articulation of voices in the private sector toward public intervention. In other words, the voice of industry will reflect the position of economic actors in global markets (Hirschman, 1970, p.15ff). This argument states that the relatively more internationally oriented economic actors will be less willing to demand substantial high technology policy and accept considerable exposure to retaliatory actions by foreign countries.

Proposition 3 *If economic actors fear retaliatory action due to their international position, they will be less inclined to demand protectionist high technology policy.*

Is there any support for this argument in the literature? In the past, European governments have supported large domestic firms (for example, support for computer and semiconductor manufacturing firms), not least when they have been considered to compete head-on with foreign producers (Flamm, 1988, 1996). 'Economic actors at the cutting edge of international competition are likely to support policies that promote open trading, not only for their own products but more generally for intensified specialization in international trade. Actors less well placed in international competition are more likely to support protection or modes of shelter or subsidy' (Gourevitch, 1986, p.56). Milner has supported this argument: 'In general, the argument is that firms with more extensive exports and multinationality should be less likely to demand protection, and more likely to resist it actively, even when facing serious import competition' (Milner, 1988, p.18).

Industries and firms that are more multinational in orientation can be assumed to be interested in a two-level game where they sometimes prefer open trade and non-intervention and other times call for intervention (Milner, 1988). The relative political strength of industrial interests, together with the degree to which they are export-oriented and have presence in other regions, shape their preferences toward public policy (Grossman and Helpman, 1994, p.835). Thus, the more economic actors depend on foreign markets and the more activities they have in those markets, the less prone they will be to ask for substantial political intervention at home that may trigger retaliatory action by other countries in which they have significant operations and market shares. Economic actors who have relatively fewer international activities and who are less competitive are likely to be less reluctant to ask for protectionism. 'The decline in the relative competitiveness of many of its manufacturing industries make the US government prone to take actions to prop up domestic production in specific industries' (Price, 1992, pp.33–34).[31] For example, the political rejection in spring 2002 of standards for higher fuel efficiency in the US automotive industry is a clear example of the regulatory capture of political institutions, the lobbying strength of the car industry, and the corrupting influence of 'soft money' and financial campaign contributions. Those economic actors in the US who are less competitive in light of relatively more fuel-efficient technology and use in East Asia and Western Europe favor domestic protectionist industrial policy.

The relative competitive position of firms, sectors, and the technologies in which they have invested in the domestic and international economy thus shapes economic actors' interests. Hayward, for example, has argued that the international orientation of European aerospace industry result in a stronger focus on EU (and also transatlantic and global) politics. 'Transnational companies may not have the same loyalty and commitment to single national governments and national governments may not have quite the same degree of commitment to supporting transnational aerospace companies. This may facilitate the growth of European supranational authority in the aerospace

---

31 The import competition argument argues that the more the sector is subjected to the forces of foreign imports the higher is the likelihood that the domestic sector will demand defensive intervention (Baldwin, 1982). Industrial restructuring pressure can encourage firms to engage in political action at the EU level. The relative position of economic actors in terms of export dependence, and the degree to which they operate on an international basis, allows for both divergence and unity within the industry in terms of preferences toward public intervention (Milner, 1988, p.44).

sector' (Hayward, 1994, p.362). The examination of policy sectors and their specific institutions and relationships must be understood within the context of the international economy (Sandholtz and Zysman, 1989; Cowles, 1995; Grant, 1997; Rhodes and van Apeldoorn, 1997; Sung, 1997). An international dimension has been integrated into the organizational structure and activities of corporations, notably in industries characterized by high R&D costs, rapid pace of product innovation and short product life cycles (Magaziner and Patinkin, 1989; Howells and Wood, 1993; Wyatt-Walter, 1995). International competititon and the 'politicization' of the international division of labor play an important role (Dyson and Humphreys, 1988; Dyson, 1991).

2.3 POLITICAL ENTREPRENEURSHIP

Political entrepreneurship focuses on the internal logic of political institutions and the relationship to other political institutions. Lewis defines a public entrepreneur 'as a person who creates or profoundly elaborates a public organization so as to alter greatly the existing pattern of allocation of scarce public resources' (Lewis, 1980, p.9). However, this definition is based on individual personalities. In this study, the political entrepreneur refers to the EC, acting within the broader institutional structure and interacting with the economy and society, with the potential of influencing authoritative action.

Bureaucratic Cohesion

The condition of bureaucratic cohesion suggests that internal cohesion of the public bureaucracy is necessary for effective public policy to occur. The cohesion of the bureaucracy plays a role in bolstering effective public policy.[32] The fragmentation of the EC into departments and units with potentially different positions, as well as the principle of collegiate decision making within the EC services, may impede its potential to act decisively (Peters, 1992; Cram, 1997).[33] Moreover, there is a recurrent tension between

32 Patterson (1998) stresses in her study on EU biotechnology policy that we need to make a distinction between different forms of coordination – both formal and informal – in EU policy-making processes. Coordination is required when 'policy problems are interdependent and the actions of one actor can deliberately or inadvertently interfere with or contribute to the goal achievement of another actor' (Patterson, 1998, p.44). This view suggests that the more actors are aware of linkages between policy problems, the more administrative units and outside actors are interested in influencing the process. Hierarchical structures of coordination rather than consistent inter-service coordination have dominated EU policymaking.

33 The 'bureaucratic politics' perspective does not assume that governments are characterized by a unity of purpose and action, complete information, value-

the advocacy of policy innovation and its role as conserver of procedures (Cohen et al., 1972). Bureaucratic competition for budget can further narrow down available options and either help or reduce coordination and coherence.

Proposition 4 *If the relevant parts of the bureaucracy do not agree on high technology policy provision, it is less likely that it will be provided.*

The EC bureaucracy has been viewed as a 'multi-organization', where different Directorate Generals (DGs) may have different ways to approach policy areas and interact (formally and informally) with outside actors (Cram, 1994, 1997). Various DGs can have different conceptions of what needs to be done, why it needs to be done, and how it should be done. This may open up potential for tensions (as well as forms of administrative cooperation and interaction) that resembles the processes of national public policymaking (Menon and Hayward, 1996). Also, an actor may believe that taking on a responsibility will hurt, rather than help, the current power position or the prospect of strengthening its position.

The bureaucratic cohesion argument suggests that the capacity and willingness of the bureaucracy to intervene is greater when it is supported by internal cohesion within the bureaucracy. As was argued previously, the EC has been accredited institutional and 'connection power' in that it can influence public and private sector behavior that would not emerge in the absence of EU intervention (Sandholtz, 1992a, 1992b). It may be able to 'massage' the direction of industry behavior and strategy toward intra-EU cooperation and enjoy certain flexibility in approaching private and public sector organizations (Cram, 1994). The EC has at times been willing and able to encourage coordination among economic actors, which also 'facilitated the growth of the Commission authority and has encouraged the transfer of some

---

maximization, or are centrally coordinated (Allison and Halperin, 1972). Peters (2000, p.45) states that 'government is rarely a unitary actor, and rarely has a single coherent and integrated set of preferences for policy'. The basic unit of analysis is government action as the result of compromise and conflict among officials/units with diverse interests and unequal influence through bargaining along regularized channels. The EC consists of several levels of administrative units, on the one hand, and the executive EC, on the other. The latter is 'responsible for the final phrasing and timing of the publication of policy initiatives, for establishing the opportune moment for action, and for the final selection of the instruments with which policies should be implemented' (Cram, 1994). Parts of a larger organization may attempt to use a certain issue for their own benefit, for example, DGs within the EC services. Hence there may be conflicts. Policymaking evolves within a context of the existence of competing preferences and processes, bargaining games, and interpretations of the same problem among bureaucratic entities and civil servants who represent 'a department or agency along with the interests and constituencies their organization serves' (Allison and Zelikow, 1999, p.256).

corporate lobbying to the Union level' (Hayward, 1994, p.349). The EC is assumed to be interested in expanding its functions. 'Over the last decade, the European Commission began to intervene in more aspects of aerospace – mainly in trade and competition issues. More recently, its has expanded its R&D sponsorship functions to aerospace. These developments led European firms to increase their lobbying activity at the European level, and to a broader corporate and political interest in creating more efficient structures for European industrial and technological support' (Hayward, 1994, p.347).

The capacity of the EC to govern effectively is conditioned by its ability and willingness to cajole and twist the arms of other actors to enhance cohesiveness, to bargain for influence in policymaking and implementation, and thereby extend its administrative reach and control. The bureaucracy has a strategic position between the private and public sector due to accumulated experience and access to specialized preferential information and expertise, which both facilitates its tasks and challenges it through information overload and interest conflicts (Mosher, 1978; Olsen, 1983). 'The powers and skills of the officials who occupy positions in these institutions allow them to aid one side or another in policy quarrels and to take the lead in mobilizing pressure and shaping opinion' (Gourevitch, 1986, p.61). Disunity among economic actors can offer civil servants (and legislators) greater opportunities to play out the demands of competing interests against one another to increase the probability of reaching benefits (Peltzman, 1974). The opportunity for entrepreneurship is here assumed either to increase when the support for policy proposals is balanced among those actors that are important for the bureaucracy in terms of political support, or when the bureaucracy benefits from disunity within the industry (Baldwin, 1985, p.14).

External Support

The external support condition for the provision of high technology policy suggests that in the face of expected political opposition from EU Member States (that is, a large majority), the bureaucracy will be less willing to risk confrontation and promote public intervention proactively. In the EU context, the expectation of a large majority of EU Member States opposing high technology policy, the more reluctant the EC will be to risk its reputation and resources, whereas in the case of an expected potential to win the support of the majority in the future despite initial opposition may warrant promotion.

Proposition 5 *If there is an expectation of considerable external opposition, the public bureaucracy will be reluctant to promote high technology policy.*

The capacity and willingness of the EC bureaucracy to intervene in support of high technology policy will be conditioned by external political pressure, for example the degree of independence from centralized political control and leadership. The mandate for political intervention in the market to assist the implementation of technology ultimately rests with the EU Member States. Even if the EC manages to mobilize a variety of actors and encourages collective action, the outcome of coordination will still be uncertain. In some areas we may expect the EC to be able to play a greater role while in others it will be more constrained or not able (even if willing) to play any significant role (Ross, 1993; Wyatt-Walter, 1995, p.439; Harcourt and Radaelli, 1999).

The EC bureaucracy can couch the need for public intervention in terms of competitiveness, independence or other collective goals to raise the support for intervention among EU Member States. The EC may use the EU as an organizing political-economic concept engaged in cooperative or competitive relations with non-European countries and regions. It may find it politically advantageous to promote proposals in the name of economic prosperity or other collective goals. In other words, the need for external support is central for the development and execution of an effective public policy. The support for the European Parliament (EP) has also become increasingly important. The clearer the industrial and macroeconomic benefits for EU Member States, the more likely it is that proposals will gain momentum and be acted upon collectively by the EU institutions. 'By justifying its intervention in terms of bolstering the competitive position of European industry *vis-à-vis* Japan and the US, the Commission has, to some extent, succeeded in selling intra-EU collaboration to reluctant EU governments. Defending and promoting European industry has become, therefore, the EU equivalent of national protectionism at the level of the nation state. As such, the policy instruments were relatively familiar, and acting collectively in the European interest provided a legitimizing concept' (Cram, 1994, pp.202–203).

Finally, the effectiveness of government depends in part on the range and effectiveness of the instruments that give effect to policy. Public intervention occurs through the use of governing instruments, that is, 'the actual means or devices which governments have at their disposal for implementing policies, and from among which they must select' (Howlett and Ramesh, 1995, p.80). In general, we can assume that the requirements of direct public spending will impact on the probability of political intervention. In other words, the 'treasury function' will impact on public intervention, that is, the possession of fungible money-like stocks and checkbook resources that can be used to influence outsiders as a part of political intervention (Hood, 1986, pp.5–7). The EU has scarce funding resources, which has directed the policy machinery toward a market-oriented and regulatory-inclined policy regime.

2.4 A MODEL OF EU HIGH TECHNOLOGY POLICY

The intention of the more refined analytical framework has been to allow a comparative political-economic analysis of the provision of and variation in high technology policy. It will function as a roadmap for the following examination and comparison of the selected cases, thus integrating the different empirical stories into a more coherent whole through a theoretically guided analysis. A general proposition was derived from the basic incentive-demand-supply model of high technology policy. When high technology is provided, this proposition is complementary to those that are derived from the more refined model (Figure 2.2).

Figure 2.2 Analytical Framework

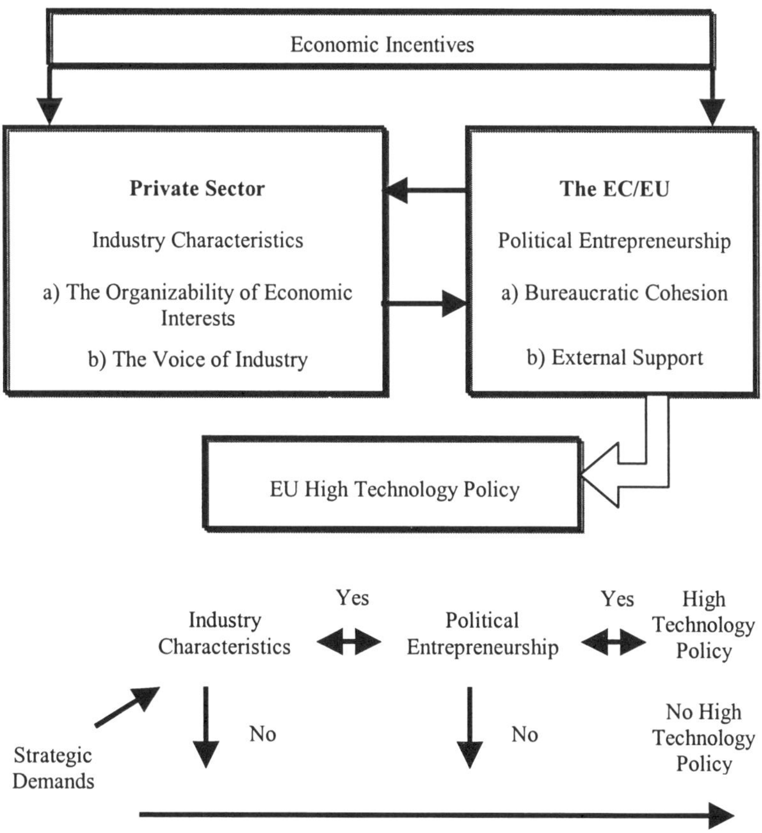

2.5 Time Frame of the Analysis

A criterion for the selection of the large-scale European-wide technology infrastructure projects was a sufficiently comparable time frame for their development. The focused time frame for the study is 1995–2001 with some minor variations between the cases. With regard to satellite navigation, the EC adopted the first dedicated communication in June 1994, followed by a supporting EU Transport Council decision in December the same year. In December 2001, the Transport Council again postponed a political decision on funding for the Galileo development phase (2002–2005), but the project has been supported in general. In the case of UMTS, DG INFSO launched a report ('Green Paper') on the European mobile communications sector, which would prepare the ground for subsequent dedicated policy initiatives in the mid-1990s. In 1995, DG INFSO convened the UMTS Task Force, which was preceded by the publication of a 1994 Green Paper, to initiate a policymaking process and define a European strategy. A directive on UMTS was formally published in early 1999. The analysis of the standards development process takes the study up to 2001. Finally, regarding digital radio and the DAB project, the EC's DG EAC had touched upon radio as a budget line around 1994–1995. Around this time, the European radio industry strengthened its political representation at the EU level. In 2001, the EC informed the DAB supporters that it will not offer the same support for DAB as it had given to the other projects examined in this study.

2.6 Variation in EU High Technology Policy

Observed variation in EU high technology policy provision has been a central methodological criterion for the selection of the cases. Both Galileo and UMTS received substantial support, reflected in policy and strategy documents and endorsements by the EU Member States. At the same time, the EC did not issue any collective policy documents on digital radio (Table 2.1). A qualitative summary of the empirical results of the EU support and high technology policy will be presented in the concluding chapter.

Table 2.1 Variance in EU high technology policy

	Galileo	UMTS	DAB
EU political support	High	High	Low
EU institutional support	High	Medium	Low
EU financial support	High	High	Low

We need to find an understandable and fruitful way of measuring the theoretical construct and dependent variable (EU high technology policy), which is done here by transforming it into three categories: 'EU political support', 'EU institutional coordination', and 'EU financial support'. (The empirical results will be summarized in the concluding chapter.) The first category of the dependent variable, 'political support', indicates whether the industrial project in question has enjoyed formal political endorsement through official communications with collegiate support from the EC services dedicated to the project, with subsequent adoption through a formal document at the EU Council of Ministers level (at least through a Council Resolution). It is dichotomous (low = no endorsement, high = endorsement). If the project has been included in the conclusions of EU Council Summits of the EU heads of state and governments, it will also score 'high' in terms of political support; otherwise, it will score 'low'.

The other two categories, 'EU institutional coordination' and 'EU financial support', are rated along a non-numerical, qualitative ordinal scale (low, medium, high). 'EU institutional coordination' refers to the extent to which the EU institutions, and the EC in particular, was granted and accepted a centralized administrative and management role (and thus had institutional influence over) in the development and planning of the technology infrastructure. This category includes formal structures of interaction between the EC and the major actors associated with each infrastructure project, 'formal' referring to institutional arrangements that last for more than a week, that consist of regular meetings and in which the EC participates at least as an observer (such as the creation of a 'Forum' or 'Task Force').

The empirical examination also takes account of more informal interaction to the extent possible, such as meetings between industry and the EC, which will be used as 'supporting' information. In order to receive the score 'high', the technology infrastructure project needs to 'enjoy' a strong formal role of the EC in the day-to-day operations of the project, a strong role of national government representatives through the EU institutional framework in overall strategic planning, and at least the use of one formal working group with mandate from the EU that includes the leading private sector actors. If the EC does not have a formal operational mandate to administrate the industrial project, national governments are not formally involved in overall strategic planning, but the EC still plays a role by convening a formal working group with the major private sector actors, it will score 'medium'. Finally, when none of the conditions above apply, the industrial project will be allocated a 'low' score in terms of institutional coordination.

The third category, 'EU financial support', refers to the relative financial support the project enjoyed out of the EU budget through the EU Framework Programs (FPs). The central indicator of such support will be EU investments in support of research and technology development (RTD). The central selection criteria have been that (1) the technology infrastructure project was mentioned in the title of the project or in the statement of the 'main objective' of the EU project, and (2) that the RTD projects were initiated as part of the 4th FP (1994–1998) or the 5th EU FP (1998–2002). The actual duration of the project may stretch outside the specific time frame of this study since the EU FP have slight different time frames and projects often stretch beyond such overall program schedules. In order to score 'high' in terms of EU financial support, the technology infrastructure receives Euro 50 million or more in EU RTD funding (project costs are not directly relevant for this study but will be included as a reference), to score 'medium' it receives between Euro 25 and 49.99 million, and lower amounts of RTD funding support will result in a 'low' score. Naturally, these levels are entirely arbitrary; nevertheless, they indicate the relative 'weight' of each of the technology infrastructure projects covered in this study.

Much of this information has been collected from the Community Research and Development Information Service (CORDIS) that provides information on EU-supported RTD activities, together with other EU documentation. When the technology infrastructure was referred to in the 'technical approach' or other parts of the project description but not in the statement of the objective or title, it was not included in this study. In addition, when the statement of the objective of the project refers to three or more different technologies, the project has been excluded.[34] Regarding satellite navigation, this study will also refer to investments by the European Space Agency (ESA) in the empirical chapters since this intergovernmental body co-manages the EU radionavigation program Galileo, although this funding will not be included in the concluding evaluation. Otherwise, the focus is on RTD investments directly wholly funded or co-funded by the EU and the EC. Projects that are managed through programs not funded or co-funded by the EU and by EU Member States or third countries (for example, Eureka or G-7 pilot projects) are not included in EU financial support.

---

34 Furthermore, it was decided (on an arbitrary basis) to exclude funding aimed at integrating satellite communications infrastructure with UMTS, which in Europe is referred to broadly as 'S-UMTS', not least since these ambitions have been primarily supported by the space sector and less so by the wireless 'terrestrial' sector, with some exceptions. For example, there were at least five projects under the EU 4th FP and its ACTS subprogram that targeted S-UMTS (INSURED, NEWTEST, SINUS, SUMO and TOMAS), of which data on funding was identified for TOMAS and SINUS (total project cost amounted to Euro 11.37 million and EU funding to Euro 5.04 million). Under the EU 5 FP there were at least three projects that were aimed at S-UMTS (FUTURE, SATIN and VIRTUOUS), the total project cost of which amounted to Euro 16.49 million and EU funding to Euro 8.21 million.

2.6 TECHNOLOGY AND EUROPEAN COMPETITIVENESS

This section will provide some data on the relative position of the sectors in the structure of international competition that are examined in this study. It has not been possible to collect data on the relative competitiveness of the future planned infrastructures, aside from some expected benefits as defined by public or economic actors, or consultants and the trade press. A starting point for the European space sector is a comparison between the volume of space activities in Europe and the US in 1992, just prior to the launch of the first dedicated EC report on satellite navigation in 1994 (Table 2.2). It illustrates the significant difference in sales and budgets between Europe and the US. Table 2.3 provides estimates of European industry market share in different segments over the period 1991–2000.

Table 2.2 Comparison of space activities (US/Europe), 1992 (US$ billion)

	US	Europe
Sales of space equipment and services	$ 29 billion	$ 2.6 billion
Number of employees	350,000	35,000
Percentage of sales at export	2 %	15 %
Civil budget	$ 35 billion	$ 5 billion
Military budget	$ 21 billion	$ 1 billion
Total budget:	$ 56 billion	$ 6 billion

Source: ESA (1994a).

Table 2.3 World market estimate, 1991–2000 in Euro billion (BEuro)

	World market estimate (1991–2000)	European industry market share
Launchers	7–10 (BEuro)	60–70 %
Communication satellites	10–15 (BEuro)	15–20 %
Ground equipment	40–65 (BEuro)	10–15 %
Remote sensing data	4–6 (BEuro)	40–60 %

Source: Trade Press.

In Europe, France was clearly the leading player with over 13,000 employed in 1998 and a consolidated turnover of Euro 2 billion, followed by Germany (5963 employees; Euro 1 billion) and Italy with 5741 employed and Euro 931 million (Eurospace, 1999). In 1998, ESA was the major customer in Europe with a consolidated turnover of Euro 1.674 billion, followed by national civil programs (Euro 617 million) and military programs (Euro 512 million), and the EU (Euro 17 million). The total European public budget was Euro 2.82 billion in 1998 and a total commercial budget amounting to Euro 2.45 billion (Eurospace, 1999). The European aerospace industry in total employed around 350,000 people in 1998. In 1996, the aeronautics and space industries invested around 18 per cent of the turnover in R&D (Pompidou, 1998). The space sector is relatively more commercially oriented in Europe but does not come close to the sales volumes in the US (Table 2.4). In Table 2.5, the estimated government space budgets (1994 and 1996) are presented.

Table 2.4 Distribution of space industry sales, 1994 and 1996 (% of industry sales to government; US$ billion)

Country	US		Europe	
Year	1994	1996	1994	1996
Total sales	28.5	27.6	3.5	4.7
Government civil	46%	39%	57%	55%
Military	47%	37%	9%	11%
Commercial	7%	24%	34%	34%

Source: ESA (1997a).

Table 2.5 Estimated government space budgets, 1994 and 1996 (US $ billion)

Year	Civil 1994	Military 1994	Total 1994	Civil 1996	Military 1996	Total 1996
US	14.0	15.0	29.0	13.0	14.0	27.0
Europe	4.6	0.9	5.5	5.1	1.1	6.2
Japan	1.8	0.0	1.8	2.2	0.0	2.2
Russia	N/A	N/A	N/A	0.4	0.2	0.6
China	N/A	N/A	N/A	0.5	0.8	1.3

Includes agencies and research centers costs of the five leading space powers (billion USD). The European category includes the ESA and national programs.
Source: ESA (1997a).

Large military contracts and a captive market favor US industry. In 1996, the total US budget was almost six times as high as that of Europe. The European space and aeronautics sectors are proactively supported by a number of EP members: 'Today's challenge is to promote employment by implementing the R&D required to capture promising markets, through services rendered to citizens in terms of transport and communications, taking into account the need to ensure the greatest possible military independence for the EU'. The

US Global Positioning System (GPS) was the dominant satellite navigation system in the 1990s. The demand for GPS started to grow significantly in the 1990s (Table 2.6). In a report to the US Congress and Department of Defense (DoD), the total world GPS market in 1995 was projected at about US$2.3 billion, of which the majority was expected to constitute demand from outside the US (NAPA/NRC, 1995). A report commissioned by the EC by a study group led by Deutsche Aerospace estimated the world GPS market at around US$3.6 billion in 1996 (DA, 1994a).

> For several reasons, it is in the US interest to see GPS become widely accepted and employed around the world. First, the globalization of GPS markets provides an economic stimulus to firms in the growing US GPS industry, many of which already rely on exports for a significant share of their revenues. Second, technological preeminence is an important pillar of national power. The acceptance of GPS as the world standard for position, velocity, and timing applications enhances the position of the US and allows it to lead in one important part of the process of technological and economic globalization. Third, US national security is well served by the international acceptance of GPS [...] The international acceptance of GPS would also slow the development of alternative satellite radio-navigation systems, the adverse use of which could be more difficult for the US military to control or counter in wartime (NAPA/NRC, 1995, p.14–15).

Table 2.6 The World GPS market in 1995, 2000, and 2005 (US $ billion)

Year	1995		2000		2005	
Market	US	Non-US	US	Non-US	US	Non-US
Total	0.9	1.4	5.7	6.0	14.6	16.4
Aviation	0.132	0.137	1.074	0.873	1.982	2.212
Land transport	0.178	0.551	1.761	2.213	5.558	6.992
Maritime	0.107	0.234	0.603	0.708	1.515	1.842
Recreation	0.241	0.110	1.259	1.187	3.403	2.906
Scientific	0.117	0.190	0.514	0.594	1.152	1.303
Timing	0.116	0.159	0.375	0.472	0.998	1.183

Source: NAPA/NRC (1995), pp.14–29.

Considering that the world GPS market was projected at around $31 billion in 2005, and that the US GPS industry dominated the equipment market, it is evident that there was a strong interest in protecting the pre-eminent position of US industry. There were concerns about future international competition.

'For geopolitical or economic reasons, a number of foreign governments have adopted policies detrimental to the spread of commercial GPS. For example, the European Commission wants to implement a common European navigational transport network that will rely on radio-navigation systems [...] European governments worry about the political leverage and commercial advantages that may accrue to the US as the owner, operator, and manager of a system upon which so many other nations increasingly depend. Further, they are concerned that DoD's dominance of GPS policymaking and management may result in a lack of attention to civilian needs' (NAPA/NRC, 1995, pp.39 and 42).[35]

Any forecast related to the European Galileo system is highly uncertain. The projected cumulative total benefits over the period 2001–2020 (Euro 74 billion), according to forecasts in early 2001, can be divided as follows: producer benefits (Euro 12.2 billion), social benefits (Euro 5.6 billion) and user benefits (Euro 56.2 billion) (Astrium, 2001). The early 2001 forecast of the estimated user base of Galileo navigation services in 2015 has been estimated as follows: mass market services (404 million); professional services (5.5 million); safety-of-life services (~70 thousand); and specialist-scientific services (~70 thousand) (Astrium, 2001).

The advantages of the combined use of Galileo and GPS (and the reduction of the commercial monopoly of the GPS infrastructure at the global level), DG TREN argues, might spur a movement toward combined GPS and Galileo receivers that would overcome 'the small cost disadvantage expected in producing combined receivers' (EC, 2002b,15). This reasoning is based on the idea that GPS and Galileo are complementary systems; together they will increase the total global GNSS market. According to projected market assessments published by the EC, the global market for GNSS is expected to grow to around 800 million users by 2020. The EC forecasts that there will be around 100 million civilian GNSS receivers in use in Europe by 2010 compared to around 6 million receivers in 2000 (EC, 2002b,14). This assessment includes the expectation that there will be rapid growth in location-based information services based on wireless platforms (GSM, UMTS, and so on) and that GNSS will emerge as the selected positioning technology, for example for personal navigation and in-car navigation services (expected to constitute around 70–80 per cent of all GNSS receivers in Europe by 2020). Regulatory policy for positioning technology and emergency requirements might also play a central role in this regard. The

---

35 The report concluded that Europe would get its act together despite the concerns that the report identified. 'European apprehensions regarding US dominance and control of GPS are not confined to any particular nation, nor do they represent a united position within any country. In fact, no single nation, let alone the European Union, appears able to unite in opposition to GPS' (NAPA/NRC, 1995, p.45).

global revenues from product sales associated with GNSS are expected to grow significantly from around Euro 14 billion in 2000 to around Euro 50 billion by 2010 and to around Euro 155 billion by 2020. The EC thus expects that Galileo and its complementarity with GPS will offer significant advantages and facilitate further and faster growth of the overall GNSS world market (Figure 2.3).

Figure 2.3 Market growth opportunities for Galileo

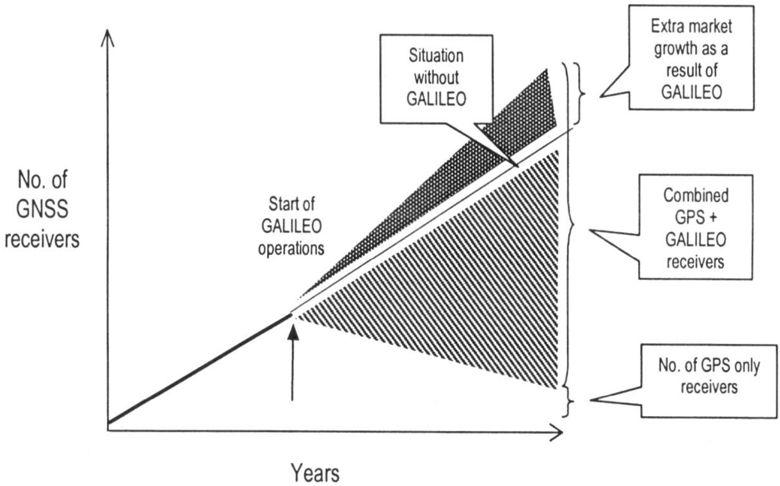

Source: EC (2002b).

When it comes to the telecommunications sector, the European mobile communications industry and market has a relatively strong international position. In 1999, Western Europe passed North America in terms of penetration (38.4 per cent and 30.7 per cent, respectively) and in 1998 when it comes to subscribers (74.6 and 89.1 million subscribers, respectively). Southeast Asia was the fastest growing market, reaching more than 100 million subscribers in 1998. In Table 2.7 background data is provided on penetration by region and subscribers per region. The data on penetration and subscribers from 1997 to 2001 should offer relatively robust assessments, whereas for the subsequent years (2002–2004) this table consists of crude projected levels of penetration and subscribers for each region.

Table 2.7 Penetration and subscribers by region

Penetration (%)	1997	1999	2001	2002	2003	2004
North America	20.1%	30.7%	46.9%	55.5%	63.9%	72.1%
Western Europe	14.1%	38.4%	69.5%	76.8%	81.0%	83.1%
South East Asia	2.4%	5.4%	9.1%	10.8%	12.4%	14.0%
Rest of World	1.1%	2.8%	6.6%	9.1%	12.0%	15.2%
Total	3.5%	7.7%	13.7%	16.4%	19.0%	21.5%

Subscribers (Millions)	1997	1999	2001	2002	2003	2004
North America	59.9	93.4	145.8	174.3	202.7	228.6
Western Europe	54.6	149.2	270.9	300.0	317.1	326.1
South East Asia	66.4	152.9	265.1	318.8	370.6	425.4
Rest of World	27.0	71.3	171.1	239.5	323.3	410.6
Total	207.9	466.8	852.9	1032.6	1213.7	1390.8

Source: Wireless Strategy Analytics (2001).

In 2001, China was the world's second largest mobile phone market by subscribers and it has been estimated that 26 per cent of the Chinese population, or about 340 million people, will use a mobile phone by 2005 (DisplaySearch, 2001). In 2001, leading wireless manufacturers (notably Ericsson, Motorola and Nokia) announced that they would significantly increase their investments in the wireless sector in China and the Asia-Pacific region and production though their subsidiaries and joint alliances. Since industry in Europe focused on GSM, the world market share of this technology relative to its competitors is a relevant measure (Table 2.8).

Table 2.8 Subscribers by technology

Subscribers (Millions)	1997	1999	2001	2002	2003	2004
Other	121.9	137.2	105.6	86.7	73.3	61.0
CDMA	7.7	51.1	131.9	184.5	248.8	318.5
TDMA	9.3	36.6	104.7	143.2	175.4	210.5
GSM	69.0	241.9	510.7	618.2	716.3	800.8
Total	207.9	466.8	852.9	1032.6	1213.7	1390.8

Subscribers (%)						
Other	58.6%	29.4%	12.4%	8.4%	6.0%	4.4%
CDMA	3.7%	10.9%	15.5%	17.9%	20.5%	22.9%
TDMA	4.5%	7.8%	12.3%	13.9%	14.4%	15.1%
GSM	33.2%	51.8%	59.9%	59.9%	59.0%	57.6%

Source: Wireless Strategy Analytics (2001).

Another indicator of the strong position of many European countries are the number of wireless subscribers. In 1999, Finland, Norway and Sweden had between 57 and 65 per cent penetration per household, followed by Italy (52.5 per cent), Japan (44.9 per cent), the UK (40.3 per cent), France (34.9 percent), the US (31.5 per cent), Germany (28.6 per cent), and Canada (22.7 per cent) (CEA, 2000).

The mobile equipment market can be broken down into two product segments: mobile terminals (handsets) and infrastructure (base stations and switches). Sales were roughly equal between the two in 1998. Equipment suppliers in 1998 earned some US$36 billion from infrastructure contracts alone. In addition, they gained large revenues from mobile handset sales (Tables 2.9 and 2.10). In 1998, an estimated 163 million handsets were sold across the world at an estimated value of US$40 billion. The telecom equipment market was estimated at around US$191 billion, of which mobiles accounted for around 25 per cent (ITU, 1999a). The world market grew by 64 per cent in 1999 and 44 per cent in 2000, measured in shipments of handsets. The world market for mobile handsets entered a period of lower growth rates around 2000 and it appeared as if the mobile phone industry would have problems maintaining growth rates. The estimated growth rates and units in 2001 in the world regions were as follows: Asia 30 per cent, 170 million units; Western Europe 19 per cent, 167 million units; the US 18 per cent, 90 million units; Latin America 17 per cent, 42 million; and rest of world 23 per cent, 38 million units (BW, 2001).

Table 2.9 Mobile equipment and handset sales by vendor, 1998

Equipment manufacturer	Handset sales	Wireless infrastructure
Alcatel	4%	N/A
Ericsson	15%	30%
Lucent	N/A	12%
Motorola	20%	16%
Nokia	23%	11%
Nortel	N/A	10%
Panasonic	8%	N/A
Others	30%	21%

Source: ITU (1999a).

Table 2.10 Wireless infrastructure suppliers, 1998

	Country	Operators	Technology
Alcatel	46	71	GSM, AMPS, TDMA, CDMA
Ericsson	100	350	GSM, TDMA (later also CDMA)
Lucent	34	N/A	AMPS/GSM, CDMA, TDMA
Motorola	80	N/A	N-AMPS/AMPS/GSM, CDMA
Nokia	37	78	GSM
Nortel	49	142	AMPS, TDMA, CDMA, GSM
Siemens	65	120	GSM

Source: ITU (1999a).

Major wireless manufacturers gradually lowered forecasts for the world handset sector, from around 500 million units in early 2001 to around 400 million six months later. Estimates on revenue streams from cell-phone advertising, commerce, and subscription fees have claimed that it would total around US$7.5 billion in 2003, of which 66 per cent (US$5 billion) would be in Asia, 22 per cent (US$1.7 billion) in Europe, and $700 million in the US (TWB, 2001a, 2001b). Ericsson, Motorola and Nokia are the leading firms (Tables 2.11 and 2.12). Table 2.13 presents data on the evolution of mobile commerce, a major application area according to European industrialists.

Table 2.11 Top mobile equipment manufacturers, 1998

	Mobile equipment revenue				
	billion $ (1998)	Change (1997–98)	% of total	% of telecom	% foreign
1. Motorola (US)	17.9	0%	61%	61%	59%
2. Nokia (Finland)	14.7	59%	94%	94%	94%
3. Ericsson (Sweden)	14.5	5%	64%	64%	95%
4. Lucent (US)	4.3	-6%	14%	16%	26%
5. Nortel (Canada)	3.7	8%	21%	22%	36%
6. NEC (Japan)	3.7	-3%	9%	29%	5%
7. Qualcomm (US)	3.3	60%	100%	100%	34%
8. Matsushita (Japan)	3.1	16%	5%	17%	51%
9. Siemens (Germany)	3.0	10%	4%	18%	69%
10. Alcatel (France)	2.7	30%	11%	13%	83%

Source: ITU (1999a).

Table 2.12 Global mobile handset market share, 2000

Manufacturer	Market share 2000	Manufacturer	Market share 2000
1. Nokia	31.3%	10. NEC	2.4%
2. Motorola	16.0%	11. Mitsubishi	2.1%
3. Ericsson	10.7%	12. Toshiba	1.7%
Total	58.0%	13. Qualcomm/ Kyocera	1.6%
4. Samsung	6.5%	14. LG	1.4%
5. Siemens	5.7%	15. Sony	0.9%
7. Panasonic	4.7%	16. Hyundai	0.8%
8. Philips	3.2%	17. Fujitsu	0.5%
9. Sagem	3.0%	18. Denso	0.5%

Source: FT (2001c).

Table 2.13 Projected market evolution of mobile commerce (US$ billion)

Year	1999	2000	2001	2002	2003	2004	2005	2006
Western Europe	0	1	4	23	41	61	86	99.8
North America	0	0	1	6	13	23	37	48.8
Asia Pacific	0	1	2	6	14	25	41	58.2

Source: Wireless Strategic Analytics (2001).

Finally, when it comes to the European radio sector, data on digital radio is almost non-existent. The major trade association WorldDAB does not provide any data base and reports it has commissioned seldom include information on market shares, revenues, or export. In 1994, the International Telecommunication Union (ITU) formally recognized the European Eureka 147 or Digital Audio Broadcasting (DAB) system as a world standard for sound terrestrial and satellite audio broadcasting. In 1996 the European market for audio products was worth Euro 14.8 billion. WorldDAB has estimated that the radio sector had an annual turnover in 1998 of around Euro 7 billion, split more-or-less equally between license fees and advertising revenues. The audio retail market with turnover of a further Euro 6 billion complemented this market (WDABF, 1999b). The European retail market value for digital radio is expected to reach over Euro 4.7 billion around 2005.

Estimates related to the radio market are often calculated based on a successful take-up of digital radio, which has been translated to at least 60 per cent of the population having digital reception. The prospective increase in market value of the consumer electronics market has been estimated for five years after launch of services (SD, 2000). Under this condition, it has been estimated that sales of home and office digital radio units would reach around 27 million units, digital radio PC cards 6.4 million units, and in-car units around 7.3 million units after five years. Regarding total employment in the European radio industry, it has been estimated at approximately 295,000 jobs, of which public and private broadcasters directly employed 55,000 and around 25,000 represented freelancing people (WDABF, 1998a).[36] In spring 2000, there were an estimated 10,000 radio stations across Europe. The European market consists largely of small- to medium-sized broadcasters with most of their revenues coming from the home market. The UK is spearheading the overall European market.

---

[36] In 1997, the EC released a report on the competitiveness of the European information and communication technologies industries, which estimated that EU employment in the consumer electronics sector had decreased from 230,000 jobs in 1980 to 110,000 in 1994 (EC, 1997d). The report stated that the European equipment manufacturing (consumer electronics) base is relatively weak, in particular in comparison with Japanese and more recently Korean producers, and that the negative trade balance in this sector had significantly worsened since 1985 and amounted to around Euro 10,000 million in 1995 (EC, 1997d).

3. EU strategy for satellite navigation

In the early 1990s, Europe had virtually no presence in satellite navigation but progress has been made since. 'In ten years Europe has made tremendous strides in satellite navigation from being completely dependent, to where we are today on the verge of launching a major project for the development of a second world constellation' (Eurocontrol, interview, 21 December 2000).

3.1 THE EVOLUTION OF SATELLITE NAVIGATION

Since prehistoric times, people have been trying to figure out a reliable way to tell where they are, to help guide them where they are going, and to get them back home again. Early techniques that were used include observing the landscape or stars on clear nights, the magnetic compass, the sextant (to measure latitude), and the chronometer (to measure longitude). A number of ground-based and inertial navigation systems were developed during and after World War II and were still being used at the end of the twentieth century.[37] Navigation satellites broadcast signals that are used by a receiver to determine precisely its position, velocity and time. In the 1960s, a significant change occurred in the field of satellite navigation with the deployment of the US Navy Navigation Satellite System, known as Transit, which was conceived in the late 1950s, entered operation in 1964, and was released for civil use in the late 1960s. The launch of Sputnik 1 on 4 October 1957, and the Cold War context, was the external trigger factor for the Transit development.[38]

---

37 In the category of ground-based systems are so-called Long Range/Medium Range hyperbolic systems (Decca, Omega, Loran-C and Chayka) and radar systems for Low/Medium Range air navigation (Automatic Direction Finder, ADF; VHF Omnidirectional Range, VOR; and Distance Measuring Equipment, DME). Among the inertial navigation systems, INS, are air and sea navigation systems.

38 The specific need that pushed the Transit development was the need for an accurate navigation system for the class of submarines that would carry Polaris missiles, which would be capable of remaining underwater for extended periods of time. The problem was that the available technology, inertial navigation systems, drifted, which made it difficult to measure the accurate position and therefore constituted a serious problem for precise targeting and launching. In early 1994,

The Transit system, developed by the Johns Hopkins Applied Physics Laboratory (APL), was the first operational satellite navigation system in the world. Two additional concepts for satellite navigation systems were examined in the mid-1960s: Timation and Program 621B (Parkinson et al., 1995). The US Naval Research Laboratory (NRL) drove the former concept, whereas the Air Force Space and Missile Organization was in charge of Program 621B.[39] The first Russian (Soviet) navigation system by satellite was the Cicada system, which was similar to the US Transit system. The development of Glonass started in the 1970s, based on the Cicada system; it was expected to be in full operation around 1995 but has been marred by economic difficulties.[40] It was intended to provide position, velocity, and precise time for naval, air, and other user groups.

---

the first successful determination of position aboard a Polaris submarine through the Transit system was computed. Transit became an important application for submarines and surface ships for many navies around the world; the Navy Astronautics Group, headquartered in California, operated it (Parkinson et al., 1995). In the late 1960s, the use of Transit extended and was first adopted by oceanographers, then by offshore oil exploration companies, and when the price of user equipment fell it was used for merchant marine, fishing fleets (deep-ocean navigation), land-survey and boat applications. The US Government planned for the replacement of Transit from the 1980s and it finally retired in late 1996.

39 The US Timation system was aimed at improving the determination of precise time and time transfer between different locations, although it could also provide navigation information. In the late 1960s, the Joint Chiefs of Staff (JCS) announced new requirements for precisely locating military forces worldwide, which subsequently led to the formation of a Department of Defense (DoD) Navigation Satellite Executive Steering Committee Group (NAVSEG). This group would evaluate different satellite navigation projects through comparative studies, involving both technical and political discussions. The Transit, Timation and Air Force 621B projects constituted the foundations for GPS. The GPS contractor team first included Collins Radio of Rockwell International, General Dynamics, Magnavox, Rockwell International, Stanford Telecommunications, and Texas Instruments (Parkinson et al., 1995).

40 In 2000, the Russian Glonass system was in decay and on the verge of collapse. There was a lack of resources to maintain the 24-satellite constellation and 10 satellites at most, and probably fewer, were operating. However, Russian authorities appeared to plan an ambitious modernization effort to upgrade the Glonass constellation, owned and operated by the Russian Defense Ministry, and new satellites were to be launched to have around 12–15 satellites operating through 2002. The Glonass modernization program was approved by a Russian presidential decree in February 1999. According to Koslov, head of satellite communications and navigation at the Russian Aviation and Space Agency, Rusaviakosmos, a second-generation Glonass system would then be put into orbit. 'By 2003, under this plan, we should be able to restore the Glonass constellation all the while we are modernizing it and using cheaper launch vehicles' (SN, 2000h). Koslov stressed the need to set up a private-public structure (also debated in Europe) in order to keep the struggling Glonass constellation from collapsing.

In 1973, the US Authorities approved the US Global Positioning System (GPS), also called Navigation Satellite Timing and Ranging Positioning System (Navstar/GPS). The US Air Force was assigned the executive responsibility for GPS and a Joint Program Office (JPO) was established in 1973. The JPO, headed by the Air Force, would enjoy strong participation from other services, not least the US Navy community, but also the Army, Marine Corps, Defense Mapping Agency, Coast Guard, Air Logistics Command and NATO (Parkinson et al., 1995, p.129). The JPO thus extended the previous monopoly of the Air Force and allowed the participation of numerous agencies in the GPS conception process. Its initial development was a more public and collective enterprise than the previous systems.[41] The first GPS satellite was launched in 1978, which was a developmental satellite called GPS Block I.[42] As a result of the crash of Flight 007 of Korean Airlines in 1983 due to lack of accurate navigational technology, former US President Reagan allowed the use of the so-called SPS signal for use in aircraft and other transportation modes (road, rail, and maritime).

By August 1993, GPS had 24 satellites in orbit, initial operational capability was established in December the same year, and the Federal Aviation Agency (FAA) declared GPS ready for aviation use in February 1994. In 2001, the GPS system operated from the Shriever Air Force Base in Colorado. It would take considerable time, however, for the added value of GPS to be more recognized and then first in the military domain, for example, for minesweeping, aircraft landing, and infantry operations. 'Its impact was not fully understood by the operational forces of the military until the Desert Storm battles showed its value as a force multiplier. The tactical

---

41 The fate of GPS was far from determined. In the 1970s, the supporters of the Navy's Transit system asked for funds to upgrade the system with the aim of providing accurate tracking technology for the Trident (submarine-launched missile) booster during test firings into broad oceans. New funding for the Transit system could result in a reduced opportunity for the GPS concept. However, promoters of the latter concept appeared to have been able to convince the parties involved that the GPS system would be able to solve the Trident problem if a few more satellites were authorized.

42 The GPS constellation emitted two types of signals. The military signal could determine location within 10 to 20 meters (a meter equals 3.3 feet). A second and unencrypted signal was intentionally distorted to degrade the accuracy of the signal to within 100 meters to, among other things, protect the use of the signal from being used by enemies to improve the accuracy of weapons targeted at the USA. The GPS infrastructure consists of three main elements: the space segment, the user segment, and the control segment. The space segment consists of the operational satellites hovering 11,000 nautical miles above the Earth. Ground stations (the control segment) located across the world continuously monitored the GPS satellites to make sure the satellites were working properly; the satellites transmitted signals that could be detected by GPS receivers (the user segment).

commanders were finally able to experience the power that comes from precise knowledge of position in a common coordination frame' (Parkinson et al., 1995, pp.152–153). It had taken more than 20 years to establish the GPS system from the previous systems. A large number of everyday life applications have emerged and continue to develop (construction, mineral and oil exploration, vehicle tracking, emergency response, mapping and surveying, precision agriculture and mining, electronic data transfer, exploration and hiking, and so on).

The World Civil Aviation Community Sets the Tone

In the early 1980s an interest gradually emerged in the international civil aviation community in using satellite-based navigation aids for critical applications, for example to enable landing in bad weather and at night. In the light of the increasing limitations of the current air navigation systems, efforts intensified in the early 1990s to elaborate a strategy of transition from conventional ground-based navigation systems to Global Navigation Satellite Systems (GNSS) in an incremental manner. The International Civil Aviation Organization (ICAO) spearheaded these developments. The international aviation community was searching for ways to increase airspace capacity, provide substantial economies in flight operations and further enhance the regularity of air transport.[43] The Future Air Navigation Systems (FANS) Committee of ICAO, which had been set up in the early 1980s, proposed the Communication, Navigation and Surveillance/Air Traffic Management (CNS/ATM) concept (ICAO, 1998a, 1998b).[44] In 1991, the ICAO adopted

---

43 In 1997, total world airline schedule passenger traffic in terms of passenger kilometers was expected to grow at an average annual rate of 5.5 per cent over the period to 2005. The total freight traffic over the same period was estimated to be even stronger and amount to 7.0 per cent annually in terms of freight ton kilometers. Finally, the annual total number of domestic and international aircraft departures on scheduled services was estimated to grow more than a quarter and the number of aircraft-kilometers flown by more than a half. The international routes that would experience fastest growth in passenger traffic would be the Transpacific and Euro-Asia/Pacific ones (ICAO, 1997).

44 The ICAO was created in 1944 to promote the safe and orderly development of civil aviation worldwide and is a specialized agency of the United Nations (UN). Headquartered in Montreal, the ICAO develops international air transport standards and regulations and serves as the medium for cooperation in all fields of civil aviation among its 185 Contracting States (as of the year 2000). This global undertaking required an unprecedented level of collaboration between aircraft operators, service providers, manufacturers, governments as well as user and other interested organizations, all of which are working together with the goal of building the safest possible global aviation system. The objective of the world aviation community has been to create an integrated global system of air traffic management linking together CNS/ATM systems.

and recommended the implementation of GNSS-1 around 2005 (completed with a geostationary overlay system) and GNSS-2, relying on a civil satellite constellation, around 2010. As a first step, the ICAO recommended the world aviation community to rely on the US GPS and Russian Glonass systems.[45]

In the early stages of the international debate on satellite navigation, there were ideas about creating a truly global seamless system. A rationale behind the push for an internationally controlled satellite-based navigation system was that it would be a disadvantage to rely solely on one system like GPS, which was controlled by one country, particularly its military branches. This was particularly important in terms of safety-of-life applications. The same concerns were voiced with respect to the Glonass constellation. 'Some people were not particularly pleased with the idea of relying on any of these two systems for large-scale civil implementation. Attempts were made to elaborate a joint system with some global organization, and people talked about doing something together under the aegis of the United Nations (UN). The US reflected on this but concluded that it was not in their interest. They did not want to give up control over their own system' (EU national government advisor, interview, 25 January 2000).[46] The US was clearly not interested in running more than one system, which helped to stall the implementation of a global system under civilian control.

---

45 The ICAO had set up a range of guidelines for institutional issues and for satellite navigation in general. The so-called ICAO FANS-11-4 report included the institutional issues and the ICAO FANS-4-8 report the general issues with regard to satellite navigation aids. The ICAO FANS-11-4 report stated that service had to be available to all parties without discrimination, that no modification was acceptable of the rights and responsibilities of States concerning the areas under their jurisdiction, or no modification of the responsibility of the ICAO in the establishment of standards and procedures. The report also recommended that the creation of new international organizations should be avoided, the modification of existing legislation kept to a minimum, and the legal responsibilities of all concerned must be clear. Moreover, if elements of the CNS/ATM system failed, there would be a need for joint technical reconfiguration of the system, while sufficient service flexibility adapted to local needs must be guaranteed at all times. The ICAO FANS-4-8 report stated that States must be ensured permanent availability of the service and protection of navigation signals against interference.

46 Official resistance, at least verbally, to the gradual move to a satellite-based navigation system on the part of the world's commercial airline community would continue in the times ahead from certain quarters. Miller, president of the British Airline Pilots Association, a powerful lobbying group, claimed that many pilots doubted that the US GPS system and its augmentations would ever be reliable enough to guide planes safely to landing. International commercial aviation authorities failed to see the value of conventional technologies. 'Our political masters tend to be lawyers and technologists. There's hardly an engineer among them' (SN, 2000g). Miller stressed that the current Instrument Landing System, based on terrestrial navigation sensors, must be maintained in operation.

In December 1995, the ICAO Council established a GNSS Technical and Legal Experts Panel. In 1998, it presented a set of draft recommendations on a future GNSS legal framework and capabilities (ICAO, 1998a). GNSS services should be considered as a public, international utility and be civilian-controlled, using existing organizational structures and with a coordinating role for the ICAO. Cooperation and mutual assistance between ICAO Member States (or regional operators on their behalf) should be strengthened. Moreover, there was no need for a worldwide and centralized operating structure. Instead, regional operating structures should be developed and coordination could be achieved through agreements among regional-based organizations. A liability regime was needed for defining whether the doctrine of sovereign immunity should apply to GNSS service provision. The ICAO was not calling for a new global governance regime but rather an inter-regional political, administrative and operational structure.[47]

The ICAO was a major stakeholder in satellite-based augmentation systems. One area where it has a particularly strong influence worldwide is in the field of frequency allocation. There are specific frequency bands allocated for the civil aviation community. 'It is the only user organization with great political power. In a way, you may say that the aviation community decides for itself on these frequency bands, although they do not have voting rights at the World Radio Conference (WRC), and this has been a hotly contested issue' (EU national government advisor, interview, 25 January 2000).[48]

---

47 In addition, it was seen as important to prevent abuse of monopoly powers on the part of GNSS signal providers, while the services should be financed by user charges, including yearly subscription rates and license fees. The issues of financing mechanisms and institutional frameworks were viewed as central for the worldwide implementation of CNS/ATM systems and for moving the projects forward. In May 1998, the ICAO agreed on a Declaration that stressed the need for international cooperation, procedures and interoperability of CNS/ATM systems (ICAO, 1998b). European authorities had substantially contributed to the orientation toward the implementation of a global civilian-controlled system (EC/ESA/Eurocontrol, 1998a). At the ICAO Assembly in late 1998, the EC, the ESA, and Eurocontrol together submitted a common position paper that supported the principles outlined above (EC/ESA/Eurocontrol, 1998b).

48 The frequency issue has been very important, not least for the second-generation GNSS (GNSS-2), since one of the bands claimed by the Europe and the US was a so-called 'air navigation band' over which the aviation community basically had veto power. The aviation community is the user group that has the most stringent demands and is very well organized. In comparison, road transport (which was estimated to represent around 70 per cent of all use of satellite navigation), does not have a common spokesperson or common organization. In addition, although the maritime sector also has a worldwide organization it does not appear to have expressed particularly stringent demands. Its demands exist within the larger picture but appear to have evolved from within the maritime sector, rather than taking into account other modes of transport and user groups.

Though the interests of the aviation community and public policymakers have not necessarily collided, there could be potential problems associated with such strong power held by one, albeit important, user group. 'This is a political problem in Europe since the EC stresses that satellite navigation should be a multi-modal system as they call it, that is, it should be applicable to many different modes of transport. The EC is not pleased that the aviation community continues to have such a dominant position in the discussions' (EU national government advisor, interview, 25 January 2000).

3.2 A Convergence of Activities

In the early 1990s, satellite navigation in Europe emerged mainly due to the activities of three organizations: the EC, Eurocontrol, and the European Space Agency (ESA). Satellite navigation activities evolved as a result of 'a convergence of activities' within these three European organizations.

The Eurocontrol organization was the first to establish a structured organization.[49] In June 1994, an important meeting took place in Copenhagen of what would later be called the MATSE, a forum consisting of Transport Ministers from the European Civil Aviation Conference (ECAC). The ECAC

---

49 Within Eurocontrol there was a management structure in place in the early 1990s (Committee of Management, CM), which was subsequently replaced by a different management structure. It consisted of high-level representatives of the Member States. In light of the work that had been going on in the ICAO Future Air Navigation Systems (FANS), they decided that Eurocontrol should start to look at satellite technology for communications, navigation and surveillance. This decision was related to the development of the CNS/ATM concept. In November 1991, Eurocontrol thus set up the Satellite Communication, Navigation and Surveillance (SATCNS) Working Group to look into aspects of future aviation systems (Eurocontrol, interview, 21 December 2000). The SATCNS established two expert subgroups, one on Communications and Surveillance and the other on Navigation. The Expert Group on Navigation (EGN), initiated around April/May 1992, represented the real start of satellite navigation activities in Europe in institutional, political, and strategic terms. The EGN had its first full meeting in Braunschweig, Germany, in July 1992. The head US WAAS officer was represented at the meeting and observed closely what was going on in Europe (Eurocontrol, interview, 21 December 2000). It was stated that the ultimate goal for civil aviation had to be a sole means of navigation for all phases of flight and for civil aviation, which would be global in nature. Finally, future navigation systems had to enable early benefits to users of existing systems, notably GPS. The EGN worked through early 1994 and its main work within Eurocontrol was to produce a so-called Satellite Navigation Strategy (SNS), which led to further cooperation with the FAA (which had started work activities that eventually would become the WAAS project), the ICAO and other world regions. The Eurocontrol Council approved the navigation strategy in March 1994.

Transport Ministers, who met every two or three years to set up policy objectives for the coming years, gave two clear Directives to Eurocontrol, the EC, the ESA and to the ECAC Member States (Eurocontrol, interview, 21 December 2000). Transport Ministers stressed that Europe should develop a component of the first-generation Global Navigation Satellite System (GNSS-1), and more importantly should put Europe in a position to make a contribution to GNSS-2. The GNSS-1 would be based on the US GPS and different augmentation systems, while the GNSS-2 would constitute a future generation of civil-controlled satellite navigation. The ECAC Directives gave a political mandate for formal cooperation between the three aforementioned European-based organizations. Furthermore, the US FAA decision to go ahead with what became the WAAS program in the early 1990s played a role in the stepping up of European activities.

The European Commission Intervenes in Satellite Navigation

In 1988, DG Research prepared and issued its first Communication on space, which stated that the market for position fixing services 'could make an important contribution to the development of more efficient, safer and integrated transport systems' (EC 1993b, p.19). The EC pointed out that although the US military satellite network GPS had been available for civil use, 'this arrangement is voluntary and can be halted at any time' (EC 1993b, p.19). In order to respond to US and Japanese competition, the EC announced that it would 'seek consultations with all parties concerned to examine the actions needed to ensure reliable and efficient position fixing services for European civil users' (EC 1993b, p.20). The EC argued, with regard to the integration of GPS into the civil air navigation system, that the accuracy for civil use would have to be improved (EC, 1988, 1991b, 1991c).

The discussion on satellite applications was presented in the context of the weakness of European electronics and components industries, which according to the EC represented a growing handicap for the European space industry (EC, 1991a, 1992). Though satellite navigation and positioning was largely 'neglected' in the early EC reports on space, DG Research/Joint Research Center (JRC) had already initiated work on navigation in the late 1980s and early 1990s, capitalizing on its close relations with the ESA.[50] In

[50] At EU level, the political emphasis within the domain of space had been on science, remote sensing and communications. The early initiatives on satellite navigation were restricted to maritime systems and the use of ground-based Loran-C networks (EC, 1991a, 1991b, 1993c). In 1993, the EU Council included satellite navigation in a decision. 'The Commission shall [...] pursue its work with a view to setting up a radio-navigation plan which takes into account the development of satellite navigation systems, of existing terrestrial systems and of the radio-navigation plans of the Member States' (Council, 1993a, 1993b).

1993, it commissioned Deutsche Aerospace AG (Germany) to conduct an 'Assessment of Satellite Navigation Systems and the European Dimension of their Utilization' (DA, 1994a, 1994b). In short, the study concluded that satellite navigation would evolve toward a utility with a much wider range of applications than classical ground-based radio-navigation aids. The EC started deliberations on the use of satellite technology, in particular in the context of Trans-European Networks for Transport as part of an integrated transport infrastructure for the EU. The main body responsible for these aspects within the EC was the DG Transport and Energy (TREN). Around the same time, DG Enterprise signaled that it wanted to be involved in the development of a European policy for space and space-related activities. The formal and informal interaction of the parties is illustrated in Figure 3.1.

Figure 3.1 Political interaction structure at EU level

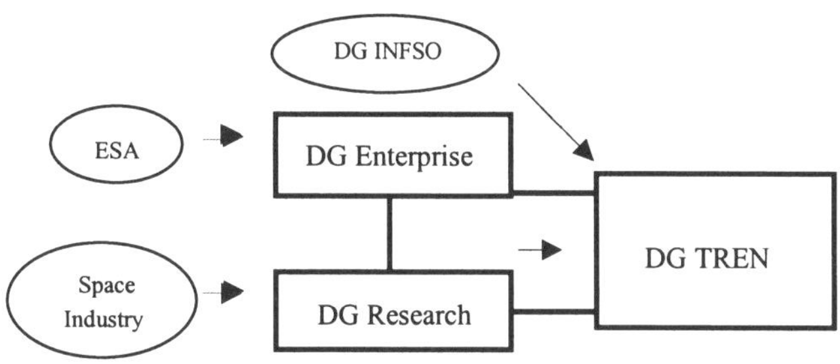

In February 1994, DG Enterprise presented a first reflection document on a European Strategy for Space and industrial policy, which was presented to the Space Advisory Group (SAP), set up by the EC in 1993. The SAP included EU/ESA Member States tasked to discuss how to strengthen Europe's investment in space. Satellite navigation was presented as a major area where the EU could encourage a market-pull development of its space industries (EC, 1993a, 1993b, 1994f, 1994g, 1994h). The GNSS dossier first landed in DG TREN Directorate for Air Transport, but was subsequently transferred to the Directorate for Trans-European Networks for Transport.[51] Though it is difficult to ascertain to what extent the transfer of the satellite

51 Through the existing Eurocontrol mechanisms (for example, the Central Route Charges Office that collects route charges from airspace users) there was a structured mechanisms for collecting fees, which was not the case in the US or in other modes of transport in Europe.

navigation dossier from the Air Transport Directorate was a result of the strong organization of air space users, it probably played a role. 'The airspace users are exceptionally vigilant to ensure that they do not pay more than their fair share of the cost' (Eurocontrol, interview, 21 December 2000). One of the big work areas that the EC has been involved in over the years is to interest other modes of transport in satellite navigation to enhance momentum and counteract the political power of airspace users (EC, interview, 26 May 1999).

The EC argued that the industrial stakes in satellite navigation would be considerable. The cost of developing and operating a satellite navigation infrastructure would amount to several billion Euros and the user segment (receivers and related products) would amount to something like Euro 5 billion. Though Japanese and US satellite manufacturers had an important lead in the high volume segments of this market, the EC believed that European industry could still gain a foothold in the high-end, system-oriented, market segments. The EC pointed to the fact that the US DoD controlled the only fully operational satellite navigation system (GPS) and recognized that the US has actively promoted it (EC, 1993c).[52] Together, these developments and interventions by government agencies around the world formed the basis of the EC's argument that it was time for Europe to step up its efforts in the satellite navigation domain.

Though satellite navigation in Europe was associated with the space sector, it was officially managed at EU level as part of European transport policy. The Single European Market (SEM) initiative included references to a common Transport Policy and provided jurisdiction for the EU to get involved in several areas (standards, patents, and so on). The EU Transport Council endorsed the initial policy efforts and set out guidelines for the stimulating Trans-European Transport Networks (Council, 1992, 1993a). Transport Commissioner Kinnock stressed that international transport required interconnected and integrated transport systems across Europe. A completely free interplay of market forces would not always provide a universal recipe. Rather, coordination at EU level was essential to construct, harmonize and rationalize Trans-European Networks and make them more 'environmentally sustainable' (Kinnock, 1995).[53]

52 Meanwhile, Russia was in the process of deploying its own satellite navigation system, Glonass, but its future remained uncertain. Moreover, Japan had announced that it would embark on a US$1 billion project to build a supplementary space segment (geostationary overlay) to GPS and Glonass, composed of two satellites and associated ground stations.

53 In December 1994, the Heads of Government at the Essen Summit endorsed a number of priority projects in terms of infrastructure, such as high-speed train lines and the Öresund Bridge between Denmark and Sweden. In addition, DG TREN pointed to the need to support inland waterways and rail freight.

The population in Europe is highly concentrated geographically, much like the northeastern and southeastern US, which gives rise to congestion and bottlenecks in transport systems. DG TREN viewed societal demands on development and integration as the fundamental drivers, which created the need to resolve bottlenecks in the transport systems across Europe (EC, 1999k). It was thus determined to foster an efficient transport chain, provide a combination of strategic and commercial services, and ensure that they were relevant to the needs of the competitive world market. This 'worldview' was the basis for the overarching policy concept of intermodality. 'An important aspect of satellite navigation in Europe is that satellite navigation was always seen as intermodal, that is, a means of navigation not only for air transport but also for other modes of transport and beyond' (Eurocontrol, interview, 21 December 2000). An overall objective was to facilitate an integrated infrastructure that was optimal (that allowed the use of the specific features of modes of transport in the most efficient and cost-effective way).[54]

New demands for effective and more precise service provision reinforced the need for advanced forms of communication, navigation and surveillance as exemplified by the concept of 'tracking and tracing'. Up to the late 1990s, all those functions had been managed with ground-based infrastructure and technology. However, this had already started to change with the introduction and expected implementation of satellite-based solutions, where the same type of services could be performed by wireless means.[55] Conventional systems had to be more efficient and accurate through new institutional and

---

54 Regarding the transport of goods, this was a type of surveillance function (to pinpoint the location of goods on the move). For those who operated the services it represented a management function (to coordinate the means of transport, for example, ships, aircraft, or trucks). Management systems could be divided into two categories, fleet management (performed by the different transport and service companies) and traffic management (performed by those who control and supervise the flow of transport and traffic). Finally, a societal-oriented function was the use of tracking and tracing by public services, such as the police, fire fighting, and search and rescue operations. DG TREN envisaged a trend toward integrated intermodal systems.

55 The most international sectors, the civil aviation and maritime, spearheaded this change, which had been subject to international rules and agreements for a long time compared to road and railway sectors. The maritime community could use satellite navigation particularly where it is difficult to navigate, such as the English Channel and the Stockholm Archipelago. The aviation community represented historically the most demanding users – due to the need for safety requirements – while other sectors, such as high-speed railways, slowly started to get organized. Inland waterways could also be an area for satellite navigation services. A particularly problematic sector was the aviation sector, where the intensity of flights (not least with the increasing traffic in point-to-point routes and small aircraft resulting from new demands) had now resulted in a crisis in Europe, and was experienced in southeastern and northeastern US as well.

technological solutions. DG TREN admitted that the push for satellite navigation was not well perceived by those who produced conventional ground-based navigation equipment: 'Those systems become all of a sudden obsolete, completely obsolete, with the new types of technologies, which implies that their whole business case has disappeared, which naturally is viewed as a huge threat. They naturally do everything in their power to delay the development of satellite-based systems as well as applications of new technology' (EC, interview, 6 March 2000).

The crisis in the area of air traffic management interacts with the proposals for satellite navigation systems.[56] Future satellite navigation had to be (for Europe at least) intermodal in nature and this overall strategy resulted in closer cooperation between the EC and the ESA in the 1990s.[57] Furthermore, the political and organizational framework in which air traffic management operated in Europe was in evolution with the commercialization of services, the growing separation of service providers from regulators, and in some instances even a transfer of competence from a national to a European level (Pierre, 2000). This process had accelerated with the liberalization since around 1993–1994. DG TREN acknowledged the opportunity to rationalize the European infrastructure (also as a means to expand its administrative control through harmonization and institutional coordination).[58]

56 Aircraft have been forced to navigate through (that is, fly over) ground-based beacons to know where they were, which forced them into specifically designed airspace corridors, which in turn created bottlenecks, although many aircraft already had advanced on-board intelligence that mitigated such problems. The GPS system had recently been accepted in Europe through a formerly military system, the so-called RNP-5 (Required Navigational Performance 5, where 5 stands for 'with an accuracy of 5 nautical miles'). 'However, GPS could not be used as a sole means of navigation since it was a military product and the US does not make available the required accuracy. The European view is that GPS is not optimal. The functionality is acceptable but the accuracy is not sufficient and can be improved' (EC, interview, 6 March 2000).

57 European organizations and the EC pushed for using Eurocontrol as the institutional basis for harmonization of air traffic management, which led to the launch of the so-called European Air Traffic Management, Harmonization and Implementation Program (EATCHIP) that lasted to the end of the 1990s. One of the important achievements, according to DG TREN, was the setting up of a Central Flow Management System in Brussels, through which route plans were coordinated between the participating countries.

58 DG TREN stressed that EU Member States had to comply with the treaties they had signed. Moreover, it pointed out that the traditional borderline between sectors and issues could change, including military and defense issues. 'New concepts are constantly evolving on how the military and civil traffic can be coordinated, such as the concept of "flexible use of airspace", etc. The High Level Group on Air Transport set up in the late 1990s was the first time to my knowledge that the EC invited both transport and defense ministers in an air traffic management issue. This group, which also includes airlines, service providers and other interest

The Space Technology Community

In the late 1980s, work had also been going on within the European Space Agency (ESA) and the national space agencies, in particular the French Centre National d'Études Spatiales (CNES). Parallel with the emerging efforts within the ICAO, the ESA started to realize the opportunity for a global civil satellite navigation system. In the early 1980s, it launched a program of technical studies, which resulted in a preliminary definition of a satellite navigation system (ESA, interview, 25 May 1998). The definition of a European contribution would concentrate on the transition from the first-generation to a second-generation civilian system, GNSS-2 (Ferrazzani, 1998 [ESA], p.5).[59] GNSS-2 would give Europe 'independence from foreign national/military satellite systems and control over its own element within a global civil navigation satellite system' (Barbarance et al. [ESA], 1996).

In addition, European producers gradually became more interested, although they were more skeptical than the European organizations. In 1993, DG Research/JRC tasked Deutsche Aerospace AG to assess the potential for a European contribution to satellite navigation.[60] The assessment and set of recommendations, reached with the flanking support of DG Research/JRC,

---

organizations, will present conclusions in June 2000 that will be the platform for the EC's working program for the long-term' (EC, interview, 6 March 2000).

59 The ESA studies and R&D work within its program for Advanced Research in Telecommunications Systems (ARTES), which was approved in December 1993, would complement work by the EC on telematics and transport infrastructures. In late 1994, the ESA Ministerial Council invited the ESA Director General to define the program proposal in close consultation with the EU on a European contribution to a satellite navigation system (Barbarance et al., 1996). The ESA would promote R&D activities (under the so-called ARTES Element 9) for the first generation of a global navigation satellite system (GNSS-1). This infrastructure system would be complementary to the existing GPS and Glonass and be based on the lease of Inmarsat III satellite transponders (Eurocontrol, interview, 21 December 2000).

60 In addition to Deutsche Aerospace AG, the study team consisted of Aerodata (Braunschweig), Dassault Electronique (Paris), Walter Blanchard (Dorking), The Russian Institute of Radionavigation and Time (St Petersburg and GOSNII Aeronavigatsia (Moscow). The task consisted of several aspects. First, the study team was to assess views on the current and longer-term capabilities, potential and affordability of satellite navigation systems in relation to user requirements. Second, it was to analyze constraints and obstacles in the wider and more rapid European adoption of satellite navigation systems. Third, it was to identify goals and strategy options for establishing consistent European policies in this area. Moreover, the study team identified needs for further research, technology development, and/or pilot demonstrations to support important future potential markets. It would assess the potential, the possible cost, and the advantages and disadvantages of future options for the provision of the space systems needed.

would be incorporated in EU policy for satellite navigation.[61] First, satellite navigation would become indispensable for future traffic and wireless infrastructures. Second, it would be inadequate to regard satellite positioning as a subordinate application of satellite communications. Finally, in order to achieve the goals set forth by the EC, European space policy had to be reconsidered (DA, 1994a, p.298).[62]

The report provided an interesting political analysis on (1) the problems for the application of satellite navigation in Europe, and (2) strategy options for Europe with respect to the space segment. Within the current institutional situation (a US military-controlled GPS system), aeronautical and maritime users could only use satellite navigation as a supplementary means of navigation. It was thus likely that these user groups would prefer a civil navigation satellite system that was under international control. In addition, the European equipment industry needed political assistance to achieve a stronger position compared to its US and Japanese competitors. 'While

---

61 The utilization of satellite navigation would reduce operating costs by simplifying the current radio-navigation infrastructure in Europe and the variety of aeronautical navigation aids. In addition, it would also enable more efficient road infrastructure. There were clear industrial benefits to gain in markets for satellite navigation equipment. 'In total, the importance of satellite navigation, which lies not only in the direct user benefits, but also in the associated socio-economic and industrial benefits, is of such an order of magnitude that it seems essential to make this a matter of policy' (DA, 1994a, p.299).

62 The report stressed that European policy should in particular be based on some specific application areas: (1) aviation, (2) vessel traffic, (3) land mobile applications, and (4) survey, exploration and offshore business. It stated that aeronautical users – for whom satellite navigation was expected to become the worldwide standard navigation system for all phases of flight – would be the main drivers toward technology development and institutional arrangements. The argument for satellite navigation with respect to aeronautical users was that one single navigation system for all phases of flight would lead to reduced complexity and weight of the avionics equipment (weight and cost reduction). Moreover, such a system would enable flight path optimization and thus further improve aviation economics and reduce fuel consumption. In the maritime domain (vessel traffic), satellite navigation technology was already in use but the maritime community used multiple sources for navigation as a standard practice. While the problem of system integrity was critical for the aviation community, this was perceived to be a somewhat less severe problem for the maritime users. Moreover, the report stated that land mobile applications, driven by user benefit, would form the largest user community of satellite positioning services in the future (for example, fleet management systems and luxury cars). 'Satellite navigation in private cars will probably become as familiar and widespread as car stereo equipment. With more than 120 million cars in Europe today, the potential market is enormous. Thus, developments in this area should be in the focus of industrial policy. Investments in the space segment should be made in a way to support not only the requirements of aviation, but also to provide an optimum basis for this very large, upcoming, user community' (DA, 1994a, p. 297).

American companies have developed their technology with substantial funding from military sources, Japanese competitors are benefiting from a comprehensive space program launched by their government' (DA, 1994a, p.300). Moreover, it was argued that US ownership of the space segment resulted in advantages for its domestic industry. Finally, a coherent European policy could enable harmonization of the policies of national agencies.

The study team identified five general strategy options for Europe in terms of the provision of the space and control segment: (1) GPS or Glonass, (2) GPS and Glonass, (3) GPS and Glonass plus overlay, (4) GPS and Glonass plus civil GNSS satellites, and (5) a civil GNSS. The use of either GPS or Glonass was viewed as having serious implications for Europe. In the long term it would result in inadequate performance for most applications, creating a need for large upfront investments in differential technology; no European influence or control; dependence of European infrastructure on foreign nations as applications grow; and persistent technological advantage for the satellite owners. 'This is therefore not an acceptable strategy option for Europe' (DA, 1994a, p.300). Regarding the second option to use both GPS and Glonass, Europe would not be dependent on a single source and integrity and availability would be improved, so it would be an option only if an acceptable level of control could be reached through negotiation. The team expressed doubts about the third option, since overlay concepts had been discussed by only one user community, the civil aviation community, and did not resolve the core political problem of dependence/independence. The study instead focused on the two last strategy options:

- The first step for Europe – GNSS-1 (GPS/Glonass + CIVIL GNSS): The use of GPS and Glonass combined with civil GNSS satellites would give Europe the required technological competence and result in an acceptable level of control for Europe. Even a small number of additional satellites would provide Europe with an independent capability for navigation.
- The long-term option for Europe – GNSS-2 (CIVIL GNSS): A European share in the installation of a civilian-controlled and international GNSS constellation would be within budgets available for space technology or infrastructure development. International feasibility studies had shown that a full constellation with 32 navigation satellites would cost around Euro 540 million (US$640 million). The launch of a civil GNSS constellation would consist of 30–40 dedicated satellites (or combined navigation and communication satellites).

The team recommended a long-term strategy where Europe would play a major role in the installation of an international and civilian GNSS. 'This is the only option really satisfying the needs of European users. It is deemed the only sensible long-term option for Europe' (DA, 1994a, p.301). A European satellite navigation program, which would not be limited to aviation requirements, should be initiated with the aim of making a major contribution to a worldwide GNSS. Europe needed to take a leading position in resolving the institutional, legal and organizational problems of installing, operating and supporting an international civil GNSS.

There were mainly two options available to enter a European program. On the one hand, Europe could use navigation transponders on the Inmarsat III satellites as an experimental tool, which could be used for the development of signal structures and payloads for future European GNSS satellites. Inmarsat, an international service provider, had announced a call for tender in June 1994. Policy initiatives within DG TREN to secure access to current existing systems were based on it (Eurocontrol, interview, 21 December 2000).[63] On the other hand, Europe could examine the possibilities of collaboration with those that developed and owned the current systems. 'The goal of such negotiations would be to achieve in the short term some influence and control of system operation, and to prepare a joint effort for the development of civil navigation satellites' (DA, 1994a, p.303). The team argued that cooperation, in particular with Russia, should be explored. 'Based on the existing Glonass technology, Europe could develop civil satellites and control technology together with Russia as part of these programs. This would assist the Russian space industry directly, and indirectly would help Russia to use satellite navigation on its own territory, which could be a contribution to the improvement of traffic infrastructure in Russia, too' (DA, 1994a, p.303).[64]

---

63 The call for proposals was expected in summer 1994 with the transponders to be allocated before the end of the same year. The announcement said that Inmarsat was searching for customers for the navigation transponders on board its planned geostationary satellites (Inmarsat III), designed to relay navigation signals to improve GPS/Glonass for civil applications, which would start operation from 1996 onwards. In June 1996, the ESA signed leases for the first two Inmarsat navigation transponders (which would be used to broadcast Egnos signals to users) together with France Telecom and Deutsche Telecom, the two signatories of the Inmarsat operating agreement that had formally submitted the Egnos proposal to the Inmarsat Council. The leases were signed in the presence of the EC and Eurocontrol (ESA, 1996). The transponders were attached to two Inmarsat III satellites (Atlantic Ocean Region, AOR, which was launched in April 1996, and Indian Ocean Region, IOR, which was launched in August 1996). Together they would cover the whole of Europe, South America and most of Asia.

64 Alcatel Espace, a leading space-hardware manufacturer, pointed to the benefits for European space industry. From a long-term perspective, work on GNSS-2 was necessary (1) to make decisions in Europe on possible technical solutions (regional

The European Commission and High Technology Policy Coordination

DG Enterprise made efforts to coordinate European industry. 'The European space industry has to overcome many national and industrial rivalries, and institutional problems such as the relationship of the ESA to the EU and national agencies, public and private actors, and so on. Because this is such a politicized industry, it is important to have the company leaders to show European consensus on priorities for Europe' (EC, interview, 13 January 1999). The EC managed to convene a number of industrialists and set up a high-level group (HLG). Alcatel Space Division headed a Working Group on Navigation, which worked closely together with DG Enterprise officials (Lembke, 2002c).[65] 'I think the members of the group were aware of the policy views of the EC, which are predominantly market-oriented and liberal. This is why even when they ask for protection or support they emphasize the support given by the US and express their aims in terms of achieving a level playing field' (EC, interview, 13 January 1999).

The HLG stressed the problem with satellite-based navigation systems under national military control. It was not acceptable for a State to depend on a foreign military system. Thus, several institutions were recommending the establishment of a civil satellite navigation system, which would constitute an important public service that would have to ensure the security of millions of people. Such a system would thus have to be under the responsibility (and control) of all the bodies that were using the system, and in particular the 'transport community' (EC, interview, 13 January 1999).[66] The HLG

---

or worldwide implementation, type of signal, and so on), and (2) to give the industry the necessary know-how through system studies, development of hardware and experimentation/validation (Alcatel Espace, 1995). In industry, satellite navigation was recognized as a key element in order to handle increases in European traffic. GNSS would be implemented in Europe regardless of the decision of European interests. On the other hand, participation in GNSS would enable Europe to (1) make a first step toward the control of the space segment, (2) ensure a good positioning of its industry and institutions into GNSS-2, and (3) give industry opportunities for export markets (interviews at Alenia Spazio, Matra Marconi, Alcatel Space and ESA headquarters 1998–1999).

65 DG Industry stressed that although there is something that could be called 'space industry', the different applications of space have very different policy aspects associated with them. Navigation was viewed as even less mature than earth observation, which was basically an R&D issue, while satellite telecom was partly regulatory, partly trade policy and partly research and development (R&D). DG Industry accordingly divided the HLG industrialists and companies into different subgroups, one of which worked on aspects related to satellite navigation.

66 The main applications of GPS that the HLG Navigation Group identified were (1) aeronautical applications, (2) maritime applications for navigation and fleet management, (3) terrestrial applications for navigation and fleet management, and (4) geodesy applications. The world satellite navigation market was estimated at

recognized that the future GNSS-2 was far from being technically defined, which posed a major problem. The European industrialists called for urgent preparations since the risk was big that Europe would be excluded from the future implementation of the system, with associated world market shares at stake. US activities, they argued, could imply an international constellation dominated by the US GPS industry.[67] The final HLG report argued that Europe had to step up its efforts to propose systems with international capability.[68] 'Otherwise, it could find itself in the absurd position of missing out on the fruits of the important effort made in space technology just when society is beginning to demand space technology, with a vast impact on industry, trade, and employment' (HLG, 1996, p.11). The report stressed the urgency needed to promote a European approach: 'The issue of Europe's active participation in the future world navigation system is both fundamental and urgent for the European space industry. It was totally absent from this major application at the sector level and only very marginally involved in terminals due to advances achieved by American industry in its military and civil markets. The European space industry could simply be eliminated as a supplier of the future international system' (HLG, 1996, p.22).

2.7 million units in 1996 and the distribution in terms of application markets was the following: in-car navigation (73 per cent), marine (16 per cent), handheld (8 per cent), aviation (2 per cent) and surveillance (1 per cent). The aviation community represented only 2 per cent but was the most demanding community in terms of technical requirements.

67 The future GNSS-2 system would (1) evolve as a globally designed system, offering worldwide coverage, or (2) as a multi-regional system where each region (Europe, Russia, Asia, North America, and so on) would develop a system offering a regional coverage. The global coverage would in the latter case be achieved through the juxtaposition of the various regional contributions. In both scenarios, the system architecture would be based on either GPS or Glonass. The international body in charge of initiating standards for civil aviation, to ensure that the same equipment could be used in every part of the world, was the ICAO and its GNSS panel, through the establishment of so-called SARPS (Standards and Recommended Practices). Such standards, in turn, would constitute the basis for regional organizations in order to establish operational performance standards for equipment design and development.

68 The European space industry was faced with heightened international competition where its main competitors, in particular the US, enjoyed advantages on which they could capitalize, not least its vast public-sector market, leaving European industry in an unfavorable competitive position. The US showed a determination to transform leadership in the military sector into a competitive international advantage for US industry. Satellite navigation had been and would remain an application at the interface of the private and public sector and would require mixed financing and efforts (Council, 1997, 1998a, 1999b; EC, 1993d, 1994e, 1995a, 1995b, 1996a, 1996d, 1996e, 1996f, 1996m, 1997e, 1997f, 1997g).

The industrialists stressed that the development and implementation of GNSS-1 in Europe on a multilateral basis would enable European industry to develop expertise in ground and user segments. In 1996, Europe had not yet committed itself to the deployment of a future European satellite navigation system (GNSS-2). The HLG argued that the EC acknowledged the importance of this issue, though.[69] 'The EC is willing to give some attention to this industry, and to try and reinforce the European position worldwide. It does not, however, want to be seen as acting in a top-down style, so it needs the initiative and inputs from industry' (EC, interview, 13 January 1999). The HLG report had a direct bearing on the 1996 Communication on space policy in Europe (Lembke, 2002c).

3.3 A European GNSS Strategy

In June 1994, DG TREN issued the first Communication dedicated to the European contribution to GNSS with the stated aim of establishing a consolidated European strategy and implementation program 'considered urgent and essential for both industry and potential users' (EC, 1994a). DG TREN based the launch of an EU program on Article 129C of the Treaty on the EU, which mandated the EC to prepare guidelines for the implementation of trans-European networks, of which satellite navigation was perceived as an integral part. DG TREN also justified the strategy proposal by calls from the European Parliament (EP) urging Europe to accelerate efforts, together with other legal documents (EC 1994b, 1994c, 1994d). 'The proposed action must allow a European contribution to the development of a world and multi-mode system of navigation by satellites. The setting-up of a navigation system by satellites aims to improve productivity of the transport systems by placing at the disposal of users a system enabling them to position geographically. This action contributes therefore to the development of the durable and sure mobility of persons and assets, which are the fundamental objectives of the common Transport policy' (EC, 1994a: annex).

Regarding economic and industrial benefits, DG TREN argued that European industry would benefit from growth in both the space and user segments and that users would benefit through the replacement of established

---

69 The HLG linked satellite navigation to the key issue of improving air navigation in Europe; in the longer term this activity would allow Europe an active part in the world satellite navigation market. Concerning the cost of an autonomous European system, the group estimated it to be less than Euro 1 billion for the initial deployment of a suitable constellation and the cost of modernization of the system to be less than Euro 100 million annually (HLG, 1996, p.22).

technology (EC, 1997l, 1998g, 1999m).[70] The global US monopoly on the provision of satellite navigation services through GPS made European efforts imperative (Allgeier, 2000; Busquin, 2000). 'If Europe does not act promptly then the control of the entire system will be done from overseas by implementing a civil American complement to the military GPS system. Those who will own and operate the upgraded system will set user requirement standards and certification schemes for equipment. The result would be a major dependence of Europe on the provision of a strategic asset for the future and a poor perspective for its industry to capture the huge associated market for user equipment' (EC, 1994a, p.11).

Europe had to act swiftly to respond to the Inmarsat tender. 'Absence of action in this frame will jeopardize the ability of the EU, and Europe in general, to contribute to the establishment of a global civil navigation satellite system' (EC, 1994a, p.4). Furthermore, DG TREN stressed that the provision of navigation signals would bring needed expertise to enable the implementation of GNSS. Eurocontrol, the ESA and transport associations had recognized the operational, industrial, political, technical and financial implications of satellite navigation and the provision of space and user services. The EC proposed a gradual approach for satellite navigation. As the first step in the short to medium term, the political and institutional issues related to US ownership and control of GPS had to be resolved to guarantee continuity of service for Europe and other countries outside the US. The first phase involved the setting up of a European complement to the existing military systems in order to give Europe sufficient control of the navigation system (EC, 1994a: annex). In the longer term, GNSS-2 would be promoted. Europe had to start as soon as possible with activities to elaborate possible

70 The total world market for satellite navigation equipment in 1996 was expected to reach around 3 million units to an approximate total value of around Euro 3 billion, with further substantial growth in the following years. Growth was also expected in the land vehicle market. Europe with its then around 120 million cars would have a significant potential. Moreover, the space segment of the current GPS, Glonass, and Inmarsat systems would have to be replaced around 2003–2005 and this constituted 'a significant potential market for European industry if it acquires the appropriate expertise and know how' (EC, 1994a, p.10). Finally, the replacement of the current and 'extensive' radio navigation infrastructure in Europe would benefit user communities (including a variety of aeronautical navigation aids (for example, maritime systems, maritime radio and visual beacons, and a range of land navigation experiments using short-range radio communication beacons and cellular mobile networks). The 1994 Communication argued that the present infrastructure was 'very costly to operate and maintain' and that a 'significant proportion of this amount might be saved if satellite navigation becomes used as the primary means of navigation' (EC, 1994a, p.11). DG TREN suggested that the civil aviation community would have to take on a large responsibility in the replacement of classical ground-based navigation aids.

definitions of GNSS-2 which would allow a civil internationally controlled service to be provided to users as an independent means on an equitable basis (EC, 1994a, p.9).[71] From the beginning, satellite navigation was heralded as an integral element of economic growth, social development, and the trans-European Transport and Telecommunication Networks. 'There is accordingly a need for the European Union to play a leading role in their development and implementation in order to maintain its independence in a strategic area' (EC, 1994a, p.13).

In December 1994, the EU Transport Council adopted a resolution on GNSS that stressed the importance of satellite navigation for the completion of the internal market, the long-term and sustainable mobility of people and goods throughout Europe, and transport safety. It was recognized that the EU and its Member States were called upon 'to give political impetus', provide regulatory frameworks, and promote appropriate transport projects 'within the context of the information society'.[72] The Council recognized the need for Europe to play a central role in its implementation and called on the EC 'to initiate or support work needed for the design and organization of a Global Navigation System (GNSS) for civil use' (Council, 1994).[73] The EC thus convened a High Level Group (HLG) of representatives from the Member States, the ESA and Eurocontrol in order to draft an Action Plan and remove relevant barriers with the aim of achieving 'the operational implementation in

---

71 The 1994 Communication proposed three specific actions: (1) definition of the operational specifications and of the architecture of satellite navigation and positioning; (2) setting up a European complement to the American GPS (Euro 25 million, 1995); and (3) allocating funds for the future navigation system (Euro 20 million, 1996–1998). The proposal said that Euro 13 million would be required by the ESA (project leader) for the first action, that is, the definition work (operational specifications and architecture). In the 1995–1998 period, Euro 13 million would be required for Action 1 and Euro 45 million for Actions 2 and 3, with a total amount for the whole period of Euro 58 million (EC, 1994a: annex).

72 Another objective was to enable European industry to rise to a level of competitiveness that would allow it to compete in the deployment of satellite navigation and in the market for user devices. The Council authorized the EC to move ahead and use Inmarsat III or any other augmentation technology required for achieving the goals set out in the Resolution. It was recognized that Community guidance would be necessary to achieve those goals (Council, 1994).

73 As satellite navigation activities evolved in Europe it became associated with strong political connotations. 'The work within the EC and the EU has been defended by the strategic and economic nature of satellite navigation through the creation of jobs and reduction of unemployment, which is a very important part of EU policy' (Eurocontrol, interview, 21 December 2000). Economic and job-creation aspects and political and strategic aspects constituted the basis for the policy arguments developed by the EC. The linkage between economic and political-strategic aspects would remain at the core 'because at the moment satellite navigation is dominated by one country, and in particular the military branch of that country' (Eurocontrol, interview, 21 December 2000).

Europe and throughout the world of a seamless multi-modal satellite-based radio navigation/positioning system' (Barbarance et al., [ESA], 1996).

The new European system for positioning and radio navigation via satellite represented a race toward a position as a valid actor in the international arena by developing a civil navigation satellite system more efficient than the US and Russian solutions (Ferrazzani [ESA], 1998). In 1996, DG Research elaborated and published the third Communication on the EU and Space.[74] The report stated that it was imperative to embark on reinforced efforts to promote the satellite navigation domain in order to 'ensure that Europe has a position commensurate with its strategic and commercial ambitions in this rapidly growing application' (EC, 1997a, p.3).[75] The industrial and political control of infrastructures and standards was viewed as crucial. 'Experience shows that only those nations and industries with a decisive influence on the system infrastructure will remain competitive in the market. The objective of the EU is to raise European industry to a level of competitiveness enabling it to participate in the development of a GNSS, in the definition of interoperability requirements and industrial standards, thereby giving industry the opportunity for an early entry in the market' (EC, 1997a, p.23).

74 The EP Committee on Research, Technological Development and Energy (CERT) agreed with the push by the EC and the ESA for more active efforts to develop and promote satellite navigation. The benefits for both air and surface transport would be considerable and the first European contribution to a satellite navigation system would be important in this respect. Substantial resources within the Fifth Framework should be devoted to navigation, among other fields. CERT asked the EC to pay particular attention to the strengthening of the industry's technical and economic potential and to promote the introduction of 'an autonomous and integrated European navigation system, with a view to a worldwide system' (EP CERT, 1997, p.7). The Committee on Economic and Monetary Affairs and Industrial Policy (EMAC), whose opinion was included in the report, stressed the longer-term importance of a European contribution to GNSS-2 fully devoted to civilian applications and managed in an international framework. 'It is a strategic issue for Europe to gain autonomy in satellite navigation systems and promote European industry participation (GNSS-2) in a second-generation system, to be operated by an international civil organization' (EP CERT, 1997, p. 17).

75 Estimated savings derived from such systems had led both the ICAO and the International Maritime Organization (IMO) to adopt satellite-based systems as key components for the future (IMO, 1997, 1998). The EC argued that the free availability of the satellite signals and the political complications associated with current military-civil systems made it difficult to develop such advanced satellite systems on a purely commercial, private basis. Rather, only 'a government-industry partnership has a chance to provide a breakthrough' (EC, 1997a, p.23). It was perceived that the current efforts by public authorities and industry around the world to commission augmentations to the dominant systems would lead to 'a seamless, worldwide augmentation system for air navigation' (EC, 1997a, p.23).

3.4 EUROPEAN GOVERNANCE

At the time when the EC launched its 1994 Communication on Satellite Navigation, 'no real organizational structure and action plan' were in place 'to set the scene for a consolidated European strategy and a subsequent implementation program' (EC, 1994a, p.3). A number of institutional studies in the mid-1990s by the EC and Eurocontrol had concluded that, at some point in time, a more coherent institutional structure had to be established.[76] It would prove difficult, however, to establish an institutional arrangement at European level. 'The idea of setting up an Agency is that it will take over the overall supervision and management. The moment in the past has not really been opportune to create an agency since there was really nothing to create it for' (Eurocontrol, interview, 21 December 2000). The HLG report argued that an operational entity for satellite navigation services had to be created to ensure Europe's active participation in the future world navigation system and markets (HLG, 1996, p.22). It stated that the creation of a new European entity (possibly including non-EU members) was urgent and would take on four functions (HLG, 1996, p.23):

- Act as the EU representative in the field of satellite navigation services, with the power to negotiate with other countries;
- Act as focal point for assessment, planning, and development and be in charge of the necessary investment in the satellite navigation system (independent European satellite system or European share in a large system developed with other international partners);
- Be in charge of operations and maintenance of the future satellite navigation system, totally or partially, depending on the nature of the system developed;
- Have missions covering all aspects of mobile infrastructures and markets: airborne, maritime and terrestrial.

The justification for an entity with such functions was to guarantee continuity from GNSS-1 to GNSS-2 and the provision of services, provide legal responsibility, and establish the necessary civil institutional arrangement: 'This unique entity, entrusted with the budgetary resources, will reinforce Europe's political strength' (HLG, 1996, p.23). HLG called for the urgent

[76] In March 1994 there was a program in Eurocontrol called the European Air Traffic Harmonization and Integration Program (EATCHIP). On behalf of the ECAC Member States, Eurocontrol managed this program, that is, the Eurocontrol Member States plus the additional states that made up ECAC. Satellite navigation was integrated into EATCHIP and Eurocontrol set up four Task Forces within the program to look into (1) research and development (R&D) aspects, (2) cost and benefit aspects, (3) operational aspects, and (4) institutional aspects.

creation of a European Agency for Navigation in order to (1) enable European independence (sovereignty), (2) safeguard the commercial interest of Europe, and (3) ensure that European industry to get all the necessary technology and experience to consolidate any GNSS-2 architecture.

The push for satellite navigation is associated with a number of people of which three persons in particular may be referred to as the 'founding fathers of satellite navigation in Europe': Luc Tytgat (DG TREN), Andrew Steciw (ESA) and John Storey (Eurocontrol). They headed a multinational and professional European network of experts, policymakers and administrators, which also stretched beyond Europe into other regions and international organizations like the ICAO. The EC, the ESA and Eurocontrol had together negotiated a Memorandum of Understanding (MoU) and official documents committing them to Egnos in the mid-1990s (ESA, 1996). The political mandate given by the ECAC Transport Ministers in June 1994 eventually led to the formal signature of the European Tripartite Agreement (ETA) in June 1998. The European Tripartite Group (ETG) – composed of the EC, the ESA and Eurocontrol – was developing Egnos (the European Geostationary Navigation Overlay Service), which would cover the broader European Civil Aviation Conference (ECAC) region.[77]

The ETG was formalized 'with the aim of establishing a satellite navigation and positioning service for Europe as a contribution to a global effort' (ESA/Eurocontrol, 1998). The joint European cooperation would strive to promote a coordinated system for telecommunications, transport and navigation for economic, political and strategic reasons. Egnos/GNSS-1 would help to improve not only navigational accuracy and performance but also communication and surveillance capabilities, and thereby increase safety and reduce time delays, fuel consumption and costs. Moreover, the organizations envisaged wider macroeconomic and social benefits in that a satellite navigation infrastructure would 'greatly contribute to improve

77 Egnos/GNSS-1 was created under the overseeing role of the European organizations. Eurocontrol had responsibility for user requirements, and validation, operational, and safety-related aspects in the civil aviation context. The ESA was responsible for the development of Egnos. Finally, the EC was responsible for political aspects and also for user requirements, validation and safety-related aspects for other modes of transport than civil aviation. While the EC develops policies and strategies for satellite navigation, it does not have the necessary competence actually to develop systems. The system development of Egnos, for example, was led by the ESA with European industry. 'In the beginning, the roles of the three organizations were relatively clear. But as we moved into the development phase of Egnos the roles of the three organizations became less clear. There are different contractual and financial rules and regulations between the ESA and the EC. This situation led to a number of frictions. Unfortunately, some of these frictions spread over into the early studies on GNSS/Galileo' (Eurocontrol, interview, 21 December 2000).

economic prosperity, industrial returns, employment and quality of life' (ESA/Eurocontrol, 1998). The ESA Director-General Rodota, UK Minister of Transport Strang, and Eurocontrol Director-General Lambert signed the ETA on the premises of the EU Council in Luxembourg.

It was announced that Egnos/GNSS-1 would be completed in 2002, while the next generation systems would start operation under civilian control by 2010.[78] The agreement was seen as politically important and urgent by those involved. 'This will give Europe independence from foreign/military satellite systems and control over its own element within a global civil navigation satellite system' (Barbarance et al., 1996, p.159). 'The majority of countries in Europe recognize that Egnos has been necessary to get to where we are today. This is true without a doubt. If we had not gone through Egnos, we would not be where we are today' (Eurocontrol, interview, 21 December 2000). Aside from the three European organizations, and the large consortium with Alcatel as the main contractor,[79] CNES (the French Space Agency) and the Norwegian Mapping Authority (NMA) actively contributed to the development of Egnos (Table 3.1). Furthermore, the Egnos Operator and Infrastructure Group (EOIG) was created as a legal entity in 2000 composed of major air traffic service providers in Europe. (The providers were DGAC, France; DFS, Germany; ENAV, Italy; NAV-EP, Portugal; AENA, Spain; Swisscontrol, Switzerland; and NATS, UK.) They joined the ETG as the Egnos operational implementation phase came closer. Efforts in Europe, Japan and the US to improve the reliability and accuracy of the US GPS system through augmentation systems lagged behind schedule in 2000.[80]

---

78 In the late 1990s and beyond there were three different regional so-called satellite-based augmentation systems (SBAS) under development. The FAA in the US managed the development of the Wide Area Augmentation System (WAAS), which essentially covered North America. Finally, the Japanese Civil Aviation Bureau was at the same time handling the implementation of the MTSAT Satellite-based Augmentation System (MSAS). The optimal international solution would be to coordinate these regional systems to the extent that they would generate a seamless worldwide navigation system. Around 2000, the mechanism for such international cooperation was the so-called Interoperability Working Group (IWG) and there had been successful tests indicating that integration was feasible.

79 The French company Thomson-CSF was awarded the contract for the development of the so-called Egnos Ranging Function in July 1995. In December 1995, Thomson-CSF was also awarded a contract for the detailed design of the entire Egnos infrastructure. Companies from Spain, the UK, France, Germany, Norway and Switzerland were likewise involved in the design phase as subcontractors to Thomson-CSF (ESA, 1996; EC, 1998b).

80 The European Egnos system, intended to boost the reliability of the US GPS system with a network of ground terminals (stations) linked to telecom satellites, used GPS data to two Inmarsat satellites (SN, 2000b, 2000c, 2000e, 2000f). In 2000, it was envisaged that it would also use the Glonass satellites as well as the ESA Artemis data-relay spacecraft, which would carry a navigation payload into

Table 3.1 The European institutional structure for Egnos

Governing institutions: The European Tripartite Group (ETG)		= European Commission, ESA, Eurocontrol	
CNES	Norwegian Mapping Authority		Egnos Operator Interest Group (EOIG)
	Main contractor: Alcatel Space Industries (France)		
Industrial contracts:			
Austria	*Germany*	*Italy*	*Spain*
SARC	Dasa, IFEN, DLR	Alenia Aerospazio	GMV, Sener
Siemens Austria	Airsys-ATM	Laben	Iberespacio
	MAN	Vitrociset	Indra Espacio
	Deutsche Telecom	Telespazio	
	Alcatel ANS	Space Switalia	
	Nortel Dasa	Alenia Marconi	
	Science Systems		
	Vega, ATSO/NATS		
Canada	*Netherlands*	*Portugal*	*UK*
Novatel	NLR	Inesc	Racal, Logica
		Edisoft	Airsys-ATM/UK
		Marconi Portugal	British Telecom
France	*Switzerland*	*Norway*	
Sextant Avionique	Contraves	Seatex	
Sodeteg/SRTI	Tekelec (TBC)	NMA	
Syseca			
France Telecom			

Source: Modified from the ESA website (www.esa.int), 1999.

geostationary orbit and provide an additional layer of reliability for Egnos. 'What we have in development now is in full compliance with requirements given to us by international civil aviation authorities. It makes Egnos more reliable, but also slightly more expensive,' said Lo Galbo, head of navigation at the ESA, which was financing the bulk of the cost for Egnos (SN, 2000a). At the beginning of 2000, European governments had still not decided where to install 47 separate ground facilities for the Egnos project (SN, 2000d).

3.5 THE EUROPEAN POLITICS OF EGNOS (GNSS-1)

Although the majority of countries agreed to contribute financially to Egnos, different opinions did exist in Europe. Sweden, for example, decided to remain outside the Egnos project and thus does not contribute financially.[81] The Swedish delegation stressed the disadvantageous technical-strategic aspects.[82] 'Sweden soon reached the conclusion that a system that is based on geostationary satellites is not appropriate for northern latitudes. It is too common that you do not have access to the signals' (EU national government advisor, interview, 25 January 2000).[83] Germany was another country that at least initially doubted the potential of Egnos but would eventually move ahead with it. Norway, which one could have expected to be as skeptical as

---

81 The national objective of ESA Member States was to ensure that the role assigned to their national industries was acceptable, that is, that investment into the ESA would result in procurement including domestic industries. Part of the bargaining game was to threaten to withdraw or reduce contributions unless the return of invested money and program development costs met expectations. For example, Italy expressed its willingness to support Egnos substantially as long as it could host a ground infrastructure element for Egnos and receive an industrial return consistent with its investment (ESA, interview, 5 March 1998). The Executive stressed, in turn, that it was important not to jeopardize the industrial edifice and planning of earlier phases and that unanimous decisions would be required.

82 Egnos could be said to consist of two major parts in terms of functionality. The first element consists of collecting information about the status and operation of a satellite navigation system, while the other consists of distributing that information to those who need it in a rapid and efficient way. The collection of information would remain an important element, to assess whether something is wrong with the system. However, some countries, such as Sweden, had argued that Galileo provided an opportunity to build a more intelligent data distribution system than one that would rely on geostationary satellites. This would change the character of Egnos and the Swedish Galileo delegation argued that the money flowing into Egnos to pay for the use of geostationary transponders could be used differently. There should be an incentive for those who were paying to follow this line of development. There was uncertainty and different opinions in Europe as to how Egnos should be related to Galileo. One important aspect for some countries was the use of geostationary satellites, while another major issue was timing. Egnos would after all start operation only a few years before Galileo due to delays. 'Changing time frames is something common in large-scale space projects when people have to address a range of complex issues, including political decision making aspects' (Eurocontrol, interview, 21 December 2000).

83 Sweden has been active in the field of satellite navigation and the use of satellite technology, although actors had been critical of Egnos in the domain of satellite navigation, in particular for civil aviation. 'The position of Sweden, certainly the Swedish aviation community, is that they accepted Egnos was necessary to get Europe into a position to where it is today. But they see Egnos more as a R&D experience gathering a system rather than a system which will one day become operational' (Eurocontrol, interview, 21 December 2000).

Sweden for technical-strategic reasons, had industrial investments in Egnos and took a more positive stance than Sweden, although it reputedly moved closer to the Swedish position in the late 1990s. 'Moreover, Denmark chose not to participate in the Egnos program for budgetary reasons.

Aviation Politics: Airspace Users and Air Service Providers

The general public in the streets in Europe and the US would probably give a blank look if asked about satellite navigation at the end of the 1990s and even beyond. Though the EC argued that the support of satellite navigation would create jobs, there was no real public debate because it was such a specialized field. Eurocontrol pointed to the benefits that satellite navigation could bring to improve safety on railways, which has been a problem particularly in the UK, over the last few years, and particularly high-speed railways. 'This is something in the public debate, but it is not a subject like taxes or the mad cow disease, which attract the attention of the normal man and woman in the street, not high technology like satellite navigation. I even wonder how much of the launch of the [Airbus] A380 in Europe really attracts ordinary public attention, apart from the fact that it is going to create several thousands jobs' (Eurocontrol, interview, 21 December 2000).

Satellite navigation, however, was not free from controversies. The satellite navigation activities in Europe raised strong concerns from the airspace users. 'We had to go through the pain barrier with Egnos to get where we are today with Galileo: experience is an expensive commodity. We had to pay the Egnos price, although not a large price, to get the experience to put Europe in a position to move forward with Galileo' (Eurocontrol, interview, 21 December 2000). The civil aviation community criticized Eurocontrol, the ESA and the EC and stressed the need to retain diverse navigation and positioning technologies. 'Although the civil aviation community undeniably wants to play a less visible role in Galileo than it did for Egnos, it can obviously not afford to stay on the sidelines of such an ambitious European project' (EC, 1999l, p.25, 1999g).

> Clearly, the Egnos-Galileo sequencing upsets the traditional timetable and this will encourage the aviation community to insist that Galileo not make Egnos equipment obsolete. Without an assurance that Galileo will not be allowed to mitigate the usefulness of Egnos, aeronautical users will naturally choose to ignore the augmentation system and wait instead for Galileo to become available, thus depriving Egnos of the largest share of its potential customer base and endangering its financial well being (EC, 1999g, p.29).

DG TREN stated that it was difficult to determine how much influence the civil aviation community would or should have on the design of Galileo. 'With Galileo, the EU will measure the respective weight of the users not by the extent of their requirements for system performance but instead in terms of their share of the potential market' (EC 1999g, p.29). Airspace users did not want to have to finance the infrastructure. DG TREN suggested that the civil aviation community could not afford to remain too distant from such an ambitious European program. 'Civil aviation does not want to be seen as the community setting the standards for fear of the financial consequences' (EC, 1999l, p.25). Modifications in the implementation schedule (a reduced gap between the operational start of Egnos and Galileo) had implications for the definition of interests. DG TREN recognized that airlines had long transition periods to switch from one system to another due to the substantial costs involved. The transition from Egnos to Galileo would upset conventional schedules. 'For the five air traffic control service providers contributing financially to the development of Egnos (AENA, Spain; DFS, Germany; DNA, France; ENAV, Italy; and NATS, UK), a rapid change-over to Galileo would make it difficult to achieve a return on investment' (EC, 1999l, p.25).

In the face of what was viewed as an 'abundance of technology' and substantial taxpayer investment in satellite navigation, the Association of European Airlines (AEA) argued that Egnos constituted a wrong business investment from a European airspace user's point of view. GNSS should be designed on the basis of end user requirements. 'If an older technology is more cost effective, airlines will prefer to stay with this rather than to move to a more sophisticated system, unless justified by significant safety improvements' (AEA, 1999). Airlines were not interested in technology itself except as a means of achieving increased safety, capacity, and schedule reliability, as well as reduced costs. The AEA had sympathy for the argument (supported by the EC and the ESA) that the foreign and military control of GPS could make complete dependence on this system potentially dangerous for European users. However, the problem here was that the development of Egnos would not function independently from GPS.

In addition, it had become clear to Europe that the US was not willing to share control and ownership of GPS, which impeded the certification of Egnos, and the US DoD had not removed concerns about questions related to basic GPS aspects. In fact, the AEA argued 'by investing a considerable amount of tax money in an overlay system to GPS, Europe is increasing its dependency on GPS' (AEA, 1999). Political support for Egnos would not remove 'the institutional deadlock of the use of GPS in Europe'. There was a stronger case to support a civil-controlled GNSS such as the planned Galileo initiative, but its implementation must be based on transparent cost-benefit analyses and a fair allocation of costs among the different user groups. 'For reasons of competitiveness and equal opportunities of the European airline industry, Galileo should be funded by European public resources as a general utility system, like GPS has been funded by US tax resources' (AEA, 1999).

Moreover, the AEA opposed claims by the EC that Egnos constituted a technical precondition for Galileo. It stressed that the latter constellation would be a completely different system. 'No technical infrastructure of Egnos will be necessary for Galileo' (AEA, 1999). AEA did not agree with the official EC statements that Galileo should be built on the precondition that it makes use of the Egnos infrastructure. 'The argument is rather that Authorities would not feel comfortable to move to Galileo without having experience of Egnos and a European contribution to the US-owned GPS system would be politically desirable. As honorable as these reasons might be, they are unrelated to user requirements and thus cannot be the basis for user's support' (AEA, 1999).

According to the AEA, Egnos would not provide any tangible operational benefits to commercial airlines. Airlines already operated in an environment of various inertial, distance measurement and flight management systems. On-board augmentation systems and conventional navigation techniques could perform the functions of Egnos. In particular, the AEA was against the idea that European airlines should finance Egnos due to its perceived benefits outside Europe in places where there was a lack of navigational aids. 'Do not ask us to fund non-European infrastructure'. The AEA reacted against the idea that policymakers may try and recover a large portion of the costs for Egnos via the so-called European Route Charges System, that is, the airline community. 'It suffices consequently to mention that it cannot be the task of European operators to finance the necessary improvement of inadequate infrastructure in other parts of the world' (AEA, 1999).[84]

---

84 The airline community did not agree that Europe should support Egnos because the US implemented the Wide Area Augmentation System (WAAS). 'The AEA believes that Europe should learn from US mistakes. WAAS is already far more deployed than Egnos and continuously under criticism because of its high costs and limited benefits. Moreover, the business case for WAAS is built on a large general aviation community which Europe does not have' (AEA, 1999). In addition, Egnos should not be implemented as a sole means of navigation, which had been the idea of transport ministers and other interests in Europe. Such a solution would be unwise and probably unsafe. The airline industry argued that it had always relied on several navigation systems. Any future satellite navigation system, therefore, would need a type of back-up system, which for larger airlines would be the so-called inertial navigation systems (INS), which together would provide both an accurate system in the long term (GNSS) and in the short term. In addition, AEA pointed to the sunk costs involved since a large part of the AEA airline fleet was already equipped with INS technology and the airline community had to protect past investments.

Finally, Egnos would have distributional implications with negative effects on the European airline community. The AEA argued that support for Egnos would benefit the European space technology industry by ensuring that it would have an opportunity to tap into the future multi-billion Euro GNSS markets. 'It is completely unacceptable to pursue a system which is benefiting one part of the European industry (for example, the European Space Technology Industry) and then to pass on the costs to another part of the European industry (for example, the airlines) deriving no tangible benefits' (AEA, 1999). It stressed that the AEA community in 1998 transported 265 million passengers, 5 million tons of freight and directly employed 329,000 people, and that the air transport sector was an essential element of the global transport system. In addition, it was a key enabler for tourism, the world's biggest industry in terms of employment and turnover. 'The competitiveness and well doing of European airlines is consequently at least as important as the competitiveness of the European Space Technology Industry' (AEA, 1999).

The AEA argued that since the US WAAS system was already more deployed than Egnos, the European industry would remain behind the US competition if full support for the Egnos activities were pursued in Europe. Instead, Europe needed to leapfrog in order to reach and potentially surpass the American global dominance in satellite navigation. 'If Galileo would be directly pursued by Europe without passing via the expensive and unnecessary Egnos step, the European Industry would have the opportunity to be before their American competitors' (AEA, 1999).[85] Unfortunately, according to the AEA, a number of 'monopolistic' air traffic service (ATS) providers in Europe had already committed to fund around 50 per cent of the development of the first phase of Egnos infrastructure (the so-called Egnos Advanced Operational Capability).[86]

---

85 The AEA wrote to the then Commissioner Kinnock in October 1997 asking DG TREN to stop the development of Egnos for aviation purposes since it was not required. The European Regional Airlines (ERA), the European Charter Airlines (EURACA), the European Business Aviation Association (EBAA) and the IAOPA-Europe (general aviation) supported the AEA position. The Chief Executive Officers (CEOs) of several AEA airlines had also communicated this position across Europe to their respective national Ministers of Transport. In May 1998, the then AEA Chairman wrote to the President of the EU Council of Transport Ministers. The letter, according to the AEA, made it 'absolutely clear that the airspace users have no requirement for Egnos and are not willing to contribute to the financing of any parts of the Egnos operations and development costs' (AEA, 1999). 'The logical conclusion would be that Egnos is developed for industrial/political reasons and that it should be entirely funded by the taxpayer (infrastructure as well as operational costs)' (AEA 1999).

86 The AEA position paper included the following data: DFS (Germany) Euro 18.9 million; ENAV (Italy) Euro 30.11 million; AENA (Spain) Euro 23.62 million;

Several ATS providers had invested financially (and politically) in the Egnos concept and development. They took back parts of their invested money for the airlines through the so-called European Route Charges Mechanism. 'This has often led to a very unpleasant debate between the two sides of the community, the navigation service providers, on the one hand, and the airspace users (airlines), on the other' (Eurocontrol, interview, 21 December 2000). A lot of heated debate, particularly on Egnos, evolved in the professional arena, among actors in 'a very restricted subset of the general public'. A similar debate had taken place in the US some years earlier on the operational benefits of the US WAAS, criticized not least by the Air Transport Association (ATA). In terms of the operation of Egnos, European ATS providers from Germany, France, Italy, Spain, Switzerland and Spain had joined forces in 2001 as a European Economic Interest Group (EEIG). This would be a step toward a more consolidated air service provision in Europe. In early 2001, China demonstrated its ambitions in satellite navigation by launching the so-called Biedou satellites. The major interest on the part of ATS providers was to rationalize the ground infrastructure, starting with the lower end (for example some types of beacons). Air navigation providers stressed that Egnos would offer not least small airfields precision approach and vertical guidance all the way down to around 200 feet as of 2001, thus increasing safety. Moreover, air traffic authorities tend to have relatively long-term strategies.

Airlines think that GPS functions well. As long as a system works well there is no widespread interest in a new system. Public authorities have traditionally assumed responsibility of investing in infrastructure and critical technology and they then sometimes hand it over to the private sector. Interestingly, public intervention is at least as common in the US as in Europe. Air traffic authorities have a higher horizon and welcome the idea of a back-up system. The idea of having two systems has always been there. If the current systems are rationalized in the future then it is good to have more

---

NATS (UK) Euro 23 million; and Air Navigation Services Portugal (Portugal) Euro 5 million. Those ATS providers had informed the AEA that they would have to recover the costs from the airspace user community rather than from users of other transport modes (road, rail, or maritime). As a result, the AEA had written to DG TREN Commissioner Kinnock 'requesting the EC's intervention with the ATS providers concerned and suggesting the use of EU R&D budgets to compensate those ATS providers for their wrong investment in Egnos'. AEA further argued that potential benefits could be derived from a so-called Ground Based Augmentation System (GBAS) in the longer term as a new landing aid technology, which would be required, according to the AEA, even with a European Galileo navigation system. 'Unfortunately – for the moment – not enough (European) research is conducted to move GBAS forward because the vast majority of tax resources have been drained into Egnos' (AEA, 1999).

than one system (LFV, Swedish ATS provider, interview, 17 November 2000). As air traffic services start to be partly privatized (notably the UK/NATS), providers will become more business-oriented. Hefty investments in infrastructure will become less popular. However, there is a perceived linkage between public investment and jobs:

> In order for Europe to match the US funding on a military basis, the EC has to have a way to putting money into the system. If this is not done then Europe can as well have GPS and that would be the end of it. This is a high technology capability industry. The mere fact that the EC is funding and Member States are putting money into the provision of Egnos and Galileo means that people like my staff and me and other people in related industries are being employed and are in a high technology front type of industry. It is very much the Boeing-Airbus idea to provide competition. The EC has to 'cultivate' these activities to make Europe as capable in a high technology front industry as the US (NATS, UK ATS provider, interview, 15 February 2001).

Altogether, air service providers had an interest in protecting Egnos due to existing investments, technical performance, employment and rationalization (and for those that were going partly private, business differentiation and profit). Despite the tensions within the civil aviation community in the late 1990s, the major actors adopted a Common Civil Aviation Position in Europe on satellite navigation needs for the future. This common position was developed jointly through Eurocontrol with the air navigation service providers in the Member States, on one side, and airspace users, on the other. 'We have tried to defuse the situation which arose over Egnos' (Eurocontrol, interview, 21 December 2000).

When it comes to maritime interests in Europe, they were also interested in Galileo. Most markets for navigation, though, were already met by other services, such as GPS and Differential GPS, but maritime administrations agreed that it would be appropriate to have more than one single source of information (for example, using a back-up system such as Loran-C). The Global Maritime Distress and Safety System (GMDSS) met the market for safety related communications. As a result of what the maritime sector viewed as lack of political influence and effective consultative arrangements with the EC, they decided to regroup and modify their strategic responses at the EU level. The European GNSS Maritime Advisory Forum (EGMAF), established in May 1998, enjoyed secretarial and logistical support from the EC's European GNSS Secretariat in Brussels. The EGMAF group worked with the EC as the Maritime User Forum (MUF). In December 1999, it changed its name to the European Maritime Radionavigation Forum (EMFR) to underline its independent interests. Service providers, representing national maritime administrations in Europe, dominated this group, although other actors were represented, such as the European Community Shipowners Associations (ECSA).

The EMFR members pointed out that while the maritime community was a leading transport sector, and had formally expressed its needs through the primary international maritime organization (International Maritime Organization, IMO), their interests had not been adequately accounted for (EMFR representative, interview, 19 November 2001). They agreed that it was necessary to improve coordination to protect common interests and to provide a single point of contact and voice for European interests and maritime radionavigation development in Europe. In addition, they were concerned (as was the European airspace users) that the manufacturing industry would define the requirements of maritime users, which would entail loss of independence for maritime users. The maritime interests were broader than the Galileo project and encompassed the optimum integration of various navigation systems and contribution to the European Radio Navigation Plan (ERNP). They pointed out to the EC that maritime users were not accustomed to paying direct user fees for radionavigation services and would opt for alternative services unless the EU would mandate the use of Galileo.

The EC discussed the option of introducing controlled access of the signal and asked the maritime interests whether this would represent a drawback for the use of Galileo in maritime applications. The maritime sector actors stressed that encryption or controlled access of a system through a public regulated service could prevent it from obtaining IMO recognition. The maritime market was mature and in many respects a regulated market, governed by long-term international conventions, which had to be recognized by the EC. The maritime industry operates on a global basis, with many ships arriving in Europe and European ships trading with countries throughout the world. A regulated service intended for national use or EU infrastructure policy applications would not be acceptable if there was a possibility that it would temporarily replace the open service under critical circumstances. In addition, they argued that the planned Galileo search and rescue service had to be fully integrated with the so-called international Cospas/Sarsat services and that the EU would have to become a party of the agreement on which this system was based. For this to be feasible, however, the Galileo services would have to be free of charge and not require a different standard of shipborne equipment to be developed and installed.

In particular, the maritime interests claimed that the EC and EU needed to take account of their work to amend the IMO Resolution A860(20), which specified the required characteristics of future GNSS services for marine users (IMO, 1999). The EU and the consultative reports it relied upon, they stressed, lacked explicit references to the existing international regulatory environment and the ongoing IMO developments concerning maritime requirements for future GNSS services (EMFR representative, interview, 19 November 2001). This ignorance could have far-reaching implications since the preparatory work at the IMO level would offer the Galileo project a unique opportunity to secure a competitive advantage over existing GNSS constellations. GPS and Glonass were recognized only for maritime use in open water areas (high sea and oceans), but not for safety critical navigation in coastal waters or harbor approaches. The IMO recognition of Galileo would give it competitive added value and it would then potentially gain a widespread and rapid acceptance in the global maritime community (EMFR representative, interview, 19 November 2001).

3.6 TOWARD AN AUTONOMOUS EUROPEAN GNSS

EU Commissioner Kinnock stressed that a practical vision for cooperation had to be promoted between the varied interests (public and private, service providers and users, and so on). Kinnock argued that the development of, and access to, technical norms and industrial standards would be as important in the satellite navigation domain to achieve commercial advantage as in other technology areas, such as the development of high-definition television (HDTV) (EC, 1997c, 1998b, 1998c, 1999a). Kinnock also signaled to international partners (particularly the US) that Europe had realized that the US was not willing to grant equal control of GPS. Europe was prepared to go ahead with an independent system.

> The most basic strategic choice is whether we in Europe should develop our own independent satellite navigation system. Frankly, my preference – and I think that of many others – would be to develop a single global system with our international partners. Obviously, however, that course could only be followed if there was genuine collective control over the system, firm guarantees that the service (on which the safety of European transport will depend) could not be withdrawn under any circumstances, and an opportunity for European industry to compete in all segments of the market. If, after careful analysis, we in Europe are not convinced that such a global system is a realistic and dependable opportunity, I think that we in Europe will have to press ahead with our own system whilst, of course, ensuring interoperability (Kinnock, 1997).

In January 1998, DG TREN proposed a strategy for ensuring an effective role for the EU in GNSS (EC, 1998a). It included three general aims for developing GNSS: an effective and cost-effective system for civil use and compatible with military needs; high levels of safety, with an adequate European control on any future system; and reassurances that European industry would be able to compete on an equal basis worldwide. Europe was determined to position itself as a key actor in the gradual and coordinated development of a non-military and civilian-controlled satellite navigation system, which would be among the most important public infrastructure projects in the 21st century. 'Europe has taken the decision to participate in the development of a next-generation global satellite navigation network. This important commitment will ensure that Europe plays a key role in the evolution of a system that has long-term industrial, strategic and political implications' (EC, 1998b).[87]

The EC stressed that satellite navigation involved a range of elements of substantial interest for the EU: strategic, political, economic, industrial, employment, security and defense interests. 'GNSS represents a strategic challenge impacting on Europe's position in the world' (EC, 1998a, p.1). It offered the potential for major improvements in the safety and efficiency of all modes of transport systems, as well as of many other applications in different markets. Moreover, since current satellite navigation systems were under military control, security and defense considerations were at stake. 'There are strategic and political interests in relation to Europe's capacity to control the positioning and navigation services for its own territory' (EC, 1998a: executive summary). Foreign control over Europe's navigation systems would have 'serious problems for both sovereignty and security' (EC, 1998a: executive summary). Europe would find it difficult to resist and bargain in a situation where one or more countries had a dominant position or virtual monopoly. If charges were imposed at a later stage, it would be

87 As long as the basic systems and technologies were outside European influence, the capacity for European industry to compete in this booming market would be seriously constrained. According to the EC, the US had clearly indicated that it would use 'the strategic advantage provided by its military positioning system (GPS) to establish a dominant position in the world market for systems and services' (EC, 1998a: executive summary). It was therefore important that the EU and its Member States 'demonstrated their commitment to a full participation in GNSS, so that European industry and potential users have the confidence to invest in its development' (EC, 1998a: executive summary). Commissioner Kinnock stated that there would be so many domestic and international applications that GNSS would be as ubiquitous as GSM was: 'In this technology, late in will mean Europe is locked out of a market estimated to be worth 50 billion dollars by 2005' (EC, 1998c).

difficult to embark on the development of an alternative system.[88] Three broad strategy options were outlined in the 1998 GNSS Communication: (1) joint development of GNSS by all the major players; (2) the EU developing a GNSS with one or more international partners; and (3) the independent development by the EU of its own system. The DG TREN strategy proposal laid down three conditions against which any international agreement would be assessed with respect to the feasibility of developing a joint global system that met European needs (EC, 1998a):

- Europe needed firm guarantees from the outset that the service (on which Europe would depend for certain vital and safety-related applications) would not be withdrawn or disrupted, with, ultimately, a full EU role in the control of the system.
- Europe needed full participation in the design, development and operation of the system.
- Europe needed assurances that European industry would have the opportunity to compete in all segments of the market and with equal access to the basic technologies.

Europe needed to continue to invest in GNSS-2 'to ensure that, if a joint approach does not prove realistic, it will be possible to go ahead with a European system keeping European industry in a position to compete effectively in this domain' (EC, 1998a, p.10).[89] Aside from the international aspects in terms of implementation, decisions would have to be taken on a legal and organizational structure. DG TREN preferred a body and institutional arrangements at EU level to regulate Egnos services and asked for further political endorsement of the ETG (EC, 1997b). The EC would be

---

88 The development of Egnos would rely on the basic US and Russian satellite signals and would leave Europe vulnerable if the operators decided to jam the signals or impose fees on services in the future. Moreover, those players who were setting technical standards would have an enormous commercial leverage. An explicit European political commitment would strengthen the credibility of European ambitions and the associated bargaining power with third countries.

89 The EC outlined the initiatives that would be launched in the domain of international relations. First, high-level strategic discussions would be held with particularly the US and the Russian Federation, including discussions on service guarantees, liability and industrial issues for GNSS-1. Furthermore, Europe would continue negotiations on interoperability with the US, Japan and the Russian Federation. Moreover, Europe would continue to explore opportunities for regional cooperation with other countries and regional blocks on an urgent basis 'with a view to reaching provisional political and technical agreements, where possible within the frameworks of existing EU-third country cooperation agreements' (EC, 1998a, p.17). Finally, the EC would promote the European strategy in relevant international organizations (the ICAO, the IMO, and the ITU) and commit to free trade and open markets.

responsible for institutional, industrial and strategic issues and research as well as defining user requirements for other users than civil aviation.[90] Satellite navigation could be a means to harmonize and rationalize current conventional ground-based navigation infrastructure in Europe and thus unleash critical mass, net welfare economic gains, and give a further boost to the internal market. From a political view, such a development would promote a European dimension with respect to transport, infrastructure, industrial competitiveness and institutional arrangements.[91]

The EC estimated the world market at US$50 billion in 2005 and stressed that European industry would need a full opportunity to contribute to GNSS. 'European industry is aware of the economic importance of this rapidly growing market and the opportunities it will offer in the future. However, it starts from an unfavorable position: for equipment, European industry's present market share is only around 15 per cent of the European market and 5 per cent of the global market' (EC, 1998a, p.15). The EC stressed that the EU needed to develop a European GNSS-2 strategy by the start of 1999 to ensure that Europeans would play a full role in the future of satellite navigation. European industry should be involved in discussions. 'Recognizing GNSS as a significant opportunity for the EU space and high technology industries and its potential dual civil/military use, a dialog on GNSS issues should be set up

90 The EC repeated that European involvement in Egnos would provide an initial foothold in satellite navigation and would be an important stepping-stone for GNSS-2. This system would cover a large geographical area, stretching from the western Atlantic to the Far East, and from South America to the ASEAN countries and western Australia, and from the Arctic to the tip of South Africa. 'This GNSS-1 approach will provide improvements in the navigation services available through guaranteeing the integrity of the GPS and Glonass signals, bringing for the first time full harmonized and integrated coverage of the whole European area. It will increase transport capacity through the ability to support reduced separation and ultimately reducing the cost of navigation aids. The early benefits will particularly assist cohesion since they will be felt most where present navigation infrastructure is lacking. The development of Egnos is allowing the EU space sector to participate, at least on a modest basis, in GNSS and will provide the transport community (the main potential beneficiaries) with a system meeting most navigation and traffic management needs' (EC, 1998a, p.8).

91 The standard GPS signal was offered free of direct user charges in the late 1990s and the US had claimed that it would remain free for the foreseeable future. This situation had created a large civilian market for GPS equipment for US industries. Moreover, the US, as stated in the so-called Federal Radio-Navigation Plan, would 'continue to promote the international acceptance and implementation of GPS as part of GNSS for navigation in all phases of flight and will progressively discontinue terrestrial systems' (EC, 1998a, p.4). The US Department of Transport (DoT) had already developed a GPS network that would provide corrections of GPS to enhance the availability of service in local areas (so-called 'differential GPS') along the US coastline and it was expected that this network would be applied to the railway sector.

with industrial interests' (EC, 1998a: executive summary). In March 1998, the EU Transport Council endorsed the communication on GNSS and instructed the EC to investigate the possible options for GNSS-2.

As a result of the 1998 GNSS Communication, the EU GNSS High Level Group established the so-called EU GNSS-2 Forum in July 1998.[92] The Forum stressed that a rapid establishment of an appropriate cooperation framework with the Russian Federation could offer Europe an opportunity for operational know-how and Europe could also benefit from cooperation on frequency issues. 'There was consensus that the future GNSS system should be a global system including both GPS and an independent, global European-led component (GNSS-2), which should be open to the inclusion of complementary contributions from third countries and organizations, subject to industrial, political, military and security considerations' (EC, 1998e).[93]

---

92 The majority of the participants in the GNSS-2 Forum activities came from user industries (50), the different European space industries (33), organizations associated with the ETG (40), and national ministries and other governmental agencies (24) (EC, 1998e). The EU GNSS-2 Forum consisted of four working groups: Institutional and Legal Issues; Defense and Security Considerations; User and Service Requirements; and Technical and Financial Issues. The first group would focus on legal questions, particularly liability and certification aspects, as well as the creation of an organizational structure for GNSS, and would look at both transitional arrangements and longer-term solutions. The second group would examine how the security and defense requirements could be met in a GNSS-2 civil system, and explore the feasibility of civil-military institutional arrangements. It would use a study prepared for DG TREN on the interface of civil and military aspects and sectors (EC, 1998d). The third group on user and service requirements would define a broad range of user requirements for all modes of transport and the value-added services that GNSS-2 could bring. Requirements for a range of non-transport applications would also be examined (for example, agriculture, fisheries, precision timing, earth observation, survey and meteorology). Moreover, it would examine industry requirements for full participation in GNSS markets, and possible options for cost recovery.

93 At the time the EU GNSS-2 Forum was established, the EC was working together with the ESA and Eurocontrol to ensure that all the basic strategy options remained open. As stated before, the options consisted of (1) joint development of GNSS by all the major partners, (2) the EU developing a GNSS with one or more international partners, and (3) independent development by the EU of its own satellite navigation system. The EC was engaged in intensive talks with its main international partners on the possibility of achieving an acceptable joint system and at the same time it was funding research into a possible fully independent European system as a strategic back-up alternative. The Forum would evaluate the potential options in order to assist in the preparation of the recommendations that would be put before the EU Transport Council in early 1999. The evaluation would be submitted to the EU GNSS High Level Group (HLG) and the EC would then consult with the HLG on the results of the evaluation with a view to making recommendations to the Council. The HLG had forwarded the nominations for participants in the GNSS-2 Forum.

The European Space Agency: A European Navigation Satellite Program

Traditionally, the ESA develops basic systems, tests and validates them, and then leaves the projects as soon as a minimum of satellites reaches full functionality. The scope of GNSS-2 was more comprehensive than Egnos and relations and structures had to be of a different kind. It was about more than just a pilot and development project. It was about setting up a fully operational system with complementary European contributions. The objectives of the program were not technical in nature and the schedule and the financial framework would actually be determined outside the ESA framework. The initiative had come from the EU, which was undertaking to raise the major portion of the program's financing.

The discussion on GNSS-1 and GNSS-2 took place within the plenary sessions of the ESA Joint Communication Board (JCB). The JCB consisted of national delegations of the ESA Member States (for example, space agencies), representatives of the ESA Executive (for example, Ferrazzani, Fromm, Lo Galbo, Pinglier, and Steciw), a Chairman (Tjaden, Germany), and the Program Director. The ESA Director General (Rodota), the EC (Tytgat, DG TREN), and Eurocontrol (Storey), were frequent participants in the JCB meetings on navigation. The novelty of the project, and the complex relationship with the EU and the EC, prepared the ground for an intense debate within the ESA and between ESA Member State delegations in late 1998 and 1999. How should the program be conceived and structured? Would the navigation project be a regional system or a European contribution to a global system that included a number of areas (agriculture, road transport and shipping) but which was not as concerned with air navigation as the Egnos system? What would be the precise relationship between the ESA and the EU? (EC, 1993b, 1994b, 1995b, 1999d, 2000d; ESA, 2000c, 2000d).

In May 1999, Euro 17 million had been committed for ARTES-9 GNSS-2 Step 1 activities (system studies, developments, simulation and validation, and support actions), while the whole budget was estimated at Euro 33 million. In the ideal budget, it was envisaged that Italy would contribute Euro 8 million, Germany 7.8 million, the UK 6 million, France 5 million, Spain 2.5 million, and Sweden and Switzerland around 1 million each. By May 1999, the UK and France had committed most money (3.6–3.7 million), followed by Switzerland 3 million, Germany 2 million, and Italy 1.6 million (interview material). The majority of ESA delegations (for example, Germany, Finland, the Netherlands, Spain and Sweden) urged the ESA to play an important role in the Galileo program (ESA, interview, 23 October 2000). As a minimum, its technical responsibility for the basic system and development work should be unambiguous (ESA, interview, 5 May 1999). GNSS-2, it was argued, was certainly a European program but not a purely EU program. The EC needed technical and financial information to make efficient judgements. It was deemed important to link the ESA activities closely with the EU since the latter would eventually determine the technical solutions. France stressed that the EU should have the primary role. Its commitment to GNSS-2 would rely on planned EU policy decisions in 1999. The French and Norwegian delegations argued that there were no compelling reasons to make commitments to the second step of the program until political decisions on those activities were taken (ESA, interview, 23 October 2000; ESA, 1999a, 1999c, 2000b).[94]

There were different opinions on the decision making process with respect to the relationship between GNSS-1/Egnos and GNSS-2. The French, Norwegian, Spanish and Swiss delegations, for example, stressed that the two were linked and Egnos needed to gain real momentum before the ESA started GNSS-2. They continued to submit proposals to ESA to foster a rapid acceleration of the Egnos program and enhance its credibility (ESA, interview, 23 October 2000). The Danish and Swedish ESA delegations were reluctant to support the Egnos efforts for different reasons (economic and technical-strategic, respectively). DG TREN intervened and stated that backward compatibility with Egnos would be a main criterion for evaluating industrial proposals.

The majority of ESA delegations supported a proposal to move ahead with the European GNSS-2 option in cooperation with the EU through a new ESA navigation program rather than pursuing it as a new element within the ESA ARTES program. The ESA Executive and DG TREN together presented a draft proposal for a new European satellite navigation program in late 1998 (EC/ESA, 1998). 'The window of opportunity is open now, it being understood that GNSS-2 needs to be operational in 2008, in parallel with the completion of the US GPS Block 2F and in time for the renewal of the Egnos/GNSS-1 space segment' (ESA, interview, 25 May 1998). Economic and strategic interests helped to set the overall timetable. 'Any later start of GNSS-2 would put at risk the ability of GNSS-2 to conquer a reasonable segment of the commercial market and to protect overall European strategic interests' (EC/ESA, 1998). The ESA and the EC recognized that the European GNSS-2 would largely rely on public funding. Coordination was therefore important with a view to aligning the planned EU satellite navigation program. The cost estimates for GNSS-2, as estimated in late 1998 (21 MEO/3 GEO satellites) when being implemented as a complement to GPS, are presented in Table 3.2.[95]

---

94 The French delegation stressed that the second step of the program was premature and should not proceed until the EU had determined European policy for satellite navigation in 1999. Coordination with national space agencies was necessary.

95 The ESA GNSS-2 program (initially referred to as the 'Eunavsat') program would basically consist of two phases. Phase 1 (Development and Validation) comprised

Table 3.2 The ESA GalileoSat program I

	PHASE 1 Development & Validation (Euro million)	PHASE 2 Deployment (Euro million)
Prime management and System engineering	60	80
Space segment (including launches and GEO)	350	720
Ground segment (including test receivers)	136	120
Certification and security	100	200
Controlled access service	100	200
Total of contract values	746	1320

Source: Based on interviews with ESA officials.

The early ESA discussions on GNSS took place in the Joint Communications Board (JCB). Aside from the ESA Executive, national delegations and a DG TREN official, the ESA Director General (Rodota) and DG TREN Director-General (Ruete) occasionally joined the discussions, which signaled the importance of EC-ESA collaboration. The ESA Executive was well informed about other European activities on satellite navigation, such as the GNSS-2 Forum, which had met between July and December 1998, and reported on those activities to the JBC. The EU decisions would be decisive and lay down the mission requirements, specifications and security requirements. The ESA Executive assured the delegations that the ESA and the EU worked closely together (ESA, interview, 5 May 1999).

A number of delegations, such as the Norwegian, viewed Europe's decision to develop an autonomous satellite navigation system as mainly strategic. The drivers for Europe were here above all political. The French

the overall system design and the development of all major hardware and software products with the objective of validating design assumptions, compatibility with GPS, and so on, through an in-orbit validation program. This phase would run from 2000 to about 2005 and be preceded by the ARTES-9 GNSS-2 Step 2 system definition phase. Phase 2 comprised any necessary refinement of the overall project design, the procurement and launch of the first generation of operational satellites, the completion of the ground segment and the operation of the Eunavsat space and ground infrastructure. While the first phase would be publicly funded, national and European public authorities hoped that the second phase would run as a public-private partnership with the ultimate aim being to transfer the responsibility for Eunavsat service provision to industry. It was believed that industry could only be expected to take a calculated risk once the major uncertainties had been removed and the concrete boundary conditions were known.

and Norwegian delegations stressed that GNSS should not be conditional on detailed cost analyses and the promise of a public-private partnership. They regarded the navigation system as primarily a public service system. The Swiss delegation compared the situation with the strategic decision taken in 1973 to develop the European launcher Ariane. The Italian delegation also stressed the political importance of satellite navigation and the crucial role that the ESA must play under the definition and development phases. In addition, the UK stressed that the envisaged European system must be placed under civilian control (ESA, interview, 5 May 1999). By stating the strategic interest of the program in terms of industry, job creation and market aspects as well as the negative implications of adopting a wait-and-see approach, ESA members would try and persuade European governments.

The French delegation therefore declined any discussion of the management structure above the project level since this was viewed as a matter for the EC and conflicts of interest should be avoided.[96] The EC, the ESA French delegation stressed, should have the task of defining, managing, and overseeing the overall program and delegating specific responsibilities. The issue of whether the ESA would have any co-management role above the project level, and how to define applications, thus raised some intense discussion within the ESA JCB.[97] The ESA Council examined the proposed program and unanimously adopted a resolution in April 1999 that would enable its initiation (ESA, 1999b). The proposed ESA GalileoSat program would allow ESA Member States the opportunity to concentrate resources

---

96 The ESA Executive and some delegations (for example, the Dutch and Italian) stated that the technical side of the development of Galileo would be the responsibility of the ESA. In addition, the ESA Program Director for Applications, the JCB Chairman and the Finnish and Belgian delegations argued that the ESA should continue to support applications. In their view, there was no formal obstacle to prevent the ESA from examining certain applications-related aspects, even though it would come close to the EU's domain, if the necessary funding were provided from within the ESA. The Dutch and Norwegian delegations expressed the view that it would be odd for the ESA not to play a significant role in defining, developing and validating the system. The ESA Executive responded that that GalileoSat would be an ESA program under the aegis of the EC, which in particular would be responsible for international relations, and the role of ESA would be restricted to the technical area (ESA, interview, 23 October 2000).

97 The ESA Executive stressed that the important issue was that the ESA Member States supported the political gestures provided by the EU. The emerging European strategy had to be unequivocal to allow Europe a stronger bargaining position *vis-à-vis* the US. The Italian and the UK delegations both stressed that Europe had to move ahead urgently in order to be able to compete in the area of applications with the US GPS, which was being modernized. The European system would have to be available before the modernized US GPS system was implemented. The ESA Director-General stressed that the program was important for Europe. It created jobs and demonstrated Europe's capacity to act as one.

into a harmonized policy under ESA management.[98] The resolution further stated that the ESA industrial policy scheme should cover GalileoSat.[99]

Summary

The EU had literally no presence in satellite navigation in the 1990s and this sector was dominated by the US industry supported by its government (with a strong presence also of Japanese industry in customer applications). The increasing use of the US GPS constellation for public and commercial use triggered activities in Europe. The combined efforts by a network of administrators, experts and policymakers in the ESA, Eurocontrol and the EC prepared the ground for the gradual movement toward a European approach and a common strategy for satellite navigation. The development of the first generation of a European capability in this field, the Egnos constellation, provided Europe with the necessary expertise for developing and launching a second-generation capability. The majority of ESA Member States supported the Egnos/GNSS-1 project, while Sweden and Denmark remained outside the project for technical-strategic and budgetary reasons, respectively. The GNSS-1 efforts prepared the ground for the second-generation global navigation satellite system (GNSS-2). The EC strategically represented the need to support satellite navigation in terms of European economic welfare, strategic-political independence and ownership, industrial competitiveness, technological development, and the need to reduce the current US global monopoly. It pointed to the benefits to local constituencies and government revenues. The next step for the EC was to foster a strategy for a longer-term capability by supporting GNSS-2 (Galileo).

---

98 The objectives and activities of the ESA GalileoSat program, within the context of the implementation schedules proposed in the EC's GNSS-2 Communication in February 1999, consisted of five points: (1) to define, as part of and in support of, the overall GNSS-2 definition activities, paying due attention through GalileoSat to the requirements derived from the objective of setting up a competitive GNSS-2 system under a public-private partnership scheme; (2) elaborate detailed specifications for GalileoSat and undertake system simulations and analyses in support of concluding the specification work; (3) support any decision of the GNSS-2 Program Management Board for an integration scenario of GNSS-1/Egnos into GNSS-2; (4) provide a complete and binding industrial proposal for the development and validation of GalileoSat including a ceiling price offer for its full deployment and operations; and (5) monitor closely the evaluation of the GNSS-2 initiative with a view to reporting to governments.

99 The geographical return targets would be based on the scale of contributions on the part of the ESA Member States that subscribed to the GNSS-2 sub-envelope. The ESA Member States expect contracts and work activities associated with the program, and roughly equal to their contributions, to return to their economies through the ESA industrial policy of fair return.

4. Galileo: Europe's response

In 1994, the EU Transport Ministers invited the European Commission (EC) to initiate the preparatory work needed for the design and organization of a European contribution to a second-generation global navigation system (GNSS-2). This infrastructure system would be under civilian control and operate according to international guidelines (Council, 1994).

> In the beginning it was not specifically based on the goal to build our own satellite navigation system in Europe. Rather, when it was time to build the next generation navigation system Europe should definitely be part of the process. The idea was that users would not always rely completely on an American system. If there were a joint system, then one of the conditions on the European side would be that Europe would help to build it. In addition, there was a prestige aspect involved in the whole development, that Europe wanted to show that they could indeed do something (EU national government advisor, interview, 25 January 2000).

4.1 MAKING HEADWAY FOR THE GALILEO PROGRAM

Initially, the momentum behind satellite navigation came from the European Tripartite Group (ETG) organizations and certain EU Member States. 'The initiative to move ahead with the GNSS-2 satellite navigation system was not least supported initially by France, but also by Germany and the UK' (EU national government advisor, interview, 25 January 2000). Europe embarked on significant investments and preparations for the future.

> It is no secret that governments with stronger satellite industries have stronger interest in the project. France, Germany, Italy, the UK – there are quite a few players who are really important. It was a political decision at that time. It was not a decision that came from the companies but from governments to say that they had a problem and wanted to solve it together. Managing such a program from the EC side is something quite unique because it is very big and expensive (ESA, interview, 23 October 2000).

According to the EC, which was brought into play around 1993–1994 and helped formulate the core of a new policy, satellite navigation would contribute to the European economy, industry and technological prowess. 'Satellite navigation is one of the key technology and strategic areas in which the EU can encourage its electronics and aerospace industries to take a market-leading development role' (EC, 1998b, p.8). The objective for the EU and Europe was to ensure that Europe would play a future role in satellite navigation. To support this ambitious European initiative, the EC and the ESA stressed the need for new institutional arrangements and mechanisms (for example, multinational ownership, operation and control).[100]

> Navigation is an infrastructure on which you will start to depend and over which you will have control. It is like [...] a road network. You want to make sure that everyone can use it and it is thus a kind of public investment. Navigation is giving you a tool, which can be used in many different ways, including military aspects. Moreover, you want to have all the intellectual property rights on the technology. There was a real point for the political side to get involved. Navigation was not a national endeavor. You really have to do it as Europeans (ESA, interview, 23 October 2000).

Between the EC report on satellite navigation in 1994 (EC, 1994a) and proposals to establish the European tripartite Group (ETG) in 1997 (EC, 1997b), the level of political activity remained moderate.[101]

---

100 The European Galileo radionavigation program was part of a broader institutional process. 'I was working on collaboration with the EC for years and we never spoke about satellite navigation until 1993. We started to ask Brussels what their plans were and about what the Americans were doing. We were under pressure in Europe by the Americans and we discussed whether we could do something together. A stronger relationship came naturally by the progressive definition of EU policies in the area of applications, which is the key concern today. A few years ago, some of our Member States even envisaged incorporating the ESA into the EU. This is not perhaps as mature today but in the future of progressive European integration there may be some type of umbrella. We need the political support of the EU. Could the ESA become an additional DG of the EC? Could it become an agency of the EC such as the European Environmental Agency? Could it be attached to DG Research/Joint Research Center as part of a technical center of support? Many options are discussed' (ESA, interview, 25 March 1998).

101 During 1994 and 1997, efforts went into securing funds for the first generation of satellite navigation in Europe (Egnos). The Inmarsat tender package, which triggered the EC 1994 communication (the decision to move ahead with GNSS-1 and the need to make an offer in October 1994), corresponded to 5–10 years of usage and around Euro 1.5 million per year/transponder (about 10 transponders). To exploit the network of satellites that the transponders constituted, additional heavy investments had to be made in ground infrastructure. It represented a substantial European investment in satellite navigation infrastructure. The initiative

> It really became truly serious in 1997–1998 [...] The Americans had indicated that they would not waive any control over GPS to any other parties, which in a way pushed, or you may say forced, the Europeans to consider instead having their own system. You need an independent system [...] if the system failed for political or technical reasons. GPS failed once or twice in the beginning. You cannot build all infrastructures on the assumption that you have access to another system. It is similar to a situation where you would rely solely on one producer of electricity (ESA, interview, 23 October 2000).

It was recognized in Europe that, to build a real satellite navigation system, there was a need to raise funding not only for a lengthy development process but also for its operation.[102] 'In principle, satellite navigation should be similar to the development of Airbus or Ariane where you needed a few good and strong promoters. However, given the political (strategic/military) dimension people felt quite strongly right from the beginning that it had to be integrated into the larger European context, that is, the EU' (ESA, interview, 23 October 2000). Kinnock, the EU Commissioner for Transport, stressed the pressing need for intensified efforts in Europe in 'a crucial period in the process of establishing a European strategy on GNSS' and regretted the degree of looming 'lassitude and indecision in some quarters' (Kinnock, 1998). 'I see my role as helping to make a bridge between the scientific and space interests, which fully understand the significance of GNSS, and the political and user community interests, which are only now beginning to grasp its potential' (Kinnock, 1998). Kinnock admitted, however, that Europe had experienced a change of gear on satellite navigation since Spring 1998.[103] EU Member States started to realize that Europe was reaching times when

---

originated from the work and efforts, a limited network of people in the EC, the ESA and some Member States, and their combined efforts prepared the ground for information on possible alternative venues.

[102] The ESA, which is traditionally a technical development agency, secured funding for space applications such as communication and broadcasting satellites without too many political constraints in the past. One exception was earth observation, where the ESA was forced to concentrate on radar observation, since optical technology (which allowed higher resolution) was considered to be too closely associated with military aspects and national programs (ESA, interview, 23 October 2000).

[103] 'Finally Europe managed to progress on a project on the technical and political level. Today, the problem is that we are facing a strategic and a substantial economic issue in terms of investments. The EC has defined a strategy in the satellite navigation area and we have defined technical activities matching this strategy. Unfortunately, the industrial and national interests of many Member States are prevailing' (ESA, interview, 25 March 1998).

important decisions had to be taken. However, relevant industries and user groups needed to step up their efforts to lobby governments to complement the planned EC report on GNSS-2 in early 1999. 'They [Member States] cannot be told too often that if we do not make a decisive effort now to establish a European strategy, we will have lost an unrepeatable opportunity' (Kinnock, 1998).

In order to exploit the unique opportunities at hand in terms of applications across economic and industrial sectors, market growth and job creation, Kinnock stressed that it would not happen 'without combined and consistent political action to support the commercial and technical efforts of industry' (Kinnock, 1998). The largest market would be in applications. It was thus important that European industries made a case to their governments that they would need support. Kinnock pointed to the role of political-strategic issues. 'There are security and sovereignty arguments that must be impressed upon decision and policymakers by inviting them to consider the reality. If active commitment is not made, within 10 years all of Europe's safety critical transport activities could well be reliant on a system that is not within European control' (Kinnock, 1998).

Kinnock also took the opportunity to announce that the contacts with the US authorities 'have confirmed that there is no prospect of joint ownership and control of the core GPS, primarily because of its continuing military role' (Kinnock, 1998). The US Administration would not allow a real transatlantic business partnership. 'European authorities have been negotiating with the US and Russia to join their systems with ours to create a single global service. However these negotiations turn out, we are increasingly confident that in 1999 a European program will be inaugurated, with or without them. We do not want to leave it to a US monopoly,' said Alcatel Espace (SN, 1998a).[104] Vigier, an expert in the cabinet of EU Research Commissioner

---

104 Alcatel Espace received a Euro 3.8 million (US$4.2 million) contract from the EC in 1998 to study a navigation constellation that would include 63 operational satellites operating at Low Earth Orbit at about 1.400 kilometers in altitude. An additional contract from CNES amounted to US$838,000 and permitted Alcatel to study specific technology options for the system. DASA's Dornier Satellitensysteme had received a five-year contract from the German Ministry of Transport valued at $42 million and also had smaller contracts with the EC. The DASA work included Russian companies and expertise. 'Europe is going to get this business; there is no question about it. The question remains how. European government authorities want only one industrial proposal, and that is more important than the technical differences in our proposals compared to Alcatel's,' said a DASA official. 'A big US electronics company came over to say that it had perfected a system to prevent GPS signal interference for commercial aircraft during landing. Everyone applauded, and then one of us asked if we could have access to the algorithms to the software. The response: 'Sorry, but this is classified. Only US companies are granted access"' (SN, 1998a).

Cresson, confirmed that the EC was pessimistic about developing a real partnership with the US. 'The question is whether the Americans are willing to enter into a real partnership. During our last round of negotiations, the US delegation was 40 per cent military, so I am not particularly hopeful' (SN, 1998b). These statements refined the European strategy options outlined in the EC 1998 GNSS report in that a joint EU–US approach had essentially been ruled out. 'It turns out that it is unlikely that the US would grant a substantial role in the control of the current GPS constellation. However, options such as the implementation of GPS-like complementary satellites by Europe, and even cooperation in the development of a new civil GNSS navigation system, remain open. Consequently, while an approach of joint control and ownership is excluded, there are sufficient common interests for a close cooperation between Europe and the US' (EC, 1999d: annex, p. 16).

Kinnock argued that the US willingness to cooperate was based on purely strategic and commercial interests. 'Obviously, the willingness to cooperate reflects the fact that the US needs European support in areas such as allocation of scarce frequencies. The navigation community must use its frequencies effectively and also defend them vigorously both from competing interests and in establishing GPS-compatible technical standards in Europe and internationally' (Kinnock, 1998). European ministers responsible for space affairs expressed similar views: 'My preference is for an independent European system, for the same reasons that we developed our Ariane rocket series,' said Yvan Ylieff in June 1998, Belgium's minister for science policy (SN, 1998c). ESA Director-General Rodota agreed: 'Depending on the US for our own future navigation needs poses too many risks. We cannot leave this market to US monopoly' (SN, 1998c).

The opportunities for cooperation with the Russian Federation remained unclear although its ambition to improve Glonass and develop an advanced GNSS would still be good reasons to explore cooperation. It had control of scarce frequencies and had an invaluable experience in operating a satellite navigation system. 'The basic political question, therefore, is whether we can reap these benefits without creating excessive risks for the European initiative, whilst the basic technical question is whether, taking Glonass as a starting point, we can develop a system that is interoperable with GPS in a cost effective way' (Kinnock, 1998).[105] Although Kinnock admitted that there

105 'Making Galileo compatible with GPS and Glonass is our way of assuring that navigation frequencies are protected in the future. Beyond that, we might benefit from some cooperation with Russia; it will be economics, not diplomacy, that drives our decision,' said Tytgat, DG TREN (SN, 1999c). The EC had concerns about Glonass' ownership by the Russian Defense Ministry (SN, 1999k). Japan was also mentioned as an important actor and discussions were under way to

was always a 'zero-option' (Europe would have no system at all and thus no substantial public money would be invested), he was 'certain that it would be immensely costly strategically, economically, and industrially in the medium to long-term' (Kinnock, 1998). Kinnock outlined three issues that would need careful assessment: exploring the requirements of the European military (the civil-military interface); the problematic issue of financing; and the potential for a public-private partnership (PPP) discussed in the GNSS-2 Forum.

In the EU Galileo budgetary strategy, the concept of a public-partnership project (PPP) was a central issue and would remain elusive for a long time. 'Significant initial public investment will be necessary for development and deployment of the European contribution to GNSS. However, it appears unlikely that there will be sufficient public resources available to cover all the aspects of the implementation of the system' (EC, 1998e). Commissioner Kinnock stressed that the project would require a public sector and private sector funding partnership.[106] 'PPP models should involve a distribution of risk and should let private partners identify further revenue streams that could be based on the quality of a European GNSS. Elements to be considered include the possibility to contractually guarantee levels of service. Further work needs to be done in consultation and with involvement of the European Investment Bank (EIB) and (in the case of cooperation with Russia) the European Bank for Reconstruction and Development (EBRD)' (EC, 1998e).

The main public concerns about Galileo were the issues of financial investments, program risks, benefits, and institutional issues, such as the working relationship between the EC and the ESA, the evolving program management, and negotiating GPS interoperability issues. Among the primary concerns in the private sector were the transfer of leadership for the system development to private managers as soon as possible and adding substantial funding for early market development that would parallel the technical development work. Products and services would have to be available by the time the Galileo satellites were deployed. 'As a downstream industry, Thomson is reluctant to take on risks before the end of the test and validation phase. But industry can help fund deployment costs under a private

---

explore the scope for cooperation with Japan, which had recently signed an agreement with the US.

106 The EU Council endorsed an initiative for a European Employment Pact in June (Cologne, Germany) that would take into account the strategic-economic importance of Galileo and the need to step up efforts to secure private funding (GPSW, 1999c). European sources had indicated that such arrangements would be open to commercial organizations from outside Europe, including the US and Japan (GPSW, 1999d).

financing investment (PFI) scheme, depending on the business plans and concession terms,' said a Thomson-CSF official (GPSW, 2000e).[107]

The EU estimate was that there was a potential market for navigation equipment in Europe of around Euro 122 billion for 2005–2023, with an expected market for services of Euro 113 billion and an export market of Euro 50 billion during the same time period. European industries involved in Galileo pushed for a decision on the financial structure for Galileo and for speedier implementation in order to reduce 'the time to market' of future Galileo services and applications. 'We cannot invest money into a system if we are in an unfair situation [relative to] our US competitors. Nobody would put one dollar or one Euro in a program where first we have to spend money to bring us to the same level as our competitors. There's no chance,' said an Alcatel Espace official (GPSW, 2000e). In November 1999, Alcatel had already been tapped as lead contractor for the EC's GALA and such industry statements were clearly intended as signals to the EC and EU Member States. The EU would have to spend government funds to bring Galileo to the point where it was equivalent to GPS before European industry would invest.

By this time, the main market focus had shifted to the promises of land transportation, which was viewed as being 77 per cent of Galileo's potential market. This sector includes cars, trucks and fleet management and assumes the integration of Galileo capabilities with mobile phone services (GPSW, 2000a, 2000d). A consultant at the French Development Council, who was very close to policymakers in DG TREN, stressed that sea and air markets would not be the largest customers. 'Sea and air applications now make up a mere five percent and one percent, respectively, of the total European satellite navigation market. Expenditures in these sectors are estimated to reach US $37 million by 2010, a low figure relative to other market segments due to the fact that these sectors do not replace equipment very often' (Campagne, 2000, p.18). One suggestion being studied was to have the EU mandate that each car be equipped with a Galileo receiver and a wireless communications device to convey the car's position to a control center. The justification for requiring satellite navigation might come from a need to control traffic on Europe's increasingly congested highways.[108]

---

107 Parish, Director-General of the influential UK Federation of Electronic Industries, gave a strong endorsement of Galileo in 2000 and urged the UK government to support it at EU level. Infrastructure, he argued, must be funded in large measure by government. Private investment in systems and services will exceed the infrastructure costs many times over. The UK will have to contribute its share of the EU funding if Galileo goes ahead. If the UK Government does not play its full part, the UK will gain very limited benefits and still have to contribute substantially to the development costs of the system (GPSW, 2000e).

108 Mandating satellite navigation in cars as the financial foundation for Galileo would have a number of potential ramifications. For European users there was the

> From the point of Galileo, the EC looks at aviation as only being one per cent of the real user base. What they are looking at in terms of benefits to the EU at large are macroeconomic benefits. However, the billions really come from those selling applications that use the signal, building it into mobile phones, putting them into cars through a little Galileo chip to measure speed, and use as automatic road tolls instead of having toll booths. Just what the EC always wanted – total control and Orwellian 1984, I am afraid, but that is the way the EC thinks (NATS, UK ATS provider, interview, 15 February 2000).

The cost of building Galileo was compared to high-speed train infrastructure, which was a hot topic among transport ministers in the EU. DG TREN (Tytgat) stressed that the cost of Galileo would correspond to the development cost of about 100 kilometers of High Speed Train (Tytgat and Campagne, 2000). This was further clarified: 'Public funding during the initial project phases should not be a real problem given all the important political, strategic, and social advantages that Galileo will provide Europe. For example, the construction of the Innsbruck-Fortezza high-speed train sections (55 kilometers) amounts to Euro 3.8 billion, while the development of Galileo is estimated at around Euro 3 billion' (Campagne, 2000).[109]

Within the GNSS-2 Forum (July to December 1998), which consisted of senior officials from both national and European level, and from the private and public sector, a special emphasis was placed on the Working Group on

---

potential that a mandatory on-board system would enable tracking of individual movement, account payments and service enforcement, and so on. A problem was that the average European citizen, however, might not be too enthusiastic about the associated extra cost and the potential impact on personal privacy that Galileo would entail. There was an argument out there, though, that more efficient intelligent transportation and better traffic control could lower costs, and improve safety and environmental aspects. Interviews with industry and EC officials indicated that having land transportation as the financial basis for Galileo could undermine some of the arguments for changes to the proposed system on the part of US interests, for example that safety-of-life service should be free. Land navigation mandates would probably not impact on international operations in the same way that mandates for Galileo receivers on board ships or airplanes might. The former would be a more domestic European-oriented measure that would affect US citizens and industries less.

109 As GNSS technology would become integrated with other services, the market for satellite navigation information and equipment was also becoming more geared toward integrated systems, for example, advanced asset tracking systems. As such, the future of the Galileo service may be determined by the choices companies, particularly automobile equipment manufacturers, make when they decided which receiver to integrate into their products. 'The key to the success of Galileo is to find partnership with the car industry and [with] the electronic suppliers of cars. If we are not successful with that, there could be a serious problem,' said a consultant to the EC (GPSW, 1999g).

Technical and Financial Aspects. DG INFSO Director (Telematics R&D), Richonnier, chaired the group with assistance from EC officers and the head of the ESA Communication Systems Department (Fromm). As an overall guideline for the analysis, the group linked a geo-strategic perspective (the need for Europe to be a major player in an independent and global space segment) and an economic perspective (to allow industry and service providers in Europe a significantly better position in the expected vast market). The Technical and Financial working group suggested that Europe should proceed rapidly toward the development of GNSS-2 and guarantee at least the independence of Europe for positioning, navigation and timing (EC, 1998e).[110] Not investing in a European-led GNSS would put European industry (space, manufacturing and service providers) and the whole European economy at a disadvantage in the longer term.

In addition, this infrastructure system had to be global since (1) safety-related applications in the civil aviation or shipping sectors using transport networks integrated across boundaries demanded a global system, and (2) manufacturers wanted to market interoperable user equipment directly to global markets. Galileo would 'promote opportunities for European industry to participate in all areas of the GNSS market and to develop the European market for GNSS added-value applications' (EC, 1998e).[111] The EU Space High Level Panel met at several occasions in 1999 to discuss aspects related to Galileo (for example, public-private partnership for the Galileo program and potential revenue streams), among a number of other strategic issues.[112]

---

110 The participants identified a number of overall limitations with respect to Egnos (EC, 1998e). First, GNSS-1 would be heavily reliant on the US military-controlled GPS. Second, industrial return and employment opportunities would be limited by Europe not being involved in developing its own global navigation system. Egnos was used jointly with GPS and/or Glonass to provide a GPS navigation-like signal. Egnos was also vulnerable to jamming and interference. Finally, certification for many important, and particularly safety-critical, applications was difficult if not impossible due to the lack of guarantee on (GPS) service continuity (the liability issue). The group concluded that more work had to be done on how Egnos could be integrated with GNSS-2 and how an Egnos operational structure could be adapted to fit in with the requirements for GNSS-2.

111 The overall Panel consisted of certain Commissioners, the ESA and industrial firms. The EC had established it to define European space policy. Concerning the financing strategy for Galileo, DG TREN had proposed a combined approach in its 1999 Communication (EC, 1999b).

112 The Panel concluded that different funding mechanisms and sources could be used for different phases. While the definition phase would be largely financed out of public funds (ESA/EU), the Panel members suggested that the development phase could be financed through public funds plus loan guarantees. The public funds would come from the EU Member States and the loans, which would be paid under guarantee, by institutions such as the European Investment Bank (EIB). Interest rate subsidies could be used for the validation phase under the EU Trans-

The Panel discussed in particular two possible revenue streams: a general levy on receivers and controlled access services (also called 'public regulated services', PRS). The debate on the role of encrypted public regulated services would stir a controversial political debate in Europe (Lembke, 2002c).

4.2 EUROPE'S PLACE IN SATELLITE NAVIGATION

Prior to the official release of the awaited EC GNSS-2 report in early 1999, the EU GNSS-2 High Level Group held a meeting to debate the draft communication. DG TREN Director-General Ruete chaired this restricted high-level meeting together with Director Richonnier (DG INFSO) and representatives from the EU Member States participated. Regarding the overall funding of GNSS-2, Germany insisted that it would not accept extra EU spending above the amount foreseen in the existing programs, and the project had to be co-funded by the EU Member States through the ESA. Ruete and Mastracci (ESA) gave assurances that EU funding would come from within existing budgets, that efforts would be undertaken to clarify further the financial aspects and to explore whether current cost estimates could be reduced (EC, interview, 21 December 2000). DG TREN announced that it would seek to add GNSS to the list of TEN-T (Trans-European Networks for Transport) priority projects. It would also encourage substantial private investment. Germany and France suggested that the EC should not hesitate to suggest a levy on GNSS receivers to strengthen the GNSS-2 business plan (EC, interview, 26 May 1999).

In February 1999, the EC then launched the report dedicated to the development and implementation of GNSS-2 (EC, 1999b). The European GNSS-2 concept was renamed 'Galileo' at the end of January 1999; this initiative originated from the circle headed by DG TREN Director-General Ruete and senior policy advisor Tytgat (ESA, interview, 29 January 2000). A senior EC official remarked that the Galileo communication enjoyed support from powerful Commissioners. 'It was not difficult at all. We had a very strong leadership from Kinnock and the Commissioner for Industrial Affairs, Bangemann. There was no blocking within the EC. None whatsoever' (EC, interview, 21 December 2000).

European Networks (TEN) budget. It could thus be financed through a public-private sector partnership similar to the financing of the English Channel rail tunnel (the Euro Tunnel). In the deployment phase, investors could supply equity and commercial banks, capital markets and EIB could provide non-guaranteed loans with interest rate subsidies through the budget of EU TEN. In the operational phase, funding would essentially come from private operators.

The arguments about the positive contribution of GNSS-2/Galileo in terms of applications and socio-economic benefits had been further refined over the years.[113] Kinnock stated that the EU and Europe must push ahead on its own in the absence of satisfactory international agreements. 'We would prefer to cooperate, but only under the right conditions,' Kinnock said. 'That means a full European role in systems development and operation, guarantees of adequate control of satellite signals and the opportunity for European industry to compete on equal terms' (GPSW, 1999b). The EC Galileo report set out a strategy to secure a full role for Europe in the development of the next generation of GNSS. The EC argued that the initiative provided a means of achieving Europe's strategic, commercial, transport and employment objectives at an acceptable cost (EC, 1999f, 1999g, 1999h) and EU Council of transport ministers also welcomed the report.

In May 1999, an ESA Ministerial Council meeting took place in Brussels. It was the first full-scale ministerial conference on long-term space policy since October 1995. The ESA Ministers decided to proceed with a full definition program for Galileo in cooperation with the EU. The ESA Ministers appeared to support rapid implementation in order for Europe to gain comparative advantage *vis-à-vis* the US GPS modernization program. EU Commissioners Bangemann, Cresson and Kinnock attended the meeting, together with Allgeier, the EC Space Coordinator.[114] Kinnock welcomed the

---

113 Galileo would be an integral part of European traffic policy. It would enable the implementation of a harmonized traffic management system based on satellite navigation and positioning service and complementary data communication. In the civil aviation sector, the application of satellite navigation, combined with a suitable communication system, would offer shorter routes, quicker access to airports through simplified navigation procedures, and enable final approach and landing at most airports, thereby improving passenger safety and reducing the cost of terrestrial infrastructure. It was claimed by the EU institutions that it could provide a more efficient use of the existing aircraft fleet and airport infrastructure. In the maritime sector, satellite navigation is applied to safety navigation, fishery control, container tracking, fleet management and maritime distress systems. In the railway sector, satellite-based rail applications may be used for fleet management, signaling and train control. In addition, it is used for many non-transport applications. In the agricultural sector, satellite navigation can be used to minimize fertilizer and pesticide usage and to maximize yield, a technology called 'precision farming'. Satellite positioning plays a central role in offshore exploration (for example, for exploration work, the operation and servicing of platforms and the precise allocation of claims). There are also applications for surveying (geodesy) purposes, for example, to help determine deformations of dams at detailed level. Finally, satellites provide global time reference, mainly for the synchronization of communication systems and as a frequency standard for power plants.

114 In spring 1999, all the Commissioners (including the staunch Galileo supporter Kinnock) announced that they would resign collectively due to corruption scandals. The EC was accused of cronyism and mismanagement in a report published by an independent panel of the European Parliament.

commitment to Galileo. 'Navigation in land, sea, and air transport is an excellent example of that, and the potential benefits for users and producers in Europe are massive' (GPSW, 1999a). The ESA Executive had asked for Euro 40 million through 2000 for the ESA GalileoSat program, followed by another Euro 460 million in early 2001. In fact, ESA governments committed Euro 60 million to the end of 2000, 50 per cent more than ESA's original request (SN, 1999c, 1999e, 1999g).[115] In total, the ESA had asked for Euro 3.2 billion and received commitments of Euro 2.4 billion (Table 4.1).

Table 4.1 ESA space policy and priorities (Euro million)

Program	ESA Proposal	ESA Ministers' Commitment
Navigation, 2000	40	60
Earth observation	759	593
Ariane 5 rocket upgrade	600	533
International space station, 2001	344	298.5
Vega rocket development	317	Decision delayed
Satellite multimedia technology	309	260
Microgravity experiments, 2003	98.4	48
Guyana Space Center operations, 2000	87	Decision delayed
Future launcher		
Technology program, 2003	70	50
Adv. telecommunications research	50	30

Source: SN (1999g).

The allocation of public money depended on the political decision to be taken by the EU Transport Council planned in June 1999. 'Galileo is an expression of European autonomy, but on an extremely far horizon. We are going to try

115 The UK government signaled that it was ready to commit to the planned ESA satellite navigation program and to take a leading role. 'We have experience in air traffic management and safety-critical software. This is going to be a very big market in the future,' said Lord Sainsbury, head of the UK delegation and chair of the ESA Council (SN, 1999f). 'The commercial potential for value-added products and services resulting from satellite navigation are enormous. To fully take advantage of this potential, we need to be involved in the space segment as well. This was the feeling of the broad majority at the conference,' said Sainsbury (SN, 1999f). The UK had taken a back seat in ESA/European space affairs in the late 1980s and 1990s because the ESA, in its view, focused too much on major programs of less 'utilitarian' character, for example, a manned space capability, the Hermes spaceplane project and the Ariane-V rocket programs (Reay, 1991).

to accelerate this schedule,' said Claude Allegre, French Minister for Research, Education and Technology (SN, 1999g).

On 17 June 1999, the EU's Transport Council convened to approve the creation of the proposed Galileo proposal. Heads of State and Government met in Cologne prior to the Transport Council meeting and they endorsed Galileo as a formally approved EU program. Kinnock again stressed that a European GNSS-2 component was needed to ensure that the EU was not locked out of the market for satellite-based services, which could be worth US$50 billion by 2005 (GPSW, 1999c, 1999g). 'I have to say that we have only a short span of opportunity to join the league of global players that is currently dominated by the US,' said Kinnock. 'Of course, the US naturally wants to consolidate and strengthen its position for the same reasons the EC is recommending the development of Galileo' (GPSW, 1999b). At stake were 'thousands of jobs in high-tech industries, stimulus to economic growth, increased world influence, greater regional security, and substantially improved transport efficiency and safety. In addition, the development of a European system would give the world an alternative to monopoly' (GPSW, 1999b). The promotion of satellite navigation in Europe was primarily a question of strategic independence and ownership (GPSW, 2000b, 2000e; SN, 1999a, 1999d). Allemand, CNES Galileo Manager, agreed:

> Access to positioning data and to a common time reference is vital to our economy [...] and to effective military action [...] The EU cannot remain indefinitely reliant on a system that gives the US military a monopoly in such a crucial area. Indeed, the system not only conditions essential economic infrastructures but also Europe's capacity to play an independent role in the international arena (Allemand, 1999, pp.14–15).

DG TREN Director-General Ruete stressed that the proposals presented by the EC were the result of many years of day-to-day activities (EC, 1999h). The EU Transport Council unanimously approved the EC's proposal and gave its backing to the initial definition phase of the Galileo project, with the aim of allowing the EU to enter the world satellite navigation market. The EC was instructed to come up with a detailed cost-benefit analysis of project funding and to pave the way for a public-private partnership. The EU ministers for transport urged the EC to start the definition phase of Galileo, which would run to the end of 2000, and prepare the Galileo administration (a steering committee featuring EC and ESA representatives) to decide what services could be supplied (EC, 1999c). A mandate was proposed for the EC to enter into talks with the US and Russia, while exploring the opportunities for cooperation with third countries (such as Japan and Canada). Figure 4.1 gives an overview of the various elements in the Galileo definition phase.

Figure 4.1 Management of the EC/EU/ESA Galileo activities

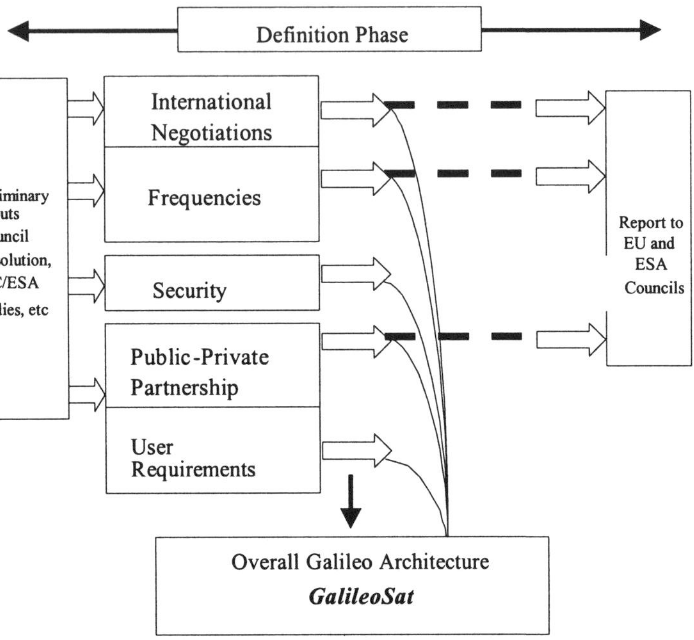

Source: EC, 2000a.

The French Director of Space and Aeronautics, Ministry of Education, Research and Technology, stated that Europe must move ahead quickly to avoid remaining 'under the yoke of an American monopoly'. 'Military location capabilities and air transport safety would be subject to American standards and entirely in the hands of the US. Given the weakness in Europe's communication, navigation and observation systems already highlighted by the recent conflict in Kosovo, this would be unacceptable' (Salomon, 1999). In July 1999, the EU Transport Council authorized the Galileo definition phase, which would address governance, operational management, system design, security and services (GPSW, 1999e). The political agreement on a European satellite navigation system, which the EC had unveiled in February 1999, was welcomed as a major step forward. 'This gives us the green light to go ahead with the different studies needed in the definition phase,' said Ruete (GPSW, 1999d). Based on the investment and operating cost of Galileo, the EC expected that more than 100,000 new and high technology jobs would be created between 2005 and 2025. Final approval from the 15 EU Member States, however, would be required before the US$3.2–3.4 billion project ever gained momentum. Ruete acknowledged that the crucial milestone, the issue of whether a full commitment would be given to build Galileo, would arrive at the end of 2000 (GPSW, 1999d). The political-economic dimension of Galileo was translated into the goal to move from 'dominance of the US industries to a fair international competition in mass markets for user equipment and services' (EC, 1999c) (Figure 4.2).

Figure 4.2 Legitimizing the European Galileo program

Political	*Key words:* European independence; strategic ownership of critical infrastructure; industrial politics.
Economic	*Key words:* Global market shares; global competitiveness throughout the value chain; job creation and protection of high-skilled jobs; efficiency of transport industry.
Technology	*Key words:* Technological advantage to European industry; critical technology development (such as atomic clocks) and know-how; disaster protection; research.
Social	*Key words:* Better and novel services for the citizens; improved safety of transport systems and environmental benefits; mobility; transport policy.

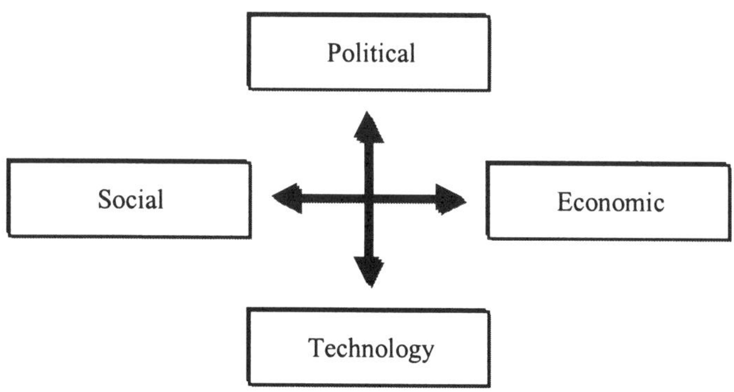

Source: EC (1999c, 2002a, 2002b); ESA (2000b). Interview material.

The EC, the EU Transport Ministers, the ESA and the space manufacturing industry supported these political concerns and objectives (ESA, 2000b).[116] In late 2000, the Galileo project was gaining momentum. Not only the EC, the ESA and EU Transport Ministers rallied behind Galileo but also navigation-related associations, professional groups and large companies in Europe's aerospace industry, which all had a stake in the project. 'Galileo is not a space program. It is an infrastructure program as defined in [EU] Treaties. In our efforts, we want to be as close as possible to users and the citizens,' said Tytgat, DG TREN (GPSW, 1999e). It had become an issue of political, professional and personal prestige for the Galileo backers.

4.3 The European Management of Galileo

A central issue regarding the management structure for Galileo was the division of labor and political status between the EC, the ESA, the EU Member States and the private sector. The definition of the relative roles of the ESA and the EC, in particular, was not entirely clear-cut. 'It is at that high level that ESA has not always been comfortable since they perceive that they are not regarded as an equal partner. At the same time there are those that think it works fine' (EU national government advisor, 25 January 2000). The ESA Executive stressed its own importance and the need for a proven leadership for the implementation process, not least since some of the political and technical constraints were not fully identified. 'The success of the ESA since its foundation demonstrates that the ESA is the most suitable body, not only because of its institutional mandate but above all because of its proven experience in taking the responsibility of large European space-based projects, protecting at the same time the interests of smaller participants. The involvement of the ESA in several projects has contributed

116 It was indispensable for Europe to secure interoperability between current systems, to reach out to the developing countries, and to install ground segment infrastructure in certain parts of the globe. By assisting countries around the world to install ground infrastructure and software, through development assistance, aid and persuasion, the result would no doubt be a stronger dependence on European exports and technology and thus revenues both to industry and governments. Interviews with industry officers, particularly in the French space sector, suggested strategic lock-in effects. If developing countries contributed financially, their incentives to seek long-term cooperation would increase. Countries such as Canada, Israel and South Africa had shown strong interest in Galileo. Japan would also be an important player. 'Galileo will probably open to other partnerships too. Indeed it would be difficult to imagine leaving Japan's engineering and industrial expertise out of the picture, and we will no doubt be working with regions for the supply of dedicated services' (Allemand, 1999, p.15).

to the creation of operational entities (for example, Arianespace, Eumetsat, and Eutelsat). The ESA has proven on such occasions its know-how and necessary flexibility to run international development programs such as Galileo' (ESA, interview, 29 January 2000).[117]

DG TREN laid out a number of 'fundamental principles' for the management structure and funding mechanisms. Firstly, it would have to be a simple, flexible and transparent structure to allow proper management of, and responsibility for, public funds, and facilitate the auditing of the use of those funds. Secondly, the operations had to be transparent to the European taxpayer and there was a need for fair distribution of costs between the different categories of users. Finally, management structures and regulation would have to accommodate the participation of, and partnership with, the private sector (EC, 2000b, 2000c, 2000d). The EU and the ESA would together constitute the Steering Committee, which would oversee all relevant implementation activities by means of a Program Management Board (PMB).[118] The Galileo Steering Committee comprised representatives from EU Member States as well as an observing ESA delegate. DG TREN argued that there was a clear need to build on the EU institutional framework as the structure that would enable political-strategic decision making, international

---

117 A new program structure for GalileoSat was established within the ESA in 2000: the Program Board Navigation (PBNAV). The ESA JCB had therefore got a sister so to speak, but it would be the same people who were delegates. Each ESA Member State could bring the people it wanted and those that had direct interests in Egnos, basically the air service providers, who participated in the national delegations to the ESA and attempted to influence the PBNAV meetings. 'In fact, they participated even in the ESA JCB meetings when satellite navigation issues were handled and were brought by each country to the meetings. This was true in particular for France, Germany, Italy and the UK; several countries brought representatives from air service providers and they could be rather vociferous' (EU national government advisor, interview, 25 January 2000).

118 The Steering Committee was responsible for securing the funding for the overall program, for the budgetary planning and overall expenditures, and for the preparations leading to the Galileo Administration. In addition, it would be responsible for the approval of the program objectives, mission requirements and implementation schedule. The PMB was responsible for overseeing development and procurement activities for the space and ground infrastructure as well as those in the areas of application and service developments and in ground-based augmentations (including the transition to the Full Operational Capability, FOC of Egnos). More specifically, it would be responsible for the elaboration and configuration control of the mission performance requirements, the architectural and system definition of Galileo in a way that would allow the initiation and management of contractual activities. Moreover, it would manage the development and validation of the Galileo infrastructure, the high-level coordination of standards, and the interfaces and interaction with operators, user groups, standards bodies, and so on. It would be composed of the EU, the ESA and various other bodies involved in navigation.

negotiations, and so on (EC, 1999e). Galileo should be run as a single European integrated project and the EU, the ESA and other potential financiers should pool their resources toward an integrated institutional center (Figure 4.3). The PMB was the highest decision making body in the Galileo program and formally consisted of only two people: the Executive Directors for DG TREN and the ESA for the respective projects (Galileo/GalileoSat). Up to mid-1999, DG TREN Director-General Ruete, who worked under EU Commissioner Kinnock, represented DG TREN in the PMB.

Figure 4.3 Galileo development and implementation: organization

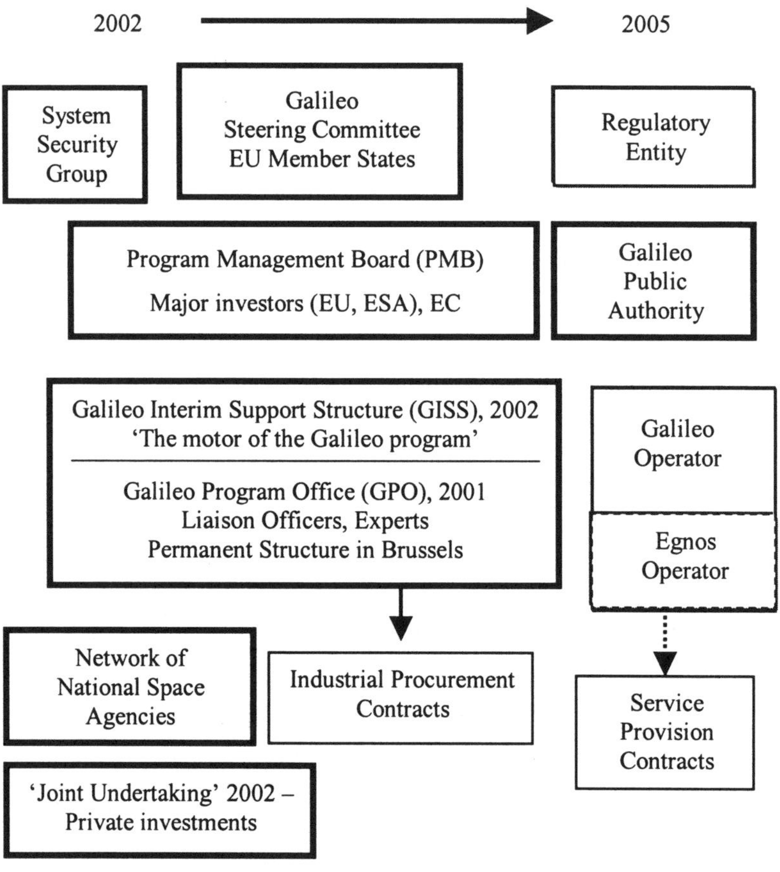

Source: EC (2000a, 2000d). Interview material.

The EC and ESA Directors brought staff members and experts with them to the PMB meetings. DG TREN and ESA established the EU Galileo Program Office (GPO) in Brussels in May 2000. Basically, the GPO would enable the Galileo Program Management Board (PMB) to execute its responsibilities.[119] The EC, the ESA and Eurocontrol argued that it was necessary to establish a unified organizational structure for the management of Galileo. The Galileo Agency would basically be the successor of the Program Management Board (PMB), which would contract out actual operations. It was seen that such an agency could boost public confidence, encourage industry to develop applications and liaise with international organizations. DG TREN viewed such an agency as a tool for administrative control and expansion. 'It could also accept responsibility for peacetime coordination and liaison, as appropriate, with defense and security organizations, including NATO and the WEU, the Conference on Security and Cooperation in Europe, Interpol and Europol, and ensure respect of arrangements related to nuclear missile and non-proliferation regimes' (EC, 1999e, p.39).[120] The ESA had explicitly

---

119 The GPO Mandate defined four tasks for the GPO. First, it should reinforce coordination of activities carried out and contribute to the management of Galileo. Second, it should facilitate the international cooperation for Galileo. In addition, it should coordinate information and public relations activities related to the Galileo initiative. Finally, the GPO was tasked to assist the Galileo PMB in the transition from the definition to the implementation phase of the program. What remained somewhat less clear was the concrete, day-to-day cooperation, for example, frequency of meetings and distribution of information. DG TREN and the ESA staffed the GPO jointly with officials and liaison officers. EC and ESA staff had worked closely together and the logistical arrangements improved in 2000 when the ESA staff moved into the GPO in DG TREN building in Rue de Mot 24. The Galileo PMB would appoint the managers of the GPO. While the ESA would appoint the Head of the GPO, the EC would appoint the Deputy Head of the office. In 2000, the Head of the GPO was the former Chair of the ESA JCB (Tjaden). The Deputy Head was from DG TREN (Landabaso). The majority of the (technical, financial, legal, and administrative) officers came from the ESA. According to the EC, this office would develop over time and should eventually take over responsibility for supervising the real development of Galileo and the operation of the system in the future. In relation to the opening of the EU Galileo office, the ESA Director-General (Rodota) stated 'this marks a step forward in one of Europe's most promising new space programs' (EP, 2000a).

120 DG TREN suggested that there could be a need to set up a GNSS Regulatory Coordinator, beyond the already existing coordination functions, to take responsibility for the development of standards required for Galileo and other related networks (for example, certification and type approval). Such standards could thereafter be submitted to recognized European and international standard bodies (for example, ETSI, CENELEC, the ICAO, the IMO, or the Eurocontrol). 'A future Regulatory Coordinator might [...], particularly when the Galileo Administration becomes established and needs regulatory support, be empowered to develop mandatory standards to be implemented by all Member States to satisfy the objective of the Trans-European Positioning and Navigation Network' (EC,

agreed to be 'subordinated' to the EU regarding the overall political decision making. The EU represented the user community and it was natural that the political steering would come from the EU level. On the other hand, it was unusual for the EU to manage such a large-scale and complex project. The ESA welcomed a new and visible technology infrastructure project.

> The Galileo program is a pilot program at full scale of what should be the future cooperation between the ESA and the EU in the field of applications benefiting from space technologies and systems. The Galileo program will materialize only if the EU takes the relevant political decision and provides part of the funding, and only if the ESA provides the other part of the public funding and its technical expertise. The key question is: how to accommodate the parallel decision making process in each organization and build up the credibility of Europe *vis-à-vis* international partners and private investors? (ESA, interview, 23 October 2000).

The administrative and political organization of the Galileo program has seen a number of changes since the Galileo definition phase. The Galileo Interim Support Structure (GISS) framework, a service contract awarded by the EC and ESA and controlled by the former, replaced the GPO, which was based in the DG TREN premises. The GISS, which was made up by people from the EC and ESA, handled the preparation of the overall Galileo development and program activities and reported directly to the Galileo PMB. It was launched in September 2001 with its own premises in Brussels and reached full operational status in early 2002. In addition, the EC's Galileo Unit, which was incorporated under DG TREN B5 in the past, was transferred to the Directorate E (Inland Transport).[121] Part of the reasoning behind this reorganization was that another dossier, Intelligent Transport Systems (ITS), was introduced and several DG TREN officials were assigned to it. There

---

1999e). DG TREN drew a parallel with the negotiations on the setting-up of a European Organization (European Aviation Safety Authority, EASA) responsible for civil aviation safety (EC, 1996b). Although the administrative and regulatory control of EASA would be restricted (since it was 'transport mode-specific'), DG TREN expected that EASA would provide important input in the definition of GNSS policy (SN, 1999e).

121 In 2002, Hilbrecht, who had acted as temporary head of unit for E4 from June 2001 when Jean Trestour stepped down, headed DG TREN Directorate 5. The Director-General for the whole DG TREN was still François Lamoreux. Other DGs that were involved in Galileo policy planning were the DGs for Research, INFSO, Agriculture, Environment, and Finance. 'An increasing interservice interest is stemming from the proper competencies. They see more and more that Galileo has an influence on their areas and that its implementation can be favorable to their politics. On the other hand, they may want to have a bigger say' (EC, interview, 21 November 2001).

was a perceived need to ensure 'synergies' between Galileo and the evolving European ITS infrastructure. EU transport ministers discussed in 2001 whether to launch the Joint Undertaking (JU). DG TREN viewed GISS as the predecessor of a possible JU; the former would eventually cease to exist. Regarding industrial and technology procurement contracts, the major initiative was the so-called Galilei Study (Table 4.2).

Table 4.2 The Galilei collaboration: industrial contracts

Project coordination: Galileo Industries SA
Technical coordination: Thales ATM (UK)
Local architecture: Alenia Spazio (I); Thales Tracs (UK)
Interoperability: Thales ATM (D)
Signals/frequencies: Astrium (D)/Alcatel (F)
Standards/certification: Thales Avionics (F)
Market Observatory: ESYS (UK)
Legal, institutional and regulatory issues: International Institute of Air and Space Law (NL)

Source: Interview material.

4.4 EU POLICY, INDUSTRY AND TRANSATLANTIC RELATIONS

The new EU Commissioner for Transport and Energy who replaced Kinnock, Loyola de Palacio, stressed that Galileo was one of the most important technological innovations in Europe. Naturally, the participation of industry was a prerequisite for the development and market implementation of Galileo. Europe, however, had no collective presence in satellite navigation in the early 1990s. The Galileo program had encouraged restructuring in European industry and the formation of a European GNSS-2 industry association designed to support the development of GNSS equipment and services.[122] In early December 1999, the EC launched the Galileo definition

[122] The Organization of the European GNSS Equipment and Services Industry (OREGIN) was launched earlier in 1999 to provide technical and marketing expertise, support efforts for standardization and certification of user equipment and services, and promote European technology, products and services. In addition, the association aimed to promote participation of universities and research organizations in Galileo-related research and development activities (GPSW, 1999e). In late 1999, OREGIN consisted of 37 European companies and

phase and the first four contracts were allocated to European industry. The research would assist European policymakers in decisions on Galileo by December 2000. The EC signed roughly Euro 40 million of contracts to define the Galileo system's overall structure. The primary contract was the GALA study (Galileo overall architecture definition) with Alcatel Space of France as the lead contractor of a team comprising more than 60 European firms.[123] A few days before the signing of the EC contracts, the ESA signed the contract for GalileoSat, worth Euro 20 million, which comprised the definition of the Galileo space segment and its ground system. 'The signature of these contracts marks a true milestone in the development of a novel program for Europe' (ESA, 1999c).[124]

In spring 1999, four large European aerospace companies (largest space hardware manufacturers) – Alcatel (Paris, France), Alenia Aerospazio's space division (Rome, Italy), Matra Marconi Space (Velizy, France) and DaimlerChrysler Aerospace/DASA (Munich, Germany) – announced a joint interest in partly financing Galileo. 'We have come to the conclusion that, given suitable institutional, political and legal arrangements, a viable business case can be constructed' (SN, 2000e). However, the size of the investment would depend on the legal structure created by public authorities to commercialize services and to generate sufficient revenue to maintain the Galileo constellation.[125] The industrial consortium that would handle the ESA

---

organizations and was coordinated by FDC, an independent French consulting firm with 10 years' experience in GNSS matters.

123 The Euro 27 million GALA contract covered mission specification, global architecture and specifications, including how it works with ground applications and organizations (TW, 2000). The Galileo Definition Phase covered a large number of technical, institutional and financial questions for which additional contracts were signed by the EC (such as GEMINUS, INTEG and SAGA). GEMINUS addressed service definition and would be undertaken by a Racal-led industrial consortium. An industrial consortium headed by Alcatel Space would work on INTEG, a contract for integrating Egnos into Galileo. An industrial consortium led by Sextant Avionique of France was awarded the Galileo Standardization (SAGA) contract. These three contracts would be fully coordinated with the GALA contract and with work being done by the ESA.

124 DG INFSO procured as well, for example through the so-called GUST project. The objective here was to support this DG in the field of satellite navigation terminals during the Galileo development. A consortium composed of two European-headquartered companies (FDC and Thomson-CSF Sextant of France) coordinated the GUST project. It further involved a number of subcontractors, of which most were members of OREGIN. Important subgroups were tasked with the elaboration of and associated recommendations for (1) preliminary specification (Adtranz, Nokia, Septentrio, and Seatex) and (2) certification (Adtranz, CNES, Ertico, and Seatex) for Galileo/GNSS receivers.

125 European companies and other interested parties submitted competitive proposals based on deadlines for Galileo contracts that began in early October 1999. The proposals covered a range of aspects, such as design contracts for the Galileo space

GalileoSat activity – headed by Alenia Aerospazio – involved more than 50 European subcontractors. In May 2000, the four largest space industry firms and conglomerates in Europe – Alcatel Espace, Alenia Spazio, Astrium GmbH and Astrium Ltd – consolidated their alliance into a joint venture, Galileo Industries (Astrium 50 per cent, Alcatel Space Industries 25 per cent, and Alenia Spazio 25 per cent) (Figure 4.4). The new Brussels-located consortium would assist the EC and ESA.

Figure 4.4 Galileo Industries SA

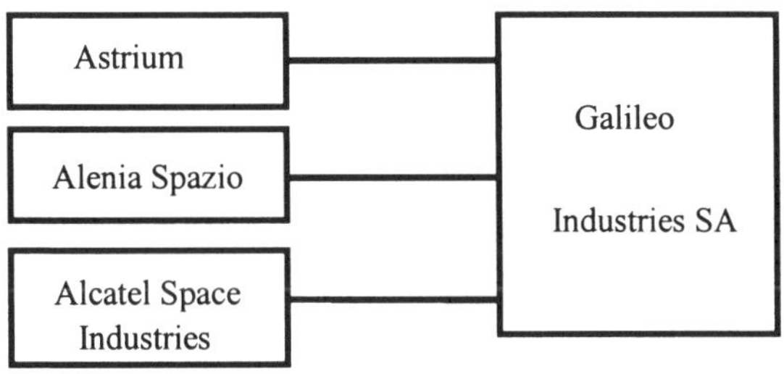

The companies stated that Galileo Industries was looking forward to have a leading role in the Galileo project and in the future Galileo Vehicle Company, which would be responsible for managing the Galileo infrastructure (Alenia Spazio, 2000; Astrium, 2000). 'The Galileo initiative is essentially a civil enterprise that include both private and public investors. It cannot be accepted that the commercial utilization of the system would be constrained by military imperatives,' stated Astrium (SN, 2000g). On the other hand, Alenia stated that the military issue should not be ruled out but would not be a central concern. 'Once the constellation will be deployed, its services will attract military and strategic security interests. However, military considerations will take a secondary-order position since the primary objective is to stimulate the European industry in the field of satellite navigation technologies' (SN, 2000g).[126]

and ground segments, signal structure, cost-benefit analyses, user and service requirements, and revenue sources.

126 Alain Gaubert, Secretary General of Eurospace (Paris) under its Swedish president Ivan Öfverholm, stressed that it was a matter of sovereignty as well as economic matters since European governments could not accept a state of dependency. 'It is undoubtedly of supreme importance for the sovereignty of Europe as a whole and for countries within Europe to have untrammeled access to reliable and unbiased information on which to base their economic development. Navigation and localization techniques using satellites are certainly essential and demanding tools

The US Administration had an incentive to maintain the US dominance in satellite navigation, for industrial and trade policy as well as for national security reasons. The US aired concerns about some aspects of the EC's plans for Galileo, which seemed to be inconsistent with the goal of compatibility and interoperability (for example, adopting common timing and survey standards as well as signal structure). This was part of the 'open system architecture' concept, proposed to the EC by the US in November 1998 (GPSW, 1999f). The definition of what an open system and open markets precisely mean, though, tends to generate differences of opinions due to public policy objectives and industrial-economic interests. 'In particular, the EC's proposal to employ controlled-access services, user fees and regulatory measures to ensure revenue streams appears to be incompatible with both the open system approach and the realization of a truly seamless, interoperable global navigation satellite system,' said a US State Department official (West, 1999, p.29).

> We have suggested that Europe could (using GPS technical characteristics) manufacture, launch and operate a functionally independent constellation of regional or global satellites that could be fully integrated with GPS [...] And European industry would have greater incentive to develop and market new applications for the service. Any suggestion that Europe needs to build its own system to avoid possible future charges by the US for GPS civil service flies in the face of this long-standing policy (West, 1999).

The EU demanded that the Galileo system must be transparent, that is the Galileo and GPS signals should be similar enough that by the time these two systems were available, a combined Galileo/GPS receiver would cost about the same as GPS-only receivers. This brought the issue back again to whether Europe and the US would use the same signal structure. Signal structure, mandated services and fees constituted key points of contention between the US and the EU. In October 1999, the US State Department (and Bialos, head of the US delegation to the GPS/Galileo consultations) indicated to the EU in a draft agreement that the US was seeking an open signal structure, with the aim of enabling manufacturers on both sides of the Atlantic to get the

---

for providing solutions to future problems such as mobility, industrialization, population density and pollution control, which are unique to European countries. Therefore, navigation and localization techniques using satellites are certainly the most conspicuous instance for strategic and security purposes, with consequences for many areas of economic activity. This dependency is not just political – it also cuts Europe off from a market for products and services with high added value' (Gaubert 1999, pp.24–25).

required information to build both Galileo and GPS receivers (SN, 2000c).[127] It expressed concerns that 'non-tariff barriers' (standards or certifications) would limit the market access of US companies to the European market, while European authorities were aware of the US GPS industrial policy. DG TREN defined GPS/Galileo interoperability as follows: 'Galileo and GPS shall be two independent (and compatible) constituents (complementary and interoperable) of the so-called GNSS-2 system. In this approach, most users rely on dual-mode receivers. This implies a coordination of both system designs' (EC, 2000a). European satellite industry was concerned, though:

> Presently in Europe, no industry for satellite navigation really exists. In other words, space-based navigation depends on the US GPS constellation. From an industry perspective, thus, it is a wholly US industry [...] This is the purpose of the Galileo project. The idea is to have a European system to also increase the share of the European industries in the receiver area. More than 80 per cent of the GPS receivers are designed and manufactured in the US. The benefits are mainly political, to catalyze political forces and to take decisions in order to proceed with a multi-billion dollar project. For the EC, this is extremely important (Alenia Spazio, interview, 16 November 2000).

Would Galileo and GPS be sufficiently compatible to allow (reasonably priced) dual receivers? Would the GPS services meet any EU mandates? Would US-manufactured receivers be allowed unrestricted access to the European market? Would European firms be allowed a level playing field in US markets and procurement contracts? Concerning the issue of US participation in the Galileo project, there appeared to be different views in Europe. On the one hand, DG TREN and Galileo Industries stated officially that they would be open to participation by non-European entities. 'The program should be open to interested investors from all countries,' said Bruns, DG TREN (SN, 2000f). 'We are designing a global system with global revenue streams. We need to have global partners,' said Dudok,

127 Under the draft agreement, each side would agree not to mandate the use of its system instead of the other. The agreement would call on existing trade agreements and organizations, such as the WTO, to settle any trade disputes. DG TREN Director-General Ruete recognized the potential problem of putting tax on GNSS receivers and that such a tax could be an issue for the WTO. 'We are looking into it. But we think if it is indiscriminately applied, like TV levies, there will be no problem' (AWST, 1999). It was difficult to see how the EU could or would be willing to accept all the provisions of the transatlantic draft agreement proposed by the US State Department. A central question was whether Europe would allow unfettered competition initially from outside the EU and resist charging user fees for certain services and still move ahead proactively with the Galileo project.

president of Galileo Industries. Others also stressed the need to support European industry. 'We believe it should be considered as a strategic European asset, and as such a wholly European system,' said Dutton, Racal Avionics of London, main contractor in Sidereus, a Galileo consortium (SN, 2000f). (Racal would later become incorporated into the Thales Group.)

European authorities and the EC were involved in tough bargaining within Europe and with third country governments about frequency bands for satellite navigation. The EC stressed the strategic political and economic importance of frequencies (EC, 1998f). The EU Member States had mandated the EC to manage the actual hands-on, day-to-day negotiations with international partners. Basically, as frequency bands become wider, higher data rates and better quality can be transmitted. The Member States intended to try and secure frequencies for Galileo at the 2000 World Radiocommunications Conference (WRC) in Istanbul. A coordinated EU policy was somewhat made more difficult due to competition between 'the safety and strategic interests of aeronautical and GNSS communications' and commercial telecommunications interests (EC, 1998f, 1998l, p.11).

The US sought a new allocation for the third GPS civil signal, while the EU wanted frequencies for Galileo. The EU had to (1) obtain access to frequencies available for satellite radio navigation and ensure interference-free utilization of spectrum sharing, and to (2) ensure additional availability of frequencies for timely and cost-efficient implementation of Galileo (EC, 2000f). 'We are open to cooperation if it makes sense and is in our interest. Such cooperation would be on the basis of these principles,' said the head of the US GPS delegation (GPSW, 2000a). It was a strategically difficult situation for the EU in that it would either have to reach an agreement to allow Galileo to share those bands with GPS and Glonass, whose supporters were not too keen on meeting such a request, or opt for an alternative approach. Despite the transatlantic differences, both GPS and Galileo supporters alike achieved success at the WRC. The final result in June 2000 was that enough frequencies for the US and the EU were allocated for their satellite navigation signals.[128] This was an important political outcome.[129]

---

128 An additional 111 MHz was added for the so-called downlink, of which Galileo would be free to use 87 MHz (the remaining was dedicated to GPS). In addition, 60 MHz was introduced for the so-called uplink and since neither GPS nor Glonass were using the new frequency range, Galileo would have the opportunity to use this in a creative way.

129 As a well-known expert in satellite navigation phrased it: 'Despite polite rhetoric on both sides of the Atlantic, the US and the EU are posturing like competitors. Undeniably, competition between the two exists, but some officials are beginning to take the maneuvers personally. For good of the users, it would be gratifying if both sides could stop taking shots at each other. Unfortunately this would require a simultaneous burst of enlightenment that is altogether unlikely' (GPSW, 2000c).

European Security Policy and NATO

DG TREN officials interviewed for this study suggested that 'security' could be defined in a number of ways. First, technical security refers essentially to the protection of plants and sensitive constructions (such as those which would be central targets in a war situation). Commercial securities referred to security measures with the aim of making it more difficult to copy or compete. Then there was the crisis management and military dimension, which is associated with the intentional degrading of the signal for military purposes (selective availability). Finally, the safety-critical or safety-of-life security aspect is concerned with ensuring a high-performance service, for example, for traffic purposes in the field of civil aviation. Within each domain there are then different levels of security (EC, interview, 6 December 2000). The development of a European Security and Defense Policy (ESDP) triggered debates on required capabilities and instruments under European control. Bialos, head of the US GPS delegation to the EU, said:

> The EU has a choice to make now. The EU has to decide whether to spend significant taxpayer dollars on Galileo, a major project, in light of other European priorities such as the NATO Defense Capabilities Initiative and in light of the existence of a very robust and modernized GPS. It is up to Europe to decide what to do and whether to go forward. That is Europe's choice. We believe, frankly, that a modernized and robust GPS can meet users needs in the twenty-first century (GPSW, 2000d).

This NATO Defense Capability Initiative was aimed at modernizing portions of Europe's military capability to support future joint operations with the US. Europe's capability needed to be upgraded in some areas. Moreover, Europe needed to improve the efficiency and coordination of coalition warfare and rationalize defense structures and budgets.[130] DG TREN apparently made efforts to explore the inclusion of Galileo into a wider defense policy and infrastructure. In March 1999, DG TREN (Tytgat) held an information session for NATO on Galileo. 'We [were] informally invited to present our activities and to identify if there is room for cooperation or not,' Tytgat said (GPSW, 1999f). '[The] European Community is mostly based in the NATO area. It is obvious NATO will be part of any type of crisis mechanism. We have to present to NATO what we are doing, as there are security aspects to it. But it will be up to EU member states to decide if we are to go further. It is

130 Bilateral and multinational efforts were under development in Europe to deepen cooperation, as witnessed for example in stronger coordination of air force squadrons to better deal with anti-aircraft missiles, and cooperation over aircraft carriers (FT, 2001a).

up to all member states to decide if NATO is the right platform to deal with security aspects of Galileo' (GPSW, 1999f). There was, however, resistance to the crisis management and security concept and the Public Regulated Service (PRS) (Lembke, 2002c, 2002d).

> There is currently a substantial opposition to the government access service within the EU Transport Council. Whether this security aspect will be integrated into or excluded from Galileo will depend on the decisions within the Council [...] No delegation has yet wanted to grapple with it. There is one Member State that has initiated it [France], which is rather insistent about keeping it on the agenda. Then there are a number of countries [...] that openly oppose the government access service. Developments in crisis management and the creation of a European rapid reaction force, and demands/requirements in this regard, may influence Galileo (EC, interview, 6 December 2000).

Aside from NATO, the EU and the EC had another option for creating a security structure for Galileo. Within the EU treaties, there is a provision for creating a common foreign and security policy, which could potentially be a starting point. Naturally, any procurement of Galileo services and receivers would result in economic benefits and thus create possible revenue streams, which were actively explored to fund the European Galileo project. The question whether Galileo should be incorporated into a wider and coordinated European defense structure or into the EU Member States individually, or whether Galileo should explicitly address military aspects at all, was apparently on the agenda in DG TREN.

> The military interest is quite clear – to have an independent and autonomous means for localization and navigation purposes for the air forces or navies of the different European countries. This is a very strong lobbying power. The military entities are extremely active in the process of definition of the Galileo system. In each panel at the different levels working on Galileo (political and even operational) there is a military presence. Space-based navigation allows the military to design new aircraft, new missiles, and new weapons with this feature. The US military has driven GPS since the beginning and built a complete set of weapons based on satellite navigation systems. The European military started to realize the potential advantages of such systems when they had to proceed to military operations in Europe or in countries nearby Europe, relying on GPS. When they operated in Kosovo they depended on GPS support and recognized that an autonomous system naturally could be helpful (Alenia Spazio, interview, 16 November 2000).

DG TREN probably reflected on what role, if any, the EC should play in this regard. 'It has been recognized that the Western European Union (WEU) might be a possible partner to deal with security aspects of Galileo. It remains to be seen. We are just now exploring,' said Tytgat (GPSW, 1999f). Within the EU Galileo Systems Security Group in the Steering Committee, there had been a debate on the role of military requirements. 'In the development of navigation there are linkages to the military and many ideas derive naturally from the military side. It is the military which has the immediate need, for example, to be independent from any navigation aids that they do not control or to be independent from any local navigation aids without having any remote-controlled or global system that would be sufficient' (EU national government advisor, interview, 25 January 2000).

There were different interests involved, for example, between those that viewed Galileo as a strictly commercial enterprise and those that wanted to explore the military dimension. Military authorities in Europe did not publicly express any requirements. 'If people all of a sudden say that Europe should have a military-controlled system then you can wonder why it is not sufficient with GPS' (EU national government advisor, interview, 25 January 2000). The looming military dimension could create problems for the justification of the entire Galileo project in Europe.

> Those Member States that oppose the crisis management and military security aspects view Galileo as a completely commercial and civilian system. The only conceivable types of security then are commercial security, protection of data, protection of constructions such as airports or sensitive civilian targets. However, when you start to approach paramilitary/semi-military areas you come outside the acceptable for those that oppose the more military-oriented security aspects. It is not the goal that Galileo should be used for military purposes. At the same time, member states have recognized that if they cannot agree about what services Galileo will provide and requirements Galileo should satisfy shortly then there is a risk that everything will be delayed (EC, interview, 6 December 2000).

While the technical discussions on satellite navigation and the transatlantic relationship moved ahead, the political climate hardened in late 2001 as European ministers were about to decide on public funding for the development and validation phase of the European Galileo program. The US government stressed that European authorities had offered no assurances that the Galileo infrastructure would be compatible with the GPS constellation and associated NATO systems, which would have security implications. The Galileo program was in the early development phase and a number of technical compatibility issues remained to be resolved.

In a letter sent out to European defense ministers in early December 2001, the US Deputy Defense Secretary, Paul Wolfowitz, took the opportunity to express concerns about the Galileo program and radio spectrum issues related to military applications, and asked those ministers to convey his concerns to European transport ministers. The US, Wolfowitz claimed, was about to modify GPS in order to meet future military and civil requirements. One of the changes was related to the so-called GPS military M-code: 'The addition of any Galileo services [...] will significantly complicate our ability to ensure availability of critical military GPS services in a time of crisis or conflict and at the same time assure that adversary forces are denied similar capabilities [...] I believe it is in the interest of NATO to preclude future Galileo signal development in spectrum to be used by the GSP M-code' (AFX Europe, 2001). Political pressure by the US government on Europe and the Galileo program is a natural result of its economic-industrial effort to maintain world leadership and of the effort to protect national military and security interests. At the same time, the EC argued that support for Galileo was of strategic importance to reduce vulnerability, to put an end to the US world monopoly, and also to increase international competititon in satellite navigation applications and services.

4.5 Political Delay and Divergence

At the end of 2000, the cost of Galileo was estimated at Euro 3.25 billion for the period 2001 to 2008 and the annual operating costs at Euro 220 million. The total Galileo investment would amount to Euro 6 billion by 2020.[131] An EC estimate of economic and social benefits up to 2020 was Euro 62 and 12 billion, respectively. However, such projections did not rule out political controversies, market uncertainties or technological challenges. Table 4.3 presents the expected total cost as of 2000 of the European Galileo radionavigation program. The estimated cost has since shifted somewhat to allow for a slightly higher total cost but remained largely in line with previous estimations and around 2002 the EC estimated the total cost of the Galileo project at Euro 3.2–3.4 billion.

131 European space industry and the future Galileo operator, it was argued, would enjoy value-added benefits of around Euro 190 million annually with an extra Euro 740 million during the deployment phase. The Galileo user's net benefits would climb steadily to amount to around Euro 8 million by 2020.

Table 4.3 The estimated cost of the Galileo program (as of 2000, per phase; in Euro million)

Cost of Galileo program including Egnos		
Definition phase	(1999–2000)	80
Development and validation phase	(2002–2005)	1100
Deployment phase	(2006–2007)	2150
Operating phase	(2008–)	200/year

Source: EC (2000c).

The EC stressed that the deployment of an autonomous European satellite system was a prerequisite for the EU's independence in the field of satellite navigation (EC, 2000b). The policy decisions that would be taken by the EU Transport Council in late December 2000 would be crucial for the future of Galileo, according to Commissioner Palacio (EC, 2000c, 2000e). DG TREN pointed to the economic-strategic interests at stake and threatened to withdraw support if EU governments did not make firm commitments. 'With Galileo, Europe will gain a foothold on the market of satellite navigation products and services, which at present is totally dominated by other players, led by the US. This will have a positive impact on the competitiveness of European industry' (EC, 2000c). The EC pointed out that a number of European Council meetings had endorsed this view.[132]

Different opinions existed, however, not only at EU level but sometimes also at domestic national level. 'We must remember that the EU Member States all have their own agendas and to get to common decisions on moving ahead in Europe is extremely difficult' (Eurocontrol, interview, 21 December 2000). In November 2000, DG TREN presented a new Galileo report, which offered an overview of the Galileo activities. On 7–9 December, the Heads of EU Governments met in Nice (European Council) and endorsed the continuation of Galileo. All together, the European Council referred to the strategic importance of Galileo in its conclusions from Cologne (June 1999), Santa Maria Da Feira (June 2000), Nice (December 2000), Stockholm (March 2001) and Laeken (December 2001) (see Lembke, 2002d).

132 However, DG TREN stressed that the continuation of Galileo beyond 2001 would be attached to three broad conditions (EC, 2000c). First, the EU would have to deploy an independent satellite navigation system. Second, guarantees for sufficient funding would be required, including investment from the private sector of up to Euro 1.5 billion for the deployment phase. Finally, an adequate legal and financial framework would have to be established to attract the private financing needed to operate Galileo, and a provisional, coordinated management structure for the project needed to be created in 2001. Based on definition-phase studies, Galileo would have a 30-satellite Middle Earth Orbit (MEO) constellation. Nine active spacecraft and one on-orbit spare would operate at 23,200 kilometers.

In late December 2000, the EU Transport Council met to decide about whether Europe should give a firm commitment to Galileo to move beyond 2001 and into the development and validation phase. In December, the ESA had committed immediate funding of Euro 50 million (US$45 million) for Galileo and another Euro 500 million in November 2001. The funding would not be released until the EU committed an identical investment. Remaining concerns within the EU Transport Council, however, resulted in the postponement of any firm political approval. The Council deferred the decision on the development phase of the Galileo program, which took the ESA, the EC and industry by surprise. In February 2001, the EC president Romano Prodi announced in Stockholm that the Galileo project had become bogged down but that the Council and the EP should reach an agreement by May 2001 (FT, 2001b). EU Transport Ministers were scheduled to meet again in Stockholm in early April 2001 under the Swedish EU Presidency. DG TREN hoped that the planned Heads of States Council in Stockholm would provide high-level political support for Galileo at European level. The Swedish Presidency identified four critical issues, which needed to be resolved: management structure, service definition, services, and the need for an independent analysis of costs and benefits. The Council called on the EC to present the financial public-private scheme that should be in place by 2003. In spring 2001, the Heads of State and Government could not reach an agreement at the Stockholm EU Summit but they reaffirmed the need for Galileo in the Presidency Conclusions:

> The European Council recalls the importance of launching the Galileo satellite navigation program without delay. In conformity with the Cologne and Nice conclusions, the private sector is required to take up the challenge with regard to participation in and financing of the project through a binding commitment for the deployment phase. The European Council notes that the private sector is ready to supplement the public budgets for the development phase. The European Council invites the Council to define the arrangements necessary for launching the next phase of the project, including establishing a single and efficient management structure before the end of 2001, be it a joint undertaking under Article 171 of the Treaty, an agency or any other suitable body (GO, 2001).

The concerns that led to the deferral of a political decision on Galileo among the EU member states lingered on throughout 2001. Those concerns are primarily associated with funding and the exact source and size of the Galileo revenue streams, the definition of security aspects and services, and Galileo's legal and management structure. In addition, the nature of the market (public/commercial) and the nature of integration of Egnos with Galileo remained to be settled. Other topics included the framework for assigning legal liabilities associated with damage or injuries resulting from a potential malfunction of the Galileo system, frequencies, and overall design of the ground infrastructure. The EC has stated that public subsidies (Euro 1.1 billion) would be indispensable for the Galileo development and validation phase (2002–2005), the financing of which would come from the EU and ESA budgets, the organizations that have shared the responsibility for the development of Galileo. Moreover, investment from the private sector (around Euro 1.5 billion) would be needed for the deployment phase (2006–2007). The total Galileo investment would amount to Euro 6 billion by 2020. EU authorities have agreed to demand early funding participation on the part of the private sector to ensure that industry is fully involved (and fully committed), while industry is reluctant to commit funds.

On the one hand, there were those who wanted to move ahead quickly with the Galileo project through a lasting commitment at the earliest date. On the other hand, there were those who wanted to generate more information for a decision on the whole project at a later date. EU Member States like France, Italy, and Finland, together with the EC and the ESA (and smaller ESA countries such as Norway and Switzerland) belong roughly in the former political grouping. Germany, the UK and Sweden belong to those that demand a clear role for and commitment by the private sector at an early stage. All the ESA and EU Member States shared the idea that the envisaged Galileo infrastructure must ultimately be developed as a civil system (that is, placed under civilian control), although some wanted to leave the door open for military communities to use the system in the future.

There were thus differences in positions among EU Member States regarding the exact nature of Galileo. France has repeatedly stressed that Europe should not hamper the Galileo project with conditions on detailed cost analyses and the promise of a public-private partnership. In their view, delays and strategic fine-tuning must not cause Europe to miss out on what was considered a unique window of opportunity. This argument has been shared by national ESA/EU delegations, such as Italy, Norway, Portugal and Switzerland, which regarded a European navigation system as primarily of a political and strategic nature and as a public service system. It was considered early on that strategic issues were driving Galileo and that those issues fell within the domain of the EU. The EC supports this view. 'Galileo will make the use of transport infrastructure more efficient, increase safety, reduce environmental pollution, positively impact on the ICT industries in particular for location-based services, and contribute to employment increase in the EU,' said Commissioner Palacio (EC, 2000f).

During spring 2001, France relaxed its desire to explore the military potential of Galileo. Together with Italy and Spain, France accepted that the system would be strictly used for civilian purposes under private sector control. Germany, though, softened its demand for a dominant private sector role early on, although the Netherlands in particular continued to call for clearer wording on the role of the private sector. In April 2001, European governments committed new public funds and industry stated that it would provide the same amount, although they would reformulate the Memorandum of Understanding (MoU) that they signed and they did not see it as binding. The difficulties in determining future market values with any accuracy are closely linked to the debate on who should pay and how much for the development of Galileo. Regarding the market size of GNSS markets in Europe, Table 4.4 breaks it down as predicted in 2001 for 2005.

Table 4.4 GNSS market segments for Europe

Market Segment	1999 (m€)	2000 (m€)	Growth (%)
Aviation – products and services	44	65	48
Commercial maritime	4	6	50
Car navigation	680	1920	182
Fleet management	40	90	125
Rail	1	25	2300
Surveying and mapping, off-shore	49	86	76
Agriculture	5	12	140
Augmentation services	46	60	33
Mobile communication (excl. services)	0	6000	...
Timing	6	10	67
Personal navigation	50	90	80
Total European GNSS market	924	8363	795

Source: EC (2001).

The largest increase was predicted to occur in the mobile communications market. The market for mobile positioning services would start from zero in 1999, since the mobile phone market for GPS positioning devices was marginal. By 2005, however, it is expected to amount to around Euro 6 billion (assuming partial integration of GNSS localization services in mobile communications) and 20 billion (assuming mandatory GNSS localization in mobile communications). There was a high degree of uncertainty involved in this regard since the take-off of satellite navigation services will depend on the future choice of positioning technology chosen for integration into mobile phones. There will probably be tough competition with network-based positioning services. Industry viewed Galileo as a central part in a future

'Global Intelligent Mobile Infrastructure' where the two fundamental components were (1) positioning and timing and (2) mobile communications. Regarding the variation in preferences among EU (and ESA) Member States, a matrix can be created to illustrate different views on the relative importance of 'economic-based' preferences (the need for profitability, low levels of public spending and early private sector commitment), on the one hand, and political or 'security-based' preferences (the need for strategic independence and ownership, political considerations, significant public funding and swift development), on the other. Table 4.5 provides a qualitative classification of the variation in preferences (see also Lembke, 2002d).

Table 4.5 Galileo: European political economy and security policy

	Economics	Economics and Security	Security
	Commercial Profitability	'Mixed'	Strategic Independence
Private Sector	Terminal vendors Users: airlines, land transportation, wireless, shipping, geodesy and surveillance		Satellite manufacturers (Galileo Industries) Service providers (The Thales Group)
Public Sector	The Netherlands, The UK, Denmark, and Germany National maritime administrations	Finland, Norway and Sweden DG INFSO?	France, Italy Spain, Portugal, Greece, Switzerland Air Traffic Service (ATS) Providers DG TREN DG Research

The explanation of the variance between EU Member States has not been a primary objective of this study and much of the information to assess this variance is largely of a confidential nature. Based on interviews with core participants and knowledge of the case and European integration in general, two explanatory factors are suggested: (1) the special relationship of certain EU Member States to the US combined with their Atlanticist orientation and their relative alignment with the 'British model of European integration' (foreign economic and security policy); and (2) the expectation of domestic political-economic returns that benefit the major domestic economic groups in each country combined with the level of already made or planned investments in GPS (military) infrastructure and the existence of preferential commercial contracts with the US GPS industry and government.

The categorization of Member States is based less on EU ministers for space affairs who largely support the Galileo project. The precise categorization of DG INFSO is tentative since there is scarce information on its preferences. In addition, parts of the German government should fall in the 'mixed' box (public funding support committed through ESA) and the Finnish and Norwegian governments have been leaning toward the 'security' box. The various government entities involved complicates the picture. The private actors in the 'security' box have been concerned with the possibility of Galileo becoming a commercially and profitable source of revenues in the shorter term. In contrast to private actors in the 'economics' box, they clearly stressed the need for public funding and security-strategic considerations. They are all, of course, concerned with sound business plans and profitability in the longer run. DG TREN (and DG Research) has seen Galileo as an institutional and policy opportunity to address the question of further incorporation of security issues within the jurisdiction of the European Community by linking economic and security considerations.

> We believe that if there was some kind of attack against GPS then we cannot do anything inside Europe [...] If there is a second source for radionavigation, that is, Galileo, it could give us better and greater confidence. Security has always been an important aspect because we think that radionavigation is becoming increasingly important and will penetrate all the different sectors of transportation (for example, railways, maritime, aeronautics applications). There is not only an economic point of view behind building Galileo but also a strategic point of view (EC, interview, 21 November 2001).

Until 2001, Galileo was largely a research and development (R&D) oriented program. Institutional arrangements lagged behind. When it came to the organization and management of Galileo, there was a consensus among European governments that there was a need for efficient governance for the project, which allowed democratic influence and political control of all participating countries and private sector organizations. A problem here has been that there was no clearly defined public customer for Galileo.

According to the satellite industry, European governments would have to spend public funds to bring Galileo to the point where it was equivalent to GPS before they would invest heavily. A political problem that has been a heated topic is whether satellite navigation will be a publicly funded market or a privately funded service market. The main actors involved appear to have different and often contradictory conceptions on whether Galileo will generate commercial revenues and be profitable. In general, industry claims that the forecasts by the EC, based on assumptions about the market in the long term, are not realistic due to the high degree of uncertainty. The size of private investment would depend on the legal structure created by public authorities to commercialize services and their plans to generate sufficient revenue to maintain the Galileo constellation. It has been suggested that industry could help fund deployment costs under a Private Financing Initiative (PFI) scheme used in some infrastructure projects across Europe.

A major question, thus, is how Galileo will fit with the market for mobile Internet services (UMTS). Customers and not politicians will choose what technology they want to use. The increased use of mobile terminals together with wireless technology will pave the way for completely new services and transformation into mobile multimedia terminals. Mobile telephony already has location and positioning service capability and this market will most likely grow considerably in the years to come. Can Galileo and future mobile telephony interact and come together? The EC reasons that added value such as better accuracy, global coverage and proven technology would increase the chances of satellite navigation technology being chosen. History, however, is evidence that not even the best technology always wins the day. The issue of how investments can be recouped (and how Galileo could run on a purely commercial basis in order to attract private investment) remained open and controversial. (This is less of a problem in the US since GPS is explicitly viewed as a national government priority.) Industry was reluctant to make a heavy investment early on in a system that would be on the market at the same time as GPS. The timing to market was a decisive factor for the European industry's ability to make a profit from the system from the start.

In the case of European aviation safety, EU Member States recognize the benefits of certification and other aspects at the European level. At the same time they say that they do not want a centralist system and are very reluctant to concede power. Galileo is somewhat different in that. the fundamental requirement is independence from the US GPS constellation, which to a large extent drives the Galileo project. Instead, the problem is that you know what GPS is, and how the Americans work, but you do not have any clear idea about who is going to control Galileo, which is a question that concerns everyone. In this regard it is stated that Galileo should be oriented toward a commercial system through a public-private partnership. This could bring it down to a level where the participating countries do not feel that a certain country or general steers the whole project but rather that it is run on a commercial basis (EC, interview, 6 December 2000).

Regulatory aspects were not a major topic before 2001, but would gradually become more important. Three levels of Galileo services have been proposed: an open access service (OAS), a commercial access service (CAS) and public services (that consist of safety-oriented and regulated services). OAS will be provided to users free of charge with an accuracy of within 6 meters and service availability of around 99 per cent. The commercial service will include a charging mechanism to generate revenues. A problem is whether the safety service should be encrypted. The international civil aviation community has opposed the encryption of the safety-critical services based on the argument that it could jeopardize security. There have been different views within European industry consortia working on the definition of services; some argue that this service should be unencrypted (the GALA consortium), whereas others argue that it should be encrypted to allow for the generation of revenues (the GEMINUS consortium). There has also been disagreement at industry level about whether Galileo should include a communication payload to transfer navigation information. The GEMINUS consortium argued that a communications payload would generate additional sources of income through management of truck, taxi and bus fleets.

From a political-institutional perspective, it boiled down to the extent European governments were prepared to delegate power to the EC. A related problem was how the program should proceed if no flexible and firm commitment could be reached to move ahead with Galileo in a way that was satisfactory for those governments that view Galileo as an urgent priority for the EU. What would happen if the reluctance on the part of some European governments to proceed rapidly and actively does not comply with the desire of staunch Galileo supporters? Would the whole program be taken out of the EU framework to a less politicized, bureaucratic governance mechanism? In November 2001, ESA ministers approved Euro 527.9 million for the ESA GalileoSat program, which is concerned with the Galileo satellites and the associated ground segment (for example, control centers and networks), and the initial development of the Galileo constellation. The President of the EC, Romano Prodi, welcomed the funding. On 6–7 December 2001, the EU Council of transport ministers decided to again defer a decision on a Euro 450 million payment (~US$394) for the development and validation phase of the Galileo program (2002–2005). A number of European Parliamentarians distributed a letter to the heads of government in support of Galileo before the European Council meeting in Laeken in mid-December 2001:

> Heads of States and Governments must seize this opportunity to demonstrate their political will and their vision of the future of Europe. Galileo is a unique opportunity today when the EU wants to prove its independence in aeronautics and space policy but also in technology and skilled jobs. More than 30 years ago, some European States began adventurous projects such as Ariane and Airbus [...] Galileo should clearly show the EU's ambition to remain at the forefront in this field (EP, 2001).

Amid the problems of the Laeken summit of EU heads of government 14–15 December 2001, an informal agreement was reached to continue work with a view to taking a decision on the funding of the development phase by March 2002 and to decide on regulatory aspects by June 2002. Central actors in the French government (President Jacques Chirac, Prime Minister Lionel Jospin, Minister of Research Roger-Gérard Schwartzenberg and Foreign Minister Hubert Védrine) urged EU heads of government to allocate funding in line with the endorsement of Galileo as a strategic infrastructure that EU ministers had offered on several occasions. Loyola de Palacio, Commissioner for Transport and Energy and Vice President of the Commission, criticized the weak conclusions of the Laeken summit and the lack of concrete decisions.

In late February 2002, the German government approved the release of EU funds for the development of Galileo, matching the financial commitment of the ESA after a deadlock in the German government between the finance ministry (led by Hans Eichel) and the transport ministry (led by Kurt Bodewig). As part of one of the 'priority actions' identified by the European Council meeting in Barcelona in mid-March 2002, the European Council again stressed the importance of the Galileo program. It asked the EU Transport Council to take the necessary decisions in spring 2002 with regard to both the funding and launching of the program and the setting-up of a Joint Undertaking in cooperation with the ESA. In late March 2002, the EU Transport Council formally endorsed significant public funding for the European Galileo program.

Summary

The European radionavigation program Galileo is a public sector initiative and represents (apart from the Euro-zone initiative) the most ambitious large-scale project ever undertaken by the EU institutions. It is supported as an effort (1) to strengthen European transport infrastructures (aviation, land transport, and maritime) and the functioning of other infrastructures, (2) to create positive macroeconomic and user benefits, industrial competitiveness and jobs, (3) to increase Europe's strategic control and ownership, and thereby also strengthen EU's position in world affairs, and (4) to bolster EU's (and the EC's) administrative and political goals. The Galileo program is promoted as a core element of the common EU transport policy and is the proposed contribution to the concept of Global Navigation Satellite Systems (GNSS). It will enable Europe to deploy a satellite navigation system by 2004–2005 and enter it into full operation capability by 2008.

EU Transport Commissioners (with support of some other Commissioners, such as Martin Bangemann, and other major figures, such as Romano Prodi and Javier Solana) have supported Galileo 2008 as one of the most ambitious technical projects ever put forward at the European level. The EU institutions stressed its economic importance and the requirements associated with the political independence of the EU in a high technology sector. Galileo would be placed under civilian control, which would meet user requirements and provide user benefits while facilitating the access for European industry into the global markets for systems, equipment and services. The EU Member States, the EC, the ESA and industry agreed that Europe could not allow itself to become increasingly reliant on a single-source and non-European controlled navigation system. In addition, Galileo has been discussed in the context of the emerging Common European Security and Defense Policy (CESDP), at the same time as the EU institutions have aimed for a model of civil ownership and management. The European Galileo program will have implications for European security policy and relations with NATO.

Among the EU Member States, France, Italy, Spain and Belgium wanted to strengthen Europe's economic and strategic independence, political sovereignty, and commercial competitiveness. Together with the EU Commission, they viewed Galileo as a public service and political project. Germany, the Netherlands, the UK and Sweden wanted a commitment from industry before the launch of the development phase of Galileo, which industry argued was unrealistic. These governments wanted a strong private sector involvement and a commercial Galileo system. While European firms, and not least Galileo Industries (the consortium formed from European space companies Astrium, Alcatel Space Industries and Alenia Spazio), hope to secure contracts to build, launch and operate Galileo, European public authorities have agreed on the importance of early private sector co-funding. In 2001 the market focus had shifted from land transportation to mobile communications more broadly, and particularly toward the vision of a 'global mobile intelligent infrastructure'. A central question is how to keep the implementation of advanced wireless infrastructures intact with satellite navigation. The development of a European industrial and regulatory strategy for wireless Internet is the topic of Chapters 5 and 6.

5. The EU strategy for wireless Internet

The growth of Internet and mobile cellular communications has been spectacular. This chapter and the following will examine the creation of a European strategy for the development of wireless Internet. In Europe, this project is referred to as the Universal Mobile Telecommunication System (UMTS), and globally as International Mobile Telecommunications 2000, IMT-2000, although these terms really are 'empty shells' that had to be 'filled with life'. European industry has positioned itself strongly in mobile communications worldwide through the so-called Global System for Mobile communications (GSM), which was implemented as a pan-European, digital-based system. UMTS, which thus is a generic term for one of the third-generation (3G) mobile cellular systems, will offer higher-speed, always-on Internet access, data, full-motion video applications and CD-quality music on mobile phones. It represents the convergence between two fast-growing markets: the Internet and mobile telephony.

5.1 BACKGROUND TO MOBILE CELLULAR COMMUNICATIONS

In the 1980s, representatives of the European telecommunications industry and regulatory community started to explore the idea of an integrated solution for a second-generation (2G) wireless infrastructure. The first-generation (1G) analog systems had been deployed within national industrial policy frameworks, monopoly provision of mobile services, and embedded contractual and institutional links between private sector actors (Redl et al., 1995). Before 1990, a disparate group of analog technologies – TACS (an American-derived technology, which was not least deployed by countries with no strong domestic manufacturers) and NMT (the Nordic Mobile Telephone system), to name but a few – had come to dominate Europe's cellular landscape. Equipment manufacturers promoted different radio frequency (RF) standards, which gave rise to divergence across the world. Table 5.1 offers an overview of the various mobile cellular standards and their footprint around the world.

Table 5.1 Mobile cellular standards in the world

Standard	Current Frequencies	Initial Service Area	Initial Year of Deployment
Analog			
Proprietary (various)	200-800 MHz	Japan/Europe	1979
NMT	450/900 MHz	Europe	1982
AMPS	800 MHz	United States	1983
TACS	900 MHz	Europe	1985
J-TACS	800 MHz	Japan	1989
Digital			
GSM	900/1800/ 1900 MHz	Europe	1992
PDC	800/1500 MHz	Japan	1993
TDMA/IS-54/ IS-136	800/1900 MHz	United States	1993
CDMA/IS-95	800/1900 MHz	Korea	1995

Source: ITU (1998); Trade Press.

In order to overcome the fragmentation, the Conference Européenne des Posts et Télécommunications (CEPT), which in the early 1980s consisted of 26 countries, took two major decisions in 1982: (1) frequency bands in the 900 MHz area would be reserved for mobile cellular systems; and (2) a committee, GSM, Group Special Mobile, was created to generate standards for a future unified European infrastructure (GSMA, 1995a, 1995b, 1996a, 1996b).[133] In June 1987 in Madeira, the CEPT GSM Committee agreed upon

[133] During the period from 1982 to 1986 a lot of discussions took place within the group about whether an analog or a digital infrastructure should be used for the mobile radio system. One driving force was the need for more capacity. At the beginning there was a close cooperation between the state-owned telecom companies/regulators in France and Germany (Deutsche Bundespost and France Telecom) on one side, and the Nordic NMT Group (incumbent operators in Sweden, Norway, Denmark and Finland) on the other. The collaboration concretized a little in 1985 when the French and German PTT Administrations decided to plan and establish research projects for cellular radio systems, soon followed by other administrations. The Permanent Nucleus based in Paris, which consisted of qualified experts working permanently together (later transformed into Project Team 12), had an important role in developing GSM. In early 1987, eight trial/pilot systems from different countries were compared. In June 1987, CEPT invited the manufacturers to participate in CEPT standardization work as advisors, which resulted in a considerable increase in competence and experience of the CEPT working groups. In September 1987, 15 network operators from 12 different countries signed a GSM Memorandum of Understanding. In the beginning of 1989 the GSM group was transferred from CEPT to ETSI with open participation in

the basic characteristics of a cellular radio standard, which would be based on TDMA (time division multiple access) technology. This decision spurred joint multinational efforts to further define the GSM concept and basic requirements.[134] At the same time, the EU Telecoms Council published two Directives, which covered the coordinated introduction of a public cellular digital mobile system in the EU and the reservation of the 900 MHz band for the GSM system in all CEPT member countries.[135] The achievements in Europe resulted from political support together with the coordinated implementation of pan-European standard setting and a frequency strategy (EC, 1993e, 1993f; Paetsch, 1993, p.316).

> We had a patchwork of analog systems on the continent and it was a very ambitious target even to dream of a pan-European system. Several players viewed it as totally unrealistic and many believed the Group GSM would fail. So nobody thought about the possibility of a world system. There was a clear differentiation between what we wanted to do in Europe and what we thought could be done in the world since we knew that the world market was dominated by the American analog systems (ETSI, interview, 16 June 1999).

European governments, industry and standards bodies aimed to create a completely unified system all over Europe in order to generate a common European market. GSM would spread across the world (Table 5.2).

GSM standardization activities (manufacturers, operators, administrations, service providers, and users). The ETSI Technical Committee working on GSM was labeled SMG 'Special Mobile Group'. Thomas Haug chaired the work on GSM between 1982 and 1992 succeeded by Philippe Dupuis (1992–1996) and Friedhelm Hillebrand (1996–2000).

134 The GSM project was characterized by close cooperation between a multitude of players and interests. 'You needed committed players who contributed in all major areas and also in all major countries. You needed people in industry like Dan Dornstetter in Matra, who has made most of the GSM radio interface, or people like Bernard Ghillebaert in France Telecom, who played a very strong role. You needed people in Telia, in Ericsson, in Nokia, in Italy, and so on. I think we needed a broad network of people and suddenly there were a dozen people who played a really important role' (ETSI, interview, 16 June 1999).

135 In 1987, the EU Council of Telecoms Ministers adopted Recommendation 87/371/EEC for the coordinated introduction of public pan-European digital mobile communications and it was complemented by Directive 87/372/EEC on the frequency bands to be made available for the new cellular radio system. The Council recommendation specified that the commercial service should begin in 1991 in the wake of pre-operative trials. The directive stated that the Member States would have to assign an increasing part, and eventually all, of the 890–915 MHz and 935–960 MHz spectrum to the emerging GSM networks.

Table 5.2 The growth of GSM

End of Year	Number of Networks	Millions of Customers	Number of Countries/Areas	Number of GSM MoU Members
1993	32	1	18	70
1994	69	4	43	100
1995	117	12	69	150
1996	167	30	94	200
1997	239	65	110	256

Source: GSM Association (GSMA, 1998).

The approach to mobile cellular communications was different in the US.[136] The US regulatory authorities stimulated a coordinated approach to analog mobile infrastructure. The regulatory policy, however, would change and result in a number of rivaling technologies, which was preceded by industry disputes, including both US and European manufacturers, not least Ericsson.[137] The development in North America was thus different from that

136 The US enjoyed a remarkable success with the first-generation analog system Advanced Mobile Phone System (AMPS), originally developed by Bell Labs. Its derivatives called TACS in Europe (for example, Italy and the UK) had something like 70–80 per cent share of the world market in the late 1980s/early 1990s (Harte et al., 1997). An important aspect of the US approach was the requirement that the analog and the new digital system should coexist until digital technology completely replaced the analog mobile system (Mehrota, 1994, p.15). The industry pushed for different technology proposals and the result was the development, approval and installation of competing technology standards to meet different industry and user needs. The first standard was the dual-mode (analog and digital) AMPS or D-AMPS system, the second was the so-called North American Digital Cellular (NADC) system based on TDMA and the third system standard was the IS-95 system based on code division multiple access (CDMA).

137 A public policy change occurred when the Reagan Administration succeeded the Carter Administration. An example of this shift is the standardization of stereo technology in AM car radio. The Carter regulatory FCC regime had supported coordinated solutions and endorsed a decision for one standard on AM car radio stereo across the US. This was important since FM car radio technology did have great coverage across the continent. The Reagan FCC abandoned the harmonization decision and ruled that the market players should decide on implementation through competition. This new policy resulted in a multitude of car radio AM stereo standards that made national coverage and economies of scale (production volume) more difficult. 'I think we should give a Medal of Honor to Ronald Reagan and his regulatory policy. I think that without this policy all manufactures (Ericsson, Nokia, Siemens, etc.) would have experienced a much more difficult time and situation' (ETSI, interview, 16 June 1999).

in Europe where the development went from a patchwork of technologies to the complete replacement of analog systems at an early stage.[138]

At the international level, the lack of a global standard for mobile cellular systems gave an incentive for people within the UN-based International Telecommunications Union (ITU) to start thinking about how to achieve a monolithic worldwide standard. This led to the concept Future Public Land Mobile Telecommunication System (FPLMTS) around 1985–1986 and the setting-up of a group called the Interim Working Party 8/13 (IWP 8/13). This group focused on the need for collaboration with the view of developing a unified global standard for mobile systems. However, market, technology and policy developments in Europe, Japan and the US altered the conditions. Around 1996, it had become clear to ITU officials that the rapid deployment of new mobile systems and growth of subscribers threatened the global vision. Instead, the ITU reoriented its activities toward elaborating a vision for an umbrella system known as International Mobile Telecommunications 2000 (IMT-2000). Regional standards bodies wanted to retain control over specifications, and the international division of labor, which gave regional organizations a strong political role, was further consolidated. 'The ITU is losing control of the actual standards but retains control of interworking between sets of standards' (ETSI, interview, 16 June 1999). In 1997, the ITU introduced the 'IMT-2000 Family' concept as a general framework for how to link the various regional and national emerging standards under a common umbrella. It allowed national and regional standards bodies to retain their traditional roles and bilateral cooperation links.

Initially, the process toward IMT-2000 mobile systems in the major regions of the world evolved independently from one another. Japan was most active in ITU in the late 1980s and first half of the 1990s. It was reaching a market saturation point, had shortage of radio frequencies, and needed a more efficient mobile infrastructure. Its industry and government hoped that its homegrown wireless cellular technology, Personal Digital Cellular (PDC), would become an export success (Mehrota, 1994, p.261).

---

138 Some Ericsson officials had experience of the early US developments. 'At the time of the introduction of personal mobile communications in the US there were intensive discussions about whether to have a mandated standard or to allow competition. Motorola, Ericsson and other companies argued that it would be good to have one single standard. Lucent and some other companies, however, did not think so and at the end there was a decision to allow multiple standards. These discussions are forgotten and today they say that completely free competition has always been the American model, a model that must be defended' (Ericsson, interview, 8 July 1999). 'Those who were not included in the D-AMPS initiative argued that it was not good and instead developed the alternative IS-95 and created the origin of the standards war. There was a big fight in North America to produce an alternative standard to D-AMPS' (Ericsson, interview, 22 September 1998).

The Japanese regulatory authorities differentiated PDC from other mobile standards to protect the domestic manufacturing industry and to leverage their export advantage. This industrial policy backfired, however, since a large domestic market failed to develop and no system managed to gain global market shares. 'The strongest player in the beginning was clearly Japan in the ITU for obvious reasons. They wanted a world market system and wanted to decide what that system should look like. It took some time before Europe and the US realized that an ITU-defined system based on Japanese inputs would constitute a major industrial-economic problem' (Siemens, interview, 28 June 1999). Japan announced that IMT-2000 services would be started in 2000, which was later changed to mid-2001.[139]

In the early 1990s, regulatory officials from the EU, Japan, and the US established an inter-regional, administration-led forum, dubbed Future Advanced Mobile/Wireless Universal System (FAMOUS). The EC's DG INFSO led the European activity, whereas the Ministry of Post and Telecommunications (MPT) and the FCC led the Japanese and US activity and delegations, respectively.[140] The Japanese delegation wanted to work within the ITU (where they contributed actively) rather than within a diplomatic setting. The meetings took place between 1991 and 1997. The objective of FAMOUS was to exchange information and views on the policy and regulatory environment for mobile communications in the different regions future (EC, interview, 29 October 1998). The delegations brought senior industry and standards experts to the meetings as observers.[141] In 1995, a number of resolutions were adopted in FAMOUS (on trilateral diplomatic information exchange on frequency bands and standardization for wireless

139 NTT DoCoMo Inc, Japan's dominant wireless powerhouse, formally launched its 3G mobile phones services for full commercial use on 1 October 2001 in metropolitan areas. In May 2001, President and Chief Executive Keiji Tachikawa announced that nationwide services would be available in spring 2002.

140 John Rietbrock in Commissioner Bangemann's cabinet led the EU delegation and was supported by Toscano, who was head of unit at that time. The EU delegation also included Niepold, who was responsible for relations with Japan, and Schwarz DaSilva, who was responsible for EU R&D in mobile communications. The two latter officers would play a major role in EU policy throughout the 1990s and beyond. Among the other participants was the then Chairman of ETSI SMG, Dupuis and Hillebrand (GSM Association) who became ETSI SMG Chairman.

141 In Washington in 1993, the administrations agreed on a number of principles that would guide the exchange. The administrations of Europe, Japan and the US should continue to seek collaborative approaches to the exchange of ideas to address the emerging needs of 3G wireless communications systems and services on a worldwide basis. The goal would be to maximize international cooperation on 3G for interoperability on a global basis. Moreover, there was a need to establish a broad common understanding of the requirements with a view to introducing these into the standardization process. The discussion on possible international roaming options regarding mobile cellular systems should continue (FCC, 1993).

communications) that reflected a more bottom-up approach compared to the ambition of the ITU to create a global standard (FAMOUS, 1995).[142]

> FAMOUS was important for building up the momentum for 3G in the US and was a catalyst on the way towards this IMT-2000 family of systems concept. The attention was captured by 2G mobile systems. The US industry, however, saw that the rest of the world took 3G seriously. FAMOUS was needed as an intermediate step for the recognition of interests in certain areas and a clear spelling out of such interests (FAMOUS delegate, interview, 16 June 1999).

US delegates requested that the US standardization policy and regime should also be introduced in Europe and Japan (EC, interview, 28 April 1998). The US had accepted Japanese and European standards. Therefore, US standards should be allowed full access to European and Japanese markets. According to EC and industry officials, US delegates would use the meetings as an opportunity to exert trade policy pressure and to propagate their preferences, which was not welcomed by the EU delegation, which stressed the proven openness toward US participation and contributions. It stressed that the administrations did not have the authority to require the established standardization organizations to accept specific US standards (FAMOUS delegate, interview, 16 June 1999). Ericsson and Nokia advised the EC to inform Japan that it was important to build on GSM. The EU delegation should only debate regulatory and spectrum issues. It should stay away from trade policy issues. 'Industry viewed FAMOUS as an inappropriate political intervention in the sector since the sector actors were somewhat nervous that the administrations would start something' (UMTS Forum, interview, 29 March 1999). The EC would push Japan to agree to close down FAMOUS in the light of the industry capture of the US delegation.

---

142 The US argued that the EU and Japanese delegations believed they had open procedures and no formal policy restrictions to prevent the introduction of any technologies in those regions. However, it also noted that it appears as if neither the EU nor Japan had any equivalent legislation to that of the FCC that permitted 'multiple technologies in the frequency band' (FCC, 1996). In 1996, industry representatives in the US FAMOUS delegation (AT&T and Qualcomm) requested that governments should establish a regulatory environment that would enable the uninhibited use of all regional/national standards in all regions/nations (Ericsson, interview, 15 September 1998). Wireless standards should be exchanged between regional/national standards bodies and mutually accepted. It was stressed that the US had admitted seven 2G PCS air interface standards and that the market and not standards bodies should make the selection.

> Qualcomm wished to force through a single standard that was not linked to GSM and the family concept because GSM was already well positioned in the market. In the end, this is the hard battle of global competitiveness and about being first to the marketplace. The main reason why it fell apart was simply that the Americans were pushing their own interests through FAMOUS in an unfair and unwise way. Qualcomm was already trying to pursue its interests at the time of FAMOUS. The closer it comes to the market the more strategic and commercially sensitive it becomes. It was not acceptable for Japan either and it was decided between Europe and Japan that we would not continue (EC, interview, 25 June 1999).

It was a race for commercial and technological leadership. According to Wireless Strategy Analytics, a Boston-based consultancy, there were around 850 million subscribers worldwide in 2001, approaching 1.7 billion by 2006. Asia-Pacific and South America would be leading growth markets by 2004, while Western Europe and North America were gradually moving toward saturation, although replacement sales were here expected to give a boost to the wireless market, the take-off of which, however, would depend on a turning around of the global economic slowdown. Global penetration was projected at 24 per cent by 2005.

European industry was not interested in any new infrastructure since it had invested and continued investing heavily in GSM equipment and service provision. When GSM turned out to be a world success, they did not believe the Japanese activities would constitute any real threat. It would be difficult for any major player to deploy a green-field infrastructure due to the variety of installed infrastructures. Industry pressure grew around 1996–1997 from Ericsson and Nokia to align with Japan in case the Japanese efforts were successful internationally. They had already gained a relatively strong market presence in East Asia and wanted to benefit from this advantage. Ericsson had concluded that the dominant Japanese operator NTT DoCoMo would likely opt for a revolutionary radio technology, which was similar to Ericsson's proposal and which was studied in the EU RACE CODIT project and corporate laboratories. There were a number of other similarities between the approaches in Europe and Japan. Both regions planned to use the same core IMT-2000 spectrum, had a similar time plan for commercial launch, and had similar main requirements for the future 3G mobile systems, based on existing ITU recommendations (Table 5.3).

Table 5.3 Comparison of features: Europe, Japan and the US

Region Parameter	Europe	Japan (NTT DoCoMo)	US (CDMA)
Frequency band	2 GHz	2 GHz	800/1900 MHz
Expected launch*	2002	2001	2003–2004?
Technology	W–CDMA GSM	W–CDMA (GSM?)	CDMA2000 IS–41
Base station technology	Asynchronous	Asynchronous	Synchronous

*Note: *Expected launch as of the late 1990s. Source: Trade Press.*

5.2 EUROPEAN POLITICS AND TECHNOLOGY

The EU RACE R&D program (1987–1995), to which the EU contributed Euro 1.1 billion, supported the technological development of UMTS (EC, 1995c, p.7). It aimed at promoting European industry and at deepening the European market for telecommunications equipment and services (EC, 1996g, 1996h). 'For European industrial policy, advances in communications technology and services are crucial for consolidating the Single Market, industrial competitiveness and a balanced economic development' (DaSilva and Fernandes, 1995). The Advanced Communications and Technologies (ACTS) program succeeded RACE. It was geared toward R&D support to provide the platform for internationally competitive infrastructure.[143] 'The focusing and stimulation effect, together with the pooling of resources, and sharing of information between participants, has helped to ensure that European industry retains world leadership in key strategic areas, as well as

[143] The selected technologies corresponded to areas in which European firms had to be more competitive, which were expanding and had good growth potential and where significant technological progress was anticipated (Arroyo-Fernandez, 1998). 'The intervention of the EC in telecommunications arises from the truly international nature of the industry. This implies that national actions, be they by governments in the regulatory field or by national telecommunications operators in the field of services, though absolutely necessary will not however be sufficient' (EC, 1996h, p.4).

mastering the technologies necessary for economic self-determination' (EC, 1996g: introduction).

Around 1994 the EU R&D community recognized that standardization activities related to UMTS were proceeding slowly. 'The EU spent a lot of taxpayer money on RACE and ACTS. They wanted to make sure that at some stage the research work would be adopted and accepted as standards' (Siemens, interview, 22 November 1999). This situation was discussed among the research officers and between the heads of the different EC units. Fernandes (Siemens), head of EU research on wireless communications, was asked to produce a UMTS vision paper, which concluded that a Task Force on UMTS should be established. The EC then went to the EP and Member States and stressed that something needed to be done to protect made investments. The EC was pushing for this activity.[144] At stake was long-term European strategic telecom and wireless communications policy (DaSilva and Fernandes, 1995; EC, 1994i, 1994j, 1994l). 'Europe already has a proven track record of successfully managing and exploiting the latest mobile communications technology, as exemplified by the worldwide success of GSM. Technical advances and significant changes in regulatory environments, backed by a strong political European vision and commitment from the telecommunications industry, led to the commercial emergence of 2G digital systems and a much wider telecom market' (DaSilva et al., 1995a). The EU R&D community wanted to integrate fixed, mobile and satellite components into the overall UMTS concept (Figure 5.1).

[144] The European RACE Mobile Project Assembly further elaborated and refined the vision of future mobile communications that had been initiated in the RACE I 1023 project. The objective was to generate joint views and the development of a European UMTS strategy that would gain the support of the mobile industry and customers. The ITU FPLMTS/IMT-2000 and European UMTS concepts had up to around 1995 been based on the assumption that a new global standard could be developed in the 1990s with commercial launch as early as 1998. A joint theme for the EU RACE program was the idea that UMTS represented a new, revolutionary system with respect to existing standards. Parallel to the reorientation at international level, the view promoted by the EU R&D mobile community would come to be transformed due to industry pressures and market realities.

Figure 5.1 UMTS vision

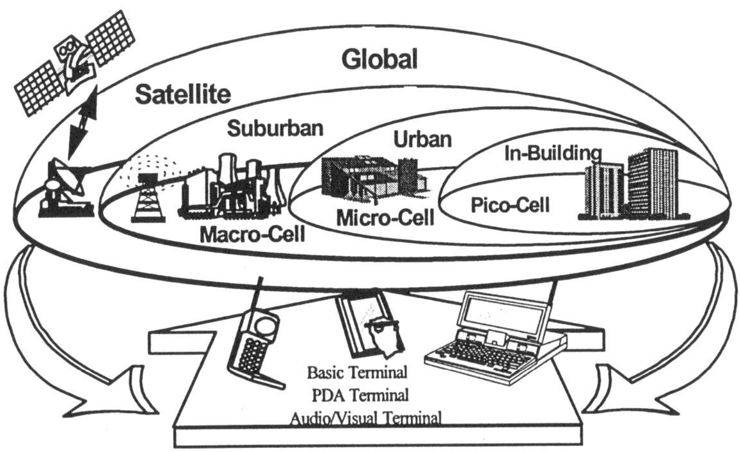

Integration with the Fixed Network

Source: DaSilva et al. (1996).

Recent market developments and initiatives taking place in Japan and North America indicate that in order to meet the challenges of globalization and international competition very significant efforts should be devoted in particular to the development of mobile/personal communications networks, systems, products and services, as a means of capturing a meaningful share of the world market. Within the EU, it is hence of strategic importance that the efforts pursued at national level be complemented on a much larger scale by EU-funded projects aimed at the development not only of a pan-European market but also of world markets (EC, 1996g, p.48).

A major strategic question was how Europe could ensure that its expertise and lead with GSM could be incorporated into IMT-2000 without alienating Japan and the US. Industry recognized the need to capitalize on the European GSM success, which could be jeopardized if Europe pushed too hard for a completely new system. 'As the global information society continues to develop, it becomes increasingly clear that the market for UMTS should not be seen as limited to the borders of the European region [...] As it is a global system, the influence of all worldwide existing systems is expected and therefore Europe must be well prepared for the challenge' (ECTEL, 1997).

In February 1995, DG INFSO initiated the UMTS Task Force to set UMTS in motion. The head of EU R&D research on mobile wireless (Fernandes, Siemens) chaired it and DG INFSO was represented through the head of its mobile R&D unit (DaSilva) and a senior officer from its policy unit (Koolen) (Table 5.4). Europe needed to speed up the development of UMTS and to elaborate a comprehensive political framework to facilitate its introduction (EC, 1995d, 1995e, 1995f). 'Europe was obviously becoming increasingly aware of the importance of winning the 3G mobile race. There were people in Europe that started to get frustrated since making standards is not enough. There is a need for regulation, spectrum, and sufficient confidence to invest. Part of the EC's remit is to advance the troika of Europe' (GSM Association, interview, 27 April 1999).

Table 5.4 Composition of the UMTS Task Force

European Organizations	Delegates	European Industry	Delegates
EC ACTS R&D Mobile	(1)	Ericsson Radio Systems	(1)
EC Mobile Policy	(1)	Nokia	(1)
ETSI SMG	(2)	Siemens (Chairman)	(1)
GSM MoU Association	(1)	Alcatel	(1)
ERO	(2)	France Telecom	(1)
ETNO	(1)	British Telecom	(1)
		Telia	(1)
		Orange	(1)
		T-Mobil	(1)
		KPN	(1)

Source: EC (1996i); interviews.

The number of delegates to the UMTS Task Force was kept limited to allow flexible and swift policy guidance to the European institutions. 'One of the motivations of the EC was to call from scratch a new European club with a balanced participation. It is always difficult to have a national club and to want to make it European-wide. This was a first fresh start and a heavyweight that nobody could easily ignore. I would say that everything was driven by industry and the EC from within Europe and there was an understanding of what needed to be done' (ETSI, interview, 16 June 1999). The EC consciously designed the group with strong representation from the major European-headquartered manufacturers (Ericsson, Nokia, Siemens, Alcatel), while for example Motorola, which had a strong presence in GSM and in the Europe market, was excluded. 'We operate according to market conditions in all the different markets we are active in. In Europe you are European and have to play the European game' (Ericsson, interview, 8 July 1999).

> The EC was 'instrumental' in setting things in motion and has been an important partner who has contributed in a constructive way. The EC really lit a fire underneath the regulators and we were very grateful for that since otherwise it would not have proceeded. The regulators were somewhat resistant since they have a certain fear that the EC and its institutions would take over. The regulators display something of a hedgehog reaction under such circumstances (UMTS Forum, interview, 29 March 1999).

The launch of the Task Force was a result of pressures from industry to capitalize on GSM to avoid the promotion of a completely new system. Huge investments had been made in GSM infrastructure. It could not be expected that industry would support UMTS financially and politically to the same extent if they could not recoup existing investments.[145] There were tensions within the EU UMTS Task Force. First, there were at this point two major radio technology proposals within the EU R&D framework (CODIT and ATDMA). Alcatel had opted for an evolutionary model of GSM for both the radio technology and network infrastructure, and had stressed the need to take the fixed telecommunications infrastructure into account (Urie, 1995; Urie et al., 1995). Alcatel's advantage and strength was primarily in cordless and satellite communications (such as DECT 1800 MHz and Globalstar).

The Japanese market was important in itself as well as being an export platform for the Asia-Pacific region for those companies with good prospects that were prepared to meet Asian competition (Berg, 1996). Ericsson and Nokia convinced EC policymakers that Europe must consider Japanese actions. The final report supported a revolutionary change in radio interface design with respect to GSM technology, which would be different from US radio technologies. DG INFSO Director for R&D Roland Huber had an interest in preserving the taxpayer money that had gone into EU research on fixed network technology (Figure 5.2) 'The researchers wanted to crush

---

145 In August 1997, the leading European telecom equipment manufacturers – Alcatel, Ericsson, Nokia and Siemens – announced that they would work together toward a common European standard viewed as the best choice for a pan-European digital wireless wideband system to succeed GSM (FT, 1997). The statement also targeted politicians in Europe and the European institutions in order to reduce criticism that industry had not succeeded in achieving one single radio technology. The coordination approach implied a general pact between the top four companies to work toward standardizing GSM-based technology and extending GSM data technology over to a UMTS platform. 'We strongly back a global and timely standardization of third-generation technology that meets with the real needs of today's operators and hundreds of millions of end-users in cellular networks all over the world. Today, there is an opportunity to base the next-generation technology on open standards that will drive the development of new breakthrough applications and the introduction of affordable multimedia services' (Nokia, 1997).

GSM. They had invested money and wanted to implement past research efforts' (Ericsson, interview, 17 September 1998). The EU R&D camp, supported by fixed telecom operators and Alcatel, wanted to build UMTS on the emerging broadband fixed infrastructure (B-ISDN).

Figure 5.2 Strategic alternatives for defining UMTS

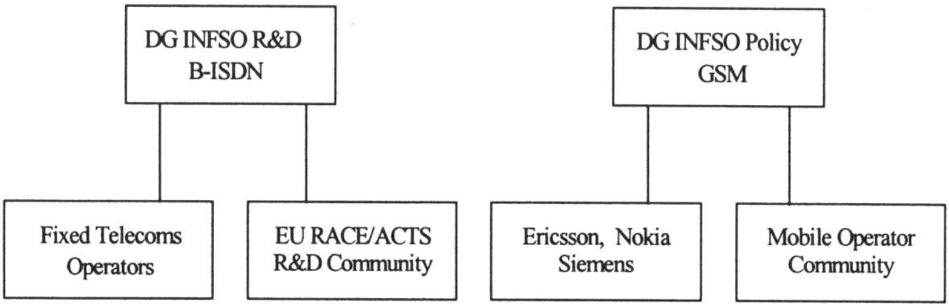

The policy branch in DG INFSO recognized the need to build on GSM and thus supported the leading wireless manufacturers. The mobile industry was less than enthusiastic about putting more money into fixed network technology and infrastructure. Ericsson, Nokia, Siemens, ETSI SMG and the GSM Association wanted to build on the footprint of the GSM core network infrastructure.[146] 'Network operators invested heavily in GSM and were not very pleased to see a new initiative starting and a new system coming across that would basically compete with their investments or with their systems. At this time this was of course a big problem. Nobody really wanted to understand what UMTS was all about and they would only try to block it. They saw it as a threat to what was being widely commercialized' (Siemens, interview, 22 November 1999).[147] From an industrial-economic perspective,

146 At the time of the task force, ideas that UMTS must be compatible with Internet Protocols (IP), the 'Internet thinking', had yet not matured. Around 1995–1996, the work on fixed network-based technology and infrastructure (B-ISDN) was reoriented to mobile network-based thinking through the so-called 'Migration Study'. This work was performed under an ETSI ad-hoc group led by Chambert (Ericsson), Ljungberg (Telia), and Söderbacka (Nokia) (ETSI, interview, 15 February 2000). A clearer shift to Internet Protocol (IP) thinking would not evolve until around 1998. Mobility management would build on Mobile Application Protocol (MAP) rather that Integrated Network Application Protocol (INAP, a fixed infrastructure solution). It was a political selection process to demonstrate which solution was superior.

147 Within the GSM Association, the Service and Regulatory Group (SERG) and then the 3G Interest Group (3GIG) aired concerns about UMTS. They argued that GSM had established a high degree of credibility. UMTS was not sufficiently based on 'leading-edge technologies', which brought into question the need for an entirely new system (GSMA, 1995a, 1995b, 1996b).

Europe was heavily oriented toward GSM that enjoyed strong growth starting around 1996 (Figure 5.3). Manufacturers largely drove the industrial and political process of defining a European strategy. Wireless network operators, however, closely followed the process and many times intervened by imposing significant pressure on manufacturers, both in Europe and at global level in the IMT-2000 standardization and requirement definition process.

Figure 5.3 Mobile cellular subscribers/penetration (EU)

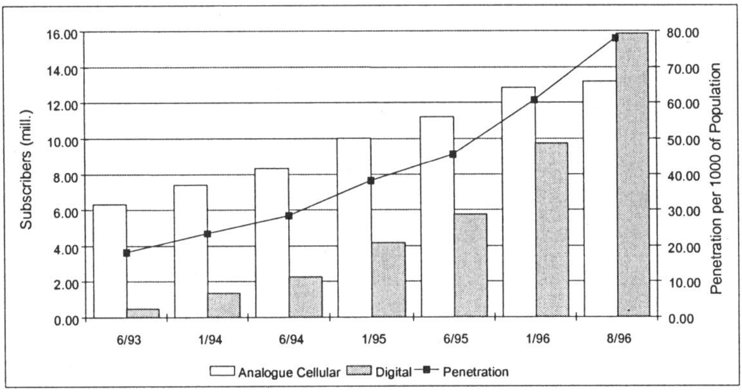

Source: DaSilva et al. (1996).

The UMTS Task Force report called for the creation of an appropriate regulatory (licensing and frequency) framework. 'A regulatory framework ensuring timely UMTS licensing will create balance between early investment and risk for prospective players. Increasing levels of investment will be encouraged when risk is reduced as time progresses' (EC, 1996i). It stressed the need for a well-functioning regulatory framework in order to achieve the 'high-priority EU targets' of growth and employment. 'The enhancement and further advancement of the European family of mobile systems is pivotal in building the European, and global, information society and infrastructures. This policy must aim at the deployment, early next century, of advanced and evolved multimedia-capable and integrated mobile telecommunications systems' (EC, 1996i). The Task Force report stressed that a 'politically determined dynamic, competitive, multi-application, and multi-operator environment' would create highly favorable conditions for 'the further strengthening of European mobile standards and technology' (EC, 1996i). Regulatory authorities needed to encourage a harmonized European-wide licensing framework. Given the triumphant success of GSM, the question was now 'what and where next?' (Huber, 1995). It was clear that industry wanted to build UMTS on the GSM trajectory (Figure 5.4).

Figure 5.4 Evolution from second- to third-generation systems

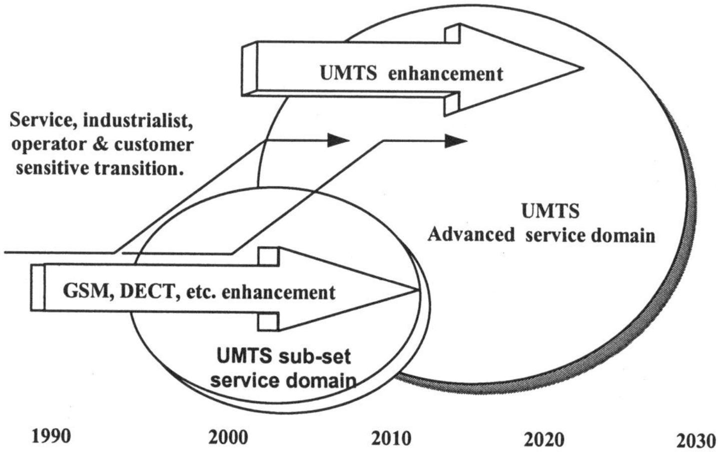

Source: DaSilva et al. (1996).

5.3 THE EUROPEAN COMMISSION AND INDUSTRY

Throughout 1996–1997, the EC and European manufacturers had preparatory meetings on the elaboration of a European UMTS strategy. The goal was 'to continue and extend the competitive advantage and the international footprint that the European industry and operators have gained with the GSM family of standards' (ENS, 1996; Brittan, 1997). The EC envisaged that UMTS would be the backbone for a 'wireless information society', driven by globalization and the pervasiveness of information and communication technologies in the private and professional spheres (DaSilva et al., 1998c, 1998d; EC, 1996k, 1996j, 1999k, 1999n).

Ericsson, Nokia and Siemens all stressed the need to speed up the development and implementation of a European strategy in the face of international competition. 'Otherwise non-European standards (for example, emerging from national arenas) may take a lead in the liberalized market areas' (ENS, 1996). Rapid allocation of frequencies and licenses were required 'to guide European standardization and to act as a global model'. The entire identified spectrum for UMTS, they argued, 'must be reserved for *only* the *European* standardized Third Generation', and additional radio frequencies would be needed in the future (ENS, 1996).[148] DG INFSO, particularly its R&D unit, engaged in a proactive lobbying campaign worldwide to promote UMTS and European industry. Leading DG INFSO officials submitted articles and organized conferences in China, Russia and elsewhere (DaSilva et al., 1997a, 1997b, 1997c, 1997d, 1998a, 1998b, 1998c, 1998d). The success of GSM as a world standard indicated the importance of concerted European action, covering R&D, political and regulatory aspects (DaSilva et al., 1998b). DG INFSO clearly linked technology development with socio-economic and political aspects. 'Specification and standardization of UMTS is expected to have an impact on many aspects other than standardization, for example competition, global trade policy and progress towards the global information society' (DaSilva et al., 1998c).

In April 1996, the UMTS Forum was set up in Europe as an independent organization with the support of the EC and funded with membership fees (Council, 1995). 'According to the political mandate given by EP and Council, the UMTS task force recommends the establishment of the UMTS Forum as the central body charged with the elaboration of European policy towards the implementation of UMTS and based on industry-wide consensus' (EC, 1996i). The forum would 'provide an ongoing high-level strategic steer to the further development of European mobile and personal communications technologies' (DaSilva et al., 1997b, p.43). The group consisted of telecom operators, manufacturers and national regulatory agencies. The first chairman was Thomas Beijer (Telia Mobile), seconded by two vice chairs (Huber, Siemens, and Wildey, Nokia); Bernard Eylert later succeeded Beijer. 'The UMTS Forum was trying to get regulators and those who allocate spectrum to coordinate their positions in order to foster a legal and uniform licensing framework' (Ericsson, interview, 15 September 1999).[149] European industry, ETSI, and the EC identified three required lines

---

148 Industry called on the EC to take into account that international agreements would be necessary 'to ensure that European system proposals would not encounter blocking actions either in development or deployment'. Actions had to be coordinated in Europe to ensure that international frequency institutions addressed the 'strategic importance to the European telecommunications industry, for example, new spectrum for mobile communications' (ENS, 1996). With the support of the EC, manufacturers stressed that 'Europe must be alert to the threat of global dominance by the US in satellite markets by 2010–2020' (ENS, 1996).

149 In December 1996, its Articles of Association were signed in Zurich by 56 member organizations and would not change until September 2000 (UMTSF, 1997b). They clearly reflected the European focus and relationship with the EU. Its first regulatory report targeted the EC and EU institutions (UMTSF, 1997a). In the original Articles, UMTS Forum members recognized the need to elaborate a 'European mobile communications policy', to address the needs of those who had invested in 'European technology', and to provide a strategy for mobile communi-

of action: regulatory framework, licensing and spectrum; basic concepts and parameters of the standard; and some type of commitment from the operators (Table 5.5).

Table 5.5 A European industry-political division of labor

The Evolution of A Coordinated European Strategy	The UMTS Vision Central Components
UMTS Vision	Consensus on UMTS concept
Dedicated regulatory framework	End 1997–early 1998
Basic parameters and technology infrastructure standards	• 'Revolutionary' radio technology • Focus on wireless Internet • Alignment with Japanese industry • Internationally oriented approach
An operator's commitment and industry-political coalition support	

Source: Interview material.

The UMTS Forum would be responsible, together with the EC and the CEPT, for the regulatory framework, including licensing and spectrum. The ETSI wireless community would be responsible for standardization, whereas the GSM Association would commit to UMTS.

The GSM experience had demonstrated the importance of obtaining European-wide agreements on fundamental policies and procedures before the creation of standards. 'Previous European experience has shown a high-level policy-forming group can generate positive direction and vigor in a complex technological development and standardization environment. UMTS needs a focused body that will bring about this top-level coordination

cations and UMTS 'for Europe' (UMTSF, 1996). Moreover, it was imperative to elaborate high-level government and industry strategies at the highest political levels in the EU, leading up to a stable regulatory framework, which in turn would create favorable conditions for investment. The signatories would advise the EU on European R&D programs. In the revised 2000 UMTS Forum Articles the references to a European strategy for UMTS, European industry, European technology, and a European mobile communications policy were officially replaced. In 2000, the signatories recognized the need 'to elaborate a consensus on regional and global mobile communications policies' and to 'provide a strategy for mobile and personal communications development' (UMTSF, 2000).

between manufacturers, operator's organizations, ETSI, CEPT, consumer electronics, content providers, the EC and users' (EC, 1996i).

5.4 EU REGULATORY POLICYMAKING

In 1994, the EC launched a Green Paper on the mobile and personal communications sector (EC, 1994k, 1994m), which was based on a 1993 EU Telecoms Council Resolution (Council, 1993c). The political intention was to launch a discussion on the future of mobile communications. In 1995, EU Member States backed the initiative by referring to UMTS 'as the principal contribution toward future personal communications' (Council, 1995). In spring 1997, the public consultation process started. 'The consultation helped to get the Member States on board at a political level. The EC received responses from all the major countries and there seemed to be a sort of unanimous opinion that it was important to pursue UMTS. I believe the Member States started thinking in the beginning of 1997 and that it was orchestrated to a certain extent by some of the companies, for example, Ericsson' (EC, interview, 25 June 1999).

In February 1997, Commissioner Bangemann met with top executives from Europe's leading manufacturers to discuss a common European strategy for UMTS (AIW, 1997a, 1997bb; RT, 1997). According to interviewees with direct access to this event, Bangemann stressed that Europe must defend its leadership (Ericsson, interview, 15 September 1998; Nokia, interview, 21 September 1998). The consultation meetings with manufacturers provided the EC with strong support for a European strategy. In March 1997, Bangemann presented the preliminary results of the consultation to the EU Telecom Council, announced that a report would be published shortly, and called for an active industrial and regulatory policy (AIW, 1997c). Around this time, a Council working group invited the UMTS Forum to inform the Council about UMTS. The invitation came from the Chair of the working group, but had been orchestrated by Ericsson and national representatives within the Council, which promoted the idea (national EU Council delegate, interview, 20 October 1998; UMTS Forum, interview, 29 March 1999).[150]

---

150 According to interview material, the Nordic countries, Germany, the UK, and also Italy, which enjoyed strong growth in mobile cellular services, demonstrated positive attitudes in the Council toward the need to grant licenses to UMTS in a timely manner. Belgium, France and the Netherlands, which had been slower on the business side, were more conservative. The CEPT/ERC had taken a decision on radio spectrum – to be allocated by 2002 – and there was some resistance. 'The Italian authorities, for example, had already allocated parts of the identified spectrum to military use. The French regulator was also resistant. The operators in

The first report dedicated to UMTS was released by the EC at the end of May 1997 (EC, 1997h).[151] The EC laid out the major strategic questions that needed to be answered. Was this the right time to define a strategy for the introduction of UMTS or would regulatory action at this point be premature? Will it be a single technology or a number of interoperable solutions based on different technologies? What should be the respective roles of the private sector and public authorities in the transition toward UMTS? How can it impact on the competitiveness of Europe's industry? How should the broader social and societal interests be secured in the development of the wireless information society? The EC pointed to the regulatory failures in the US to foster an integrated approach as a warning example. 'Europe must keep abreast of these global developments and exploit its strengths, such as the near global reach of GSM technology and its traditional capabilities in the area of communications system integration' (EC, 1997h, p.9).

> The companies that have the greatest interests together explained their views to the EC in order to be more efficient. The EC prefers that we do not come to them individually on four different occasions. If the companies can express a common position they will have significantly more impact. When companies are successful governments will contact us. The issue of European competitiveness is more independent from conservative, liberal or even socialist philosophies and people come closer together to take on a common enemy. It is about promoting European technology. There was a common European interest both internally and externally. UMTS was related to European competitiveness, employment and the introduction of new opportunities and services, which in turn could lay the foundation for additional employment (Nokia, interview, 10 December 1998).

In October 1997, the EC President Santer invited the top manufacturers and appealed to them to reach a compromise. This was followed up in November 1997 when Bangemann invited them again. Nokia and Ericsson sent their CEOs to Brussels while Siemens and Alcatel sent the heads of their mobile communications divisions. At the same time, President Ahtisaari of Finland

---

France and Italy, however, wanted the government to act' (UMTS Forum, interview, 29 March 1999).

151 The report stressed that demand for further evolution in mobile and wireless systems was emerging and that the sector promised substantial export earnings and macroeconomic benefits (EC, 1997i). 'The ICT industries are a critical component of the European economy in that they are a major and growing part of industrial activity. They are also one of the keys to future competitiveness on industrial processes, products and services and a platform for the emerging Information Society' (EC, 1997h, p.13).

discussed UMTS with the British Prime Minister Blair in London. Shortly thereafter British regulatory authorities gave their support to the Nordic Alliance headed by Ericsson and Nokia (Ericsson, interview, 15 September 1998). The European-headquartered wireless manufacturers pushed the EC to acknowledge the need to capitalize on Europe's increasing success with the GSM cellular standard. Concerted action was required to promote and defend European solutions worldwide.

> The EC has a clear role to play by setting a clear, stable agenda for the introduction of UMTS and promoting Europe's approach to multinational open standardization. There is an urgent need to move quickly to retain Europe's global influence with regard to technology and to capitalize on today's global GSM footprint that (with the right actions) promises to continue its own expansion well into the next century. GSM has granted Europe a significant advantage over other regions, for example, the US where public policies have led to competing standards and a fragmented delayed market. A technology-neutral policy must be resisted (IPP, 1997a).

UMTS would be introduced in a different environment from GSM. A major difference was that GSM had been introduced as a European standard while UMTS had to be oriented toward global export markets.[152] The European manufacturers wanted to position themselves and enter the huge Asian markets (UMTS Forum, interview, 29 March 1999). GSM was launched into an undeveloped market, whereas the market for UMTS would be strong and characterized by a relatively rapid rate of technology innovation (Table 5.6).

152 When GSM was implemented, national operators dominated the European telecommunications landscape. The operators had the role of both operator and regulator. The end of monopoly had a destabilizing effect on the operator community. In the late 1990s, the leading manufacturers had gained considerable political and market power due to their R&D expertise and to operators' reduced budgets for R&D in conjunction with deregulation. 'An important description of the changed role between manufacturers and operators is that the competence to produce specifications has moved over to the manufacturers from the operators' (Ericsson, interview, 8 July 1999). Standardization work and competence in leading-edge technologies had been passed to the ETSI where manufacturers enjoyed strong political weight (UMTS Forum, interview, 29 March 1999).

Table 5.6 GSM and UMTS market and regulatory environments

	GSM	UMTS
Introduction focus	European market	Global market
Driving actors	Operator community	Vendor community
Market structure	Monopolistic	Liberalized
Political context	National	National/European/Global
Technology	Limited innovation	Rapid innovation
Mobile market	Weak/niche markets	Strong/mass markets

According to the manufacturers, an uncoordinated European strategy for UMTS would destroy the opportunity of providing widespread pan-European coverage, interoperability, and efficient spectrum management, which were cornerstones in the GSM success. 'The manufacturing sector simply cannot afford to ignore the wider commercial and global aspects of 3G systems. This role cannot be Euro-centric' (IPP, 1997a).

> The UMTS standard should be European as minimum, but ideally it should be *the* preferred global standard. Any EU solution should reflect this aim. It is right to highlight the tremendous success story of GSM, particularly as it relates to the globalization of the standard [...] This is particularly so as Japanese and US interests both lies in moving forward the ITU global standard, with or without UMTS (IPP, 1997a).

Industry stressed that EU Member States must set a clear global example by committing to early spectrum licenses, political consensus on practical issues such as regulatory and licensing schedules, type approvals, and equitable application of competition rules. The EC should encourage other regions to implement favorable measures. 'There is no time to lose! The race is already being run' (IPP, 1997a). In the past, the adoption of a single standard had been a major blessing for the GSM industry and it had allowed manufacturers to concentrate their resources. That strategy had served the interests of the European industry well. The objective of maintaining a 'single standard policy' should be pursued vigorously (IPP, 1997b). Industry argued that while the basic technology and market would be provided there was a clear need to coordinate actions for a European strategy toward spectrum, standardization and an enabling policy and regulatory environment. Industry would be responsible for developing, deploying and operating UMTS and driving the standards process.[153] European public authorities should provide

153 There was the idea that if Europe stayed ahead and maintained its dominance then the European applications industry could get an excellent chance to get to the

political leadership and create an enabling regulatory environment. The European manufacturers presented a common front on the need for a coherent and coordinated EU strategy for wireless Internet.

> Concerning the political activity in Brussels the big four manufacturers work closely together. We have meetings with DG INFSO and have produced some joint papers. The informal cooperation is a response to the need to act rapidly. When the UMTS political issues emerged it was mostly Ericsson and Nokia that were active politically. Then the EC said that we had to try to join forces with Alcatel and Siemens. Now we act in tandem in questions where we have common views. It pleases the EC since it facilitates the opportunity to push for the issues in a stronger, more consistent manner (Ericsson, interview, 10 December 1998).

In October 1997, the EC produced a new report on UMTS that set out the results of the consultation process (EC, 1997i, 1997j, 1997k). It stated that the European mobile phone market would amount to around Euro 100 billion of annual revenues in 2005. 'Europe had a narrow window of opportunity in which to develop a clear and winning strategy' (EC, 1997j). Some EU Member States argued that UMTS should be constrained to a single technology with possible legal requirements to support interoperability at EU level. There was a broad agreement that the time was ripe for a European strategy although UMTS was not clearly defined. 'A strong home market will help European companies compete in other parts of the world and allow them to consolidate the leading position that they hold today on the world market

---

market before the US industry gained the upper hand. 'If we talk about Euro billions on the system side we can talk about Euro tens of billions on the applications side. The political dimension plays a very important role to lay the foundations for the commercialization of 3G through a rapid, consistent and harmonized approach in Europe; that is what is important' (Ericsson, 10 December 1998). A major industry debate in the late 1990s and beyond has been the international race for developing and implementing operating systems for mobile handsets. Four of the top five mobile-phone makers in 2001 (Ericsson, Nokia, Motorola Inc., and Matsushita Electric Industrial Co.), on the one hand, and Microsoft on the other, have been the main contenders. The four manufacturers were developing an operating system for mobile phones called Pearl through the Symbian software consortium based in London. Microsoft and Symbian are in a heated race to bring their rival operating systems to market first. At stake was not least control over the software inside the handsets, in addition to the hardware. Siemens, the fourth-ranked manufacturer in 2001, was using its own operating system in an attempt to differentiate its phones from those of its rivals. Interviews with DG INFSO officials indicated that the EC followed this process closely (EC, interview, 25 June 1999).

with GSM' (EC, 1997j). On implementation, the EC report supported the revealed preferences of manufacturers:

> This shows the commitment of the EU in developing a Europe-wide market for 3G mobile communications, based on a single European UMTS standard. It contributes to regulatory stability and will help in promoting innovations and investment [...] Its adoption will stimulate the development of the UMTS market as an evolution from GSM and its worldwide footprint, thus maximizing the chances for the European mobile communications industry to maintain its leading position in the field by a strong home market and early introduction of UMTS (IPP, 1998a).

A coordinated pan-European approach would be essential for securing critical mass, economies of scale in UMTS equipment and network economies in services ('network effects' or 'network externalities') (Besen and Johnson, 1986; Abbott and Snidal, 2001; Werle, 2001; Lembke, 2002a, 2002b). 'Policymakers today have an exceptional chance to help shape a seamless mobile multimedia communications environment for Europe by setting out the broader policy lines that will create a common growth basis for the development and introduction of UMTS. This will benefit European citizens and industry as a whole' (EC, 1997j).

Industry again welcomed the EC report on UMTS, which would eventually lead to an adopted decision. 'Industry was using the UMTS decision to increase their strength and position. Industry thought that if this decision could create a more favorable context, they would be pleased' (ETSI, interview, 1 June 1999). Those manufacturers with a strong presence in Asia stressed the critical significance of 'the Asian factor' in determining the future of European industry. With UMTS, market pressures came from outside Europe, specifically from Japan and Korea. Industry played on the fear of losing out in the international race. 'Europe could lose not only UMTS, but also its present position in GSM manufacturing and standards making. This danger is very real, and offers the opportunity to other non-European regions to exploit this Euro-centric weakness' (IPP, 1997c). In Europe it would take time to gain critical mass in the new industrial sector of mobile multimedia and before UMTS would be sufficiently mature to become a major export opportunity. Asian industry and governments had a less ambitious agenda (for example, a focus on voice-based communications rather than mobile data as the primary market driver). Europe therefore had to align with the regional (particularly Asian) standards bodies.[154] Europe's

---

[154] Furthermore, industry expressed concerns about developments regarding frequency auctioning and spectrum pricing. The proposal for a planned Decision

choice to focus on access to the lucrative Asian markets was a high-risk game that could either help to maintain its leading international position or lead to shifts in the international division of labor in that Japanese and US industry would gain market shares with alternative technologies. It was clear, though, that the leading wireless manufacturers, with subsequent support of the pure-play wireless operators, promoted an internationally oriented strategy toward wireless Internet and steered clear from defensive positions.

5.5 EU POLITICAL SUPPORT OF UMTS

In December 1997, the EU Telecom Council approved the proposed strategy and urged the EC to come forward with a proposal for a legislative document. The Council called on industry to take the strategic decisions required for creating a dynamic European market. 'I believe industry realized that if they would not coordinate their positions, other standards developed by Japan and the US could take over the world market' (national EU Council delegate, interview, 30 October 1998).[155] The EC proceeded in February 1998 to unveil

---

on UMTS by the EC in 1998 did not address pricing of the licensed frequency spectrum. Although this was an issue with a broader scope than UMTS (and therefore should not be included according to some of the manufacturers) the trend toward frequency auctioning in some EU Member States was of critical importance to UMTS. They argued that frequency auctioning tended to favor the financially strongest bidders, not necessarily the most qualified ones. To put more emphasis on relatively short-term economic returns could damage the prospects for a successful market development of UMTS. The argument was that UMTS would require high investments. Large investments to cover expensive licenses, which would also entail financial assistance by manufacturers and would place a disproportionate risk on them, could impede the creation of a true mass market. 'Given the long periods necessary to recoup this investment, only the biggest manufacturing corporations are able to bear both the new technology risks and the requisite vendor financing' (IPP, 1997c).

155 In January 1998, the EP Committee on Economic and Monetary Affairs and Industry (EMAC), the main responsible body for UMTS, called for European measures to ensure that Europe would remain competitive and take the opportunity of the world success of GSM. European industry had come to develop a more multi-level lobbying strategy toward the EU institutions to better ensure that their interests and views were acknowledged (Ericsson, interview, 12 December 1998). EMAC was targeted by a number of companies and organizations. All the major manufacturers had lobbying offices in Brussels with the exception of Alcatel. Ericsson was the most active manufacturer in lobbying EMAC on UMTS (together with the US manufacturer Qualcomm, which started lobbying EMAC in late 1998), while Nokia, Siemens, Lucent and Motorola were less active, and Alcatel was wholly passive (EMAC, interview, 26 October 1998). In addition, Ericsson and Nokia were most proactive in terms of stressing the need to promote UMTS

a draft Decision on the coordinated introduction of UMTS (EC, 1998i; RT, 1998a, 1998b). It came shortly after the decision by the ETSI mobile community on the radio technology for UMTS. A harmonized licensing system, a common platform for basic standards and a coordinated frequency allocation for UMTS constituted the basis for the proposed European political strategy. The proposal was intended to oblige national governments to introduce, by January 2000, rules for licensing UMTS systems and equipment. It would oblige radio authorities to allocate the 'required' spectrum frequencies in order to make the UMTS infrastructure and service implementation available by 2002.

In May 1998, the EU Telecom Council reached an unofficial political agreement to support the draft UMTS decision while waiting for the EP to give its opinion. The UK Secretary of State for Industry stressed the linkage between political activity at EU-level and the global dimension of UMTS. 'I believe it is important for Europe to go forward quickly with UMTS, building on the success of GSM, to create a global standard for future mobile services' (RT, 1998c). A spokesperson for Commissioner Bangemann said that the decision was unanimous but that Italy had expressed some concerns with regard to the allocation of frequencies. 'The majority agreed it should be possible for countries who have difficulties to ask for one year more. All except Italy said one year would be sufficient' (RT, 1998c).[156] In September 1998, the EU Telecom Council acted by a qualified majority to pass the common position on UMTS (Council, 1998c). It stated that mobile wireless communications were of strategic importance both for the development of the EU telecom industry, the information society, the economy and employment in Europe as a whole. It reiterated the need to create a coordinated and internationally competitive radio technology standard. Adopting at least one UMTS license system per country, it was argued, would ensure pan-European interoperability. Systems other than UMTS would officially be able to apply for licenses in the bands harmonized subject to national frequency

---

beyond Europe, particularly in Asia, while Lucent and Siemens took a more defensive approach with regard to lobbying positions.

156 In June 1998, EMAC welcomed the draft UMTS decision. EMAC expressed concerns about frequency auctioning processes to select winners and losers and stressed the importance of not privileging the industrial strategy of given players. Moreover, EMAC stressed the necessity for a sufficient coverage of less populated areas in Europe and full compatibility between GSM and UMTS (EC, 1998h, 1998i, 1998j, 1998m). The Economic and Social Committee (ESC) also welcomed the speed of the EC's actions, preparing and issuing three policy papers within less than a year on 3G mobile systems (RT, 1998b). In July, the EC issued an amended proposal for an EP and Council Decision on UMTS in the EU (EC, 1998k).

management and policy considerations.[157] In December 1998, the European regulatory strategy was eventually approved by the EU Telecom Council and formally went into force in early 1999 (Council, 1998b, 1999).

In November 1994, the EU Telecom Council acted on suggestions from the EC and required that a binding decision on spectrum allocation for UMTS should be taken no later than 1 January 1998. The CEPT and its European Radiocommunications Committee (ERC) concluded that industry must be given the confidence to make the necessary investment. Consequently, it was felt imperative that an adequate amount of spectrum to provide dedicated frequency bands (core bands) should be provided, which corresponded to the IMT-2000 band recommended by the ITU (ITU Resolution 212). The ERC reached a Decision in 1997 identifying those dedicated bands for UMTS and announced that the question of additional bands would be discussed. Italy and France voiced concerns about frequency allocation. Together with Greece they had not yet committed to apply the terms of the ERC Decision in late 1997 (ERC, 1997).[158] Europe and Japan had similar approaches and had

---

157 DG INFSO would continue to be active in UMTS since a number of issues had emerged in 1998 and remained on the agenda through 1999–2000, such as regulatory aspects and an emerging trade dispute with the US. It stressed the importance of three interrelated core issues for the successful introduction of UMTS. First, the conditions of national licensing procedures (which EC and industry had not been able to coordinate at EU level) would have a substantial impact on the introduction of UMTS and market developments. Second, the availability of radio frequencies was a crucial issue since radio frequencies (in particular in the range used by terrestrial mobile systems) were scarce and the availability of frequencies would determine the willingness of industry to invest in UMTS. Finally, it stressed that an open system was important when promoting a global system although it was realized that it would most likely be too difficult to achieve a single technological platform across the major regions (Argyris, 1998; Niepold, 1999). Director-General Verrue recognized UMTS as the successor of the industry success of GSM and as 'Europe's spearhead for technological innovation and growth' in mobile communications (Verrue, 1999).

158 The ECR 1997 Decision designated the frequency bands 1900–1980 MHz, 2010–2025 MHz and 2110–2170 MHz to terrestrial UMTS applications and services, and accommodated UMTS satellite component applications within the bands 1980–2010 MHz and 2170–2200 MHz. The Decision was followed by a number of other studies and planned decisions related to IMT-2000, such as on potential interference with other systems in the 2 GHz band within which UMTS would operate (ERC, 1999a) and on the licensing and global circulation of IMT-2000 terminals (ERC, 1999b). Global circulation here refers to the right of users to carry an IMT-2000 terminal originating from anywhere in the world. ERC decided that administrations should allow the global circulation of IMT-2000 terminals and their use (with the exception that administrations had the right to intervene to restrict such use if the terminals contained a non-IMT-2000 mode that was not of the receive-before-transmit type. Regarding licensing, administrations should not require individual licenses for the use of IMT-2000 terminals in a public or satellite network or outside a public network when those terminals operated on the

reserved the IMT-2000 blocks of spectrum in accordance with the ITU Resolution 212 (Chaudhury et al., 1999).

The EC had worked hard to forge a common European position and ensure that European interests would be met at the WRC conference in Istanbul in May 2000, when delegations around the world were to take decisions on global spectrum for IMT-2000. The stakes involved were substantial, as implementation, affordability, and performance of IMT-2000 services would depend on the availability of sufficient radio spectrum. The EC was clearly concerned that insufficient spectrum would jeopardize Europe's leadership. Liikanen, Commissioner for Enterprise and Information Society, stressed the weight that the EC placed on this matter.[159] 'The negotiations in Istanbul give Europe a vital chance to stake out a commanding position in the information society. If we cannot ensure sufficient spectrum availability for 3G mobile phones, this will severely handicap the leap of the Internet from being screen-based to hand-held. Europe will lose out, for the EU's big technological lead is in mobile telecommunications' (EC, 2000e). The WRC outcome would meet Europe's objectives. The EC managed to ensure a coordinated spectrum policy for UMTS, but no coordinated European-wide licensing framework was ever achieved.[160]

---

frequency band designated to that purpose by an ERC Decision (see ERC, 1996, for early UMTS overview). The EC's views of the political-economic importance of radio spectrum, viewed as the wireless infrastructure for a large number of activities, are set out in the Green Paper on Radio Spectrum Policy (EC, 1998f).

159 In the US, radio spectrum for 3G services was hotly debated. According to the wireless industry, the future of the US economy and technological prowess rested on the ability to roll out the required infrastructure. In fall 2000, the then President Clinton signed an executive order directing federal agencies to identify and make available spectrum that the wireless industry could use for new services. The FCC auction for the new frequencies was repeatedly postponed to allow bidders to figure out the value of the airwaves, and because of pressure from, among others, military authorities. In addition, the DoD controlled a large part of the most attractive spectrum and was not to eager to give it away. The Council of Economic Advisors reported that the US wireless carriers employed 150,000 people and generated US$44 billion in annual revenue. In a study, the Council depicted 3G services as a driving force for future economic growth (WP, 2001).

160 Industry had voiced concerns and argued against a purely free-market approach and auctions. 'The theory behind auctions is that market mechanisms will guarantee the most efficient use of the frequency spectrum. Some people advocate a complete freedom for those that have bought a piece of spectrum to use it for any purpose they like, provided that they do not disturb their neighbors. According to theory the development of GSM would have been a disaster. Instead of being developed in competition it was developed in consensus' (Sidenbladh, 1997). GSM had been broadly implemented and formed the basis for worldwide interoperability, long production series for equipment leading to lower prices, and a focused development of technology and services. 'If the system for selecting

The licensing process started in Finland in 1999, which awarded its IMT-2000/UMTS licenses through a beauty contest (an evaluation process aimed at selecting the proposal that best meets the requirements set up by the regulatory agency) rather than auction. It was followed by Spain, which charged what would turn out to be a modest amount compared to other countries.[161] There was a clear industrial political element in the European spectrum and licensing process. If Europe failed to coordinate and create regulatory uncertainty and thus investment and market uncertainty, European industry could be placed at a competitive disadvantage. The process was in the hands of national governments.

> A European role is key and today it goes even further than a pan-European role and this is what the UMTS Decision was for (to ensure roaming, market consolidation and that our industry can match up with industries in all regions of the world). I think we saw a pretty united Europe in 1997 because spectrum united the people. But today as the licensing process is under way and spectrum will be cleared and things come much closer to the market, competition forces prevail (EC, interview, 25 June 1999).

The auctioning of licenses in Europe generated substantial revenues for several governments and was perhaps the largest financial reallocation ever from the private to the public sector. Already around August 2000, mobile operators had invested around 10 times more for European licenses than they paid for 2G licenses. In the UK the auctioning (April 2000) raised Euro 37.5 billion while the German auctioning process (August 2000) raised Euro 50.5

---

licensees generates a profit it should be used to develop the sector, not to cover completely unrelated expenses by the state' (Sidenbladh, 1997).

161 The license process in Spain resulted in a heated debate. Like Finland, the Spanish regulator opted in March 2000 – in an attempt to show leadership in mobile technology – for a beauty contest rather than an auctioning procedure and raised Euro 520 million (US$432 million), far below the sum that would later be raised in Germany and the UK. The Spanish procedure was criticized domestically since some people viewed it as benefiting the telecommunication industry while reducing the benefits to taxpayers. In addition, when the comparatively modest sums raised in Spain with respect to the sums 'collected' in Germany and the UK became clear, the Spanish government indicated that it considered raising the fee charged to holders of UMTS mobile licenses. The four consortia which had won licenses responded that this would undermine profitability and retard investments, undermine business confidence and the credibility of the Spanish government, and that they were prepared to take the matter to the Spanish courts, and if necessary, to the EC. In September 2000, the German government announced the purposes for which they would use the revenues from the auctioning process. It would spend DM40 billion (US$17,7 billion) on infrastructure development to modernize rail networks and thereby give a boost to freight transport and passenger traffic, while the remainder would go to research and education.

billion. In 2001, the EC proposed that the EU governments should allow operators to share infrastructure to reduce the debt burden of industry.[162] By March 2001, the EC estimated that European industry had spent Euro 130 billion (US$117 billion) for 3G wireless licenses and operators had spent at least as much to roll out the new services (AP, 2001).[163]

5.6 TRANSATLANTIC TRADE AND ECONOMIC PHILOSOPHY

The EC recognized that bilateral 'relations with Japan and Japanese telecommunications interests were currently extremely positive and undergoing extension and reinforcement' (EC, interview, 12 April 1999). However, there was a looming fear among European policymakers and industry about a potential alliance with Japan and the US. The development and planned implementation of IMT-2000 wireless infrastructure would both push Europe to develop bilateral alliances and drag the EU and the EC into a

162 DG INFSO favored network infrastructure sharing between the EU's 3G wireless phone operators to soften the financial burden placed on industry. There was a need to examine further the conditions that would allow infrastructure sharing. An alternative would be to defer the collection of payments for licenses. The EC also promoted cooperative infrastructure building. New rules allowed operators to share the costs of infrastructure installment. Germany's telecom regulator (RegTP), for example, allowed the six winners of 3G licenses to share elements of the new mobile infrastructure (for example, radio base stations and towers). On the other hand, the UK promoted independent network infrastructure in order to accelerate competition between network operators. National governments acted upon concerns that the cost of auctions for the award of 3G mobile telephone licenses may cause problems within the EU's ICT sector. In a communiqué issued after the EU Council summit in Stockholm in late March 2001, the EU governments asked the EC to examine the effect of 3G licensing on European competitiveness in ICT sectors. The conclusions of such a study could attack the costs that licensing has loaded on to 3G service providers, something that the EC had strongly criticized in the past. The summit concluded that the EC should work with governments to produce a general framework law promoting 3G services (TT, 2001a).

163 Moreover, the EU institutions cooperated with the European Investment Bank (EIB) to assist indebted mobile telecom companies in the interest of preserving Europe's world lead in mobile communications. EIB has substantial weight in its capacity of being the world's largest non-governmental lending body with an annual loan distribution of some £20 billion. In early April 2001, the EIB announced it was 'very keen to keep up with the latest developments in the industry which would also include financing viable telecom projects in the 3G area' and that the EIB was in 'more or less permanent dialogue with the industry and the Commission' (TT, 2001). Over recent years, the EIB has lent the equivalent of £2.7 billion ($3.9 billion) for mobile phone projects. Again, the EIB–EU link indicates the determination of the EU institutions to support the mobile telecom sector in creating a strong home market and winning market shares.

trade rift with groups in the US. The EC was aware of the US pressure and stressed that IMT-2000 was 'an economic and political power game'. 'The US put pressure on Japan to shift over to the US development of CDMA, which would have meant that Ericsson would have lost the game. It is as simple as that. We contacted the Japanese and said that they should not give in to US pressure because they had Europe on their side. I think it helps. It is part of the political process' (EC, interview, 25 June 1999).

In early 1998, particularly those manufacturers with strong interests in Japan raised their concerns to Commissioner Bangemann, who shared the interest in developing a global standard with strong European participation. 'I believe the large manufacturers in Europe generally agree on trade policy issues' (Ericsson, interview, 8 July 1999). They were concerned about mounting US pressure on the Japanese government and standards bodies and asked the EC to support European efforts to balance US pressures. Industry argued that if the Japanese government had to bend to the US pressures, this would have a number of serious and strategic implications (IPP, 1998d).

> The next big battle will be fought over the 3G cellular mobile telephone standards. International standards for products, processes and networks are increasingly important to industry due to the quickening pace of technological innovation and the globalization of trade. Japan is key to capturing Asia, a market that by the year 2005 is expected to be larger than Europe and North America combined. Therefore the support of Japan for a future mobile communication standard is of critical importance (EC, interview, 12 April 1999).[164]

The European manufacturers pointed out that the EC must be prepared for a trade war and should further strengthen cooperation with the Japanese government. 'We consulted with industry and Ericsson and Nokia said that they did not want a trade war with the US. Because they said that, we backtracked' (EC, interview, 25 May 1999). First, industry argued that if the US succeeded in forcing harmonization ('a compromise standard') on to

---

164 'The EC has been concerned that Japan would be too dominated by the US. They have had a tradition of high-level agreements in the trade policy area where the US has imposed certain things through trade agreements that they very consciously use as an industry policy weapon. Europe had concerns that Japan could follow the US preferences or at least be closely associated with the US and then European industry would not be allowed. Our desire regarding a common European policy has been to balance up the US influence to create a level playing field in order to allow competition on commercial grounds' (Ericsson, interview, 8 July 1999). Ericsson had developed a mutual relationship with the EC. 'Ericsson is the most active in Japan among the European companies and the EC has direct contacts with our office in Japan to get information and that works well' (Ericsson, interview, 12 December 1998).

Japan it would reduce the integrity and performance of the European/ETSI technology proposal and would jeopardize the development of a truly mobile multimedia infrastructure. Moreover, it would threaten the opportunity for UMTS to gain a stronghold in Asian markets. In addition, it would disrupt the close strategic cooperation between European and Japanese industry and standards bodies and impede global economies of scale. More generally, a compromise solution would challenge the overall European objective to strengthen its ICT industry through UMTS. The balance of economic and political power would shift to the US, diminishing the chance of European technology becoming a strong world technology and potentially threaten thousands of jobs in Europe.[165]

Unofficially, the EC adopted an international strategy to 'consolidate the leading position of Europe' and to avoid 'the repartitioning of the world market' (EC, interview, 26 April 1999). The first element of that approach was to keep a low-level political profile, which the leading European manufacturers had demanded. The EC should support bottom-up activities through bilateral and regional activities, and broad political platforms in support of the development of a global set of interoperable standards, and thereby avoid claims that it was actively supporting 'European' standards. At the same time, the EC should consider opportunities to allocate funds for standardization that had not been committed to promote the global acceptance of standards. A statement together with Japan was considered to set up a task force within the ITU to work toward interoperable standards and a global infrastructure. The EC reckoned that such a move would (1) not isolate the US, (2) not further polarize the regional parties involved, (3) keep Japan and Europe in one camp, and (4) promote the development of a common global standard. Regarding developments elsewhere, the Chinese and Korean markets were seen as important. 'We stress that China should embrace a more cautious strategy, to find an alternative to the US-based model, and not to approximate their regulatory model. We attempt to ease the situation for European manufacturers' (EC, interview, 25 June 1999).

---

165 EC staff viewed US companies (for example, Lucent, Motorola and Qualcomm) as most of the time trying to impose their individual standards and use IPR licensing bargaining as strategies to promote corporate interests. IPR, for example, is Qualcomm's strategic asset. 'In an effort to conquer the Japanese market, they put pressure on the Japanese MPT to "harmonize" the W-CDMA standard with a modified version of an American standard; however, these alterations to the W-CDMA could carry IPR licensing obligations to US companies' (EC, interview, 26 April 1999). European manufacturers were in the driving seat with respect to the ongoing and planned 3G mobile developments and they could benefit from stronger Asian markets and reduce US influence in those markets. New innovative inter-regional (European-Japanese) linkages had been formed in order to respond to the urgent time pressures to achieve a global solution (Lembke, 2002a, 2002b).

The US had a slower start in 2G mobile communications than many European and Asian countries in terms of penetration and infrastructure capacity. While Europe had an almost completely digitized wireless infrastructure at the end of the 1990s, a large portion of the infrastructure in the US was based on analog technology, although there was an increasing deployment of digital infrastructure. In addition, the US had allocated spectrum for 2G mobile systems that had been identified for future 3G systems according to the 1992 ITU Resolution 212. As demand for mobile cellular services started to grow rapidly, this pushed the regulatory agency FCC to try to meet this situation, and its Chairman Kennard stressed the pivotal role of spectrum: 'We are running out of spectrum: it is an issue of national importance' (FT, 2000). Figure 5.5 shows the US situation in relation to Europe and Japan.

Figure 5.5 Mobile systems in Europe, Japan and the US

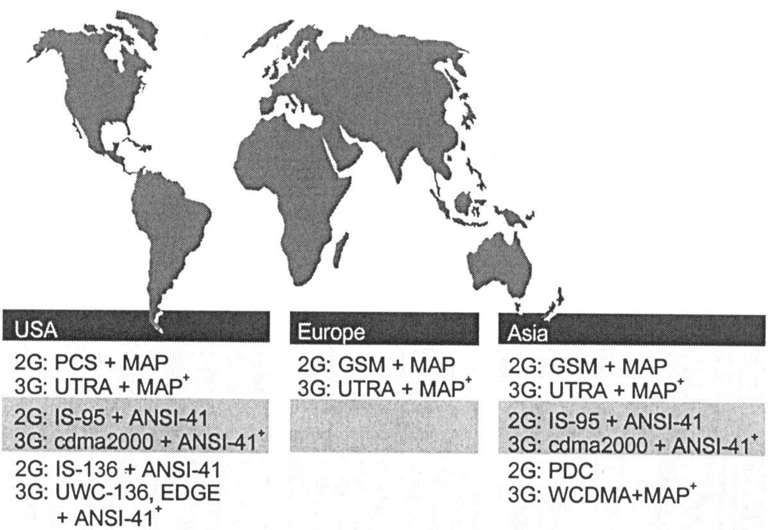

Source: Chaudhury et al. (1999).

The US had deployed three radio technologies (PCS, IS-95 and IS-136) and two network platforms (MAP and ANSI-41) for 2G mobile systems. The European approaches to both 2G and 3G mobile systems were coordinated and focused on GSM, UTRA (radio technology) and the mobile application protocol (MAP) for networks.[166] In Japan, the ambition had been to create a technology for world markets. A number of 3G technologies would be deployed, though. The US manufacturer Qualcomm, which promoted its proprietary CDMA technology, initiated an intense lobbying campaign within ETSI in 1997 and 1998 and extended it to the EU institutions.

> The main result was that it upset people like someone who is using a hammer. The reaction was that it must be something wrong and that they tried to fool us. UMTS was not regarded as a political problem until the last quarter of 1997. Politically means that it could be equally good technical solutions but the problem is that they are not compatible and that there is a choice to be made. It is political when you have valid arguments on both sides and you have to take a decision (European Parliament, interview, 13 April 1999).[167]

---

166 'In the US you have a debate on the "market determination model" that cultivates divergence while in Europe you have a "collaborative competition model" which is perceived as collusion from the former perspective. In the US, they see no difference between ETSI, the EU and industrial policy or they do not want to see it. It comes down to the way Americans and Europeans see competition. Our view does not fit with their mindset so immediately it must be a problem for them' (ETSI, interview, 1 June 1999). Moreover, 'They want to merge their standard with our standard to incorporate more patents in the latter to gain more revenues, and protect its installed base. We have the reverse intention that we do not want that market to grow but rather the GSM market. We naturally want to further build up the installed base. We have no reason whatsoever to be backward compatible to their standard; it is after all a standards war. The whole thing is very politically conditioned' (Ericsson, interview, 15 September 1998).

167 Qualcomm succeeded in getting one EP Member (Cassidy, MEP with liberal orientation) to promote an amendment to the UMTS decision (that UTRA/UMTS should not be treated in any particular way). Then the EC and some European companies actively worked to persuade in particular EMAC that the amendment was not appropriate. 'There was a need to get the politicians to understand what the attacks from the US are conditioned by and aimed at and explain how they should be conceived or at least not misconceived' (Nokia, interview, 10 December 1999). In particular the British Labour members of EP (and the leader of the British Labour in the EP, Dunnally, who until shortly before had been the coordinator of the socialists in EMAC) demanded a meeting with the EC in late 1998. The UK was in an advanced stage of the licensing process at this point in time and traditionally had a close relation to the US. The EC came and discussed with EMAC and EP members behind closed doors and strongly criticized the way Qualcomm behaved (EP, interview, 13 April 1999). According to Qualcomm, the UMTS Decision would require EU Member States to license only those systems that complied with ETSI's W-CDMA standard and would prejudge the outcome of the international standards-setting process. 'The effect of this Decision would be to cement the dominance of Ericsson, Nokia and other European manufacturers in the European market for wireless telecommunications equipment by mandating a discriminatory-exclusionary "technical regulation" for wireless communications services laid down by ETSI' (IPP, 1998b). Qualcomm's activities raised concerns

The initiation of an official political debate on Capitol Hill started in 1998 with a letter from the chair of the US House Science Subcommittee on Technology to the USTR. The letter to USTR asked for clarification on the Administration views on whether a single wireless standard or multiple standards should be promoted in the US and at global level. Moreover, the USTR was asked whether American companies had the same influence over the European standards process as European companies have over the American standards process (RCR, 1998a, 1998b).[168] Around summer 1998, US industry managed to push US legislators and government officials to discuss the European strategy for UMTS.[169] By then, however, the EU policymaking process was almost completed. A number of US legislators argued that the transatlantic marketplace was marked by an 'asymmetrical

---

in many quarters in Europe for various reasons. 'Qualcomm tried to introduce a bill that the US and all US authorities should work toward one standard backward compatible to CDMA in North America. European industry had to act quickly. We have a competitive advantage and through a common Asian-European standard we can compete on the same global market' (Ericsson, interview, 15 September 1998).

168 The same committee convened a hearing in June. Representatives from the Information Technology Industry Council (ITIC), TIA, Qualcomm, Ericsson, and a Congress member from Lynchburg, Virginia, where Ericsson has facilities, attended this. In the opening statement, the chair stressed the importance of market competition. 'I am convinced that as long as the international standard-setting process is driven by the merits of individual technologies, and not protectionist policies, US manufacturers and service providers will fare extremely well' (Morella, 1998). The Congress representative stated that wireless communications would account for a large portion of the US economy in the coming years in terms of high-tech jobs and economic growth. He stressed that Ericsson was a leading technology company with engineering and product manufacturing facilities in Lynchburg, VA, and is the leading employer in the region. ITIC stated that the US government should 'discourage other governments from using this process to promote their trade or domestic industrial policy interests' (ITIC, 1998). Qualcomm claimed that Europe used an exclusionary industrial policy in technical standard setting activities to create a protected home market base from which to export GSM technology (Qualcomm, 1998a, 1998b). Ericsson informed the committee that its US market presence dated back to 1902 and that it was in 1998 a US$3 billion company (equivalent to a Fortune 250) with over 8,000 employees in more than 100 locations throughout the US. Finally, TIA argued that the FCC should only mandate standards where voluntary action did not meet a compelling national need for interoperability in a timely fashion (TIA, 1998).

169 In addition, 3G standards proposals to the ITU had to be submitted at the end of June 1998 and this worked as a trigger for political argumentation. In late June, a number of US lawmakers from both the Congress and Senate sent a letter to Ambassador McCann (International Communications and Information Policy) at the US State Department. The letter stressed that the US proposal on IMT-2000 must refrain from favoring any one technology over another to ensure vigorous competition and innovation. 'Decreasing the number of proposed standards could stymie America's progress in the wireless telecommunications industry' (Hutchinson et al., 1998).

trading relationship' and that the global position of US wireless industry was at stake. In a letter to USTR Ambassador Barshefsky in August 1998, Senator Hollings (Senior Democrat to South Carolina, and appointed as new chair for the powerful Senate Commerce Committee in 2001) warned that US technologies could be 'locked out' of the European market. 'These technologies have been developed and actively promoted by European governments. Since European companies are able to profit from the openness of the US market, we should ensure that there is a reciprocal trading relationship that would enable US companies to profit in Europe' (Hollings, 1998).[170] The USTR and the US Administration would closely monitor the compatibility of European actions with WTO obligations (Barshefsky, 1998).

Qualcomm and Lucent, which strongly lobbied for harmonization of CDMA2000 and W-CDMA radio technology, also drove the Cellular Development Group (CDG), the leading CDMA trade association. They were successful in persuading politicians to lobby the US Administration and the latter to exert pressure on European and Japanese authorities as well as support CDMA in other regions (CDG, 1998, 1999a, 1999b). 'ETSI members, working with European policymakers, clearly hope to maintain their closed home market while enjoying the right to compete in the American marketplace with W-CDMA' (Qualcomm, 1998a, 1998b).

In December 1998, the Clinton Administration sent a first missive to Commissioner Bangemann, signed by US Secretary of Commerce Daley, USTR Ambassador Barshefsky, FCC Chairman Kennard and US Secretary of State Albright. They stressed that the EU must permit greater access for US industry to the European market. 'We are seeking specific assurances from European governments that US industry will be able to deploy competing 3G technologies and services in Europe at the same time that European-sponsored 3G technologies and services are deployed' (USTR, 1998a). The letter aired concerns about the European UMTS decision. 'The recently adopted decision of the European Commission, which appears to prohibit the operation within Europe of any 3G standard except that adopted by ETSI, would preclude marketplace consideration of rival standards and restrict consumer choice' (USTR, 1998b).

Bangemann responded to the allegations in January 1999, saying that the EU supported the ITU IMT-2000 negotiations. The EU aimed to facilitate Europe-wide services rather than prevent the competition from alternative

---

170 In Summer 1998, for example, the EU had been accused of pursuing an industrial policy that 'left unchecked may undermine the goals and objectives of the Transatlantic Economic Partnership and possibly have broader ramifications for our trading relationship with the EU' (Matsui, 1998).

technologies (Lembke, 2001).[171] 'UMTS will lead to strong competition on services and equipment to the benefits not only of consumers but also of all market players in Europe, including non-European manufacturers and operators in Europe. The requirements imposed by the UMTS Decision are considered proportionate to its legitimate purposes' (EC, 1999i). Bangemann hoped the US would understand 'the limited regulatory involvement of the EU in the process of market introduction of 3G services which is being led by industry' (EC, 1999i).

Bangemann pointed out that technology solutions were chosen by industry in accordance with the requirements and dynamics of the market. ETSI developed a particular standard for UMTS as one form of future mobile communications system through a well-defined process involving a large number of companies, not only from Europe but also from Japan, the US, and other parts of the world. The European approach was perfectly consistent with global efforts with regard to frequencies and standardization. The EU encouraged international standardization in conformity with international trade agreements.[172] 'The acute need expressed by the US for backwards

---

171 The Bangemann letter raised concerns in Europe since it appeared to leave the door open for the American CDMA technology, which was seen as a potential threat to European interoperability and as having lower performance than the European technology. 'We were very concerned about the statement from Commissioner Bangemann where he appeared to proclaim on behalf of Europe that Europe would allocate spectrum for CDMA2000-based systems, which would be suicide. The response to Ms. Albright was deplorable. The important thing in Europe has been the principle of one, and only one, system for the whole of Europe to focus resources and activities. The implication of the statement was that the pressure increased on the national level and that level is evidently completely powerless in comparison to the US. The US knows that we are concerned in Europe and play on that card. A fragmentation of the European market would weaken European industry's home market' (UMTS Forum, interview, 29 March 1999). The UK government launched an initiative within the framework of the Council of Ministers to set up a so-called 113 Committee to discuss the implications of US trade pressure and European obligations under the WTO on the introduction of UMTS in Europe. 'The French administration deplored the interpretation by EC. They were prepared for a trade war and did not think that mattered much. It was unnecessary to bend over and instead we should have resisted. The British also say that the EC's position does not hold. You cannot decide to have one standard and not to have it. You have to choose side' (Ericsson, interview, 8 July 1999). Moreover, it was argued that the US had broken the ITU Resolution 212 agreement, and implemented technology that had been developed in less democratic and open institutions than in Europe, and now the US wanted to impose their solution on others and require alignment (UMTS Forum, interview, 29 March 1999).

172 In the framework of the Transatlantic Business Dialogue (TABD), standardization and regulatory policy regarding 3G wireless services and technologies was discussed under the so-called Electronics, Electrical, Information Technology and

compatibility in this field appears to be due to the US decision to use 2G frequency bands which ITU had already identified for 3G. Moreover, it appears that the request by the US to design the new 3G standard to fit the US frequency plan would degrade overall technological performance gains and increase cost to the expense of the consumer' (EC, 1999i). The US Administration was reminded about its regulatory approach in adopting an exclusive standard for digital television in December 1996.

> There are tremendous inconsistencies in their policies and statements. It is completely opportunistic. The EU has a big market of 380 million people and we do not always have to swallow everything that the Americans say. We should not shy away from making the inconsistencies visible at the global level if the Americans put inappropriate pressure on Europe. That is the current thinking. We keep a low profile not least since our companies tell us to keep a low profile (EC, interview, 25 June 1999).

Furthermore, the political activity of the CDMA industry was opposed by a number of organizations in the US. The Computer and Communications Industry Association (CCIA) complained to the State Department 'the apparent efforts by some manufacturers of wireless equipment to use the US Government to advance their firms' parochial interests in the international standards-setting process at the expense of others' (CCIA, 1998). The North American GSM Alliance LLC, the North American Interest Group of the GSM Association, and the GSM Alliance (a consortium of US and Canadian PCS carriers), stressed the need for a multiple standards and market competition at both domestic and global levels. They supported the European GSM and W-CDMA technology. The Universal Wireless Communications

---

Telecommunications Sectors (EETSI) group. A number of principles with respect to 3G mobile infrastructure and policy were agreed in late 1998. Government regulation should be kept to a minimum, interoperability of networks and standards should be ensured, market forces should drive the process, and frequency and licensing procedures should be consistent with ITU recommendations (TABD, 1999a). This group held a specially convened meeting in Washington in February 1999 on the issue of further cooperation and dialog on 3G wireless standards and regulatory policy. 'This group strongly supports the industry-led standards-making process within ITU. There was a broad consensus that the role of governments should be to facilitate the use of spectrum and not to make technology choices in the standardization process. A single global standard is not achievable for 3G wireless. All participants in our meeting strongly agreed that the US should not on present grounds initiate a Super 301 action. This is in the interest of avoiding government action counterproductive to Trans-Atlantic progress on 3G wireless issues' (TABD, 1999b, 1999c). Motorola argued against any Super 301 action (that is, defining it as unduly restrictive to trade and thus politically punishable by unilateral trade retaliation (EETSI delegate, interview, 25 March 1999).

Consortium (UWCC), which promoted TDMA technology, joined in to urge the US Administration (and the CDMA industry) to keep a hands-off approach with regard to the ITU's 3G standards process and argued that there would be more than one global standard (EE, 1998; UWCC, 1998). The UWCC had become increasingly concerned about CDMA industry lobbying and stepped up its political and commercial efforts in late 1998 and beyond to defend TDMA technology around the world. The UWCC complained to the US DoC that the latter favored CDMA technology in bilateral talks with China and elsewhere at the expense of other US technologies. 'Due to the official statements and press reports resulting from the China trip – a trip where only CDMA was noted as a 3G technology enjoying US government support – we have encountered numerous global communications leaders who now are presuming that CDMA technology is the US government's 3G technology of choice. The Administration's absence of any public focus on TDMA in regard to its China trade agenda is creating an unintended but potentially serious problem for UWCC members that employ over 70,000 employees and have over $30 billion invested in TDMA' (UWCC, 1999).

The US Administration welcomed the EC's assurances at the same time as the signatories of the letters to the EC reiterated that they remained concerned about the previous practice of promoting a single mandatory mobile standard within Europe and stressed that it was important that Europe supported the ITU standardization process. 'While the stated intention of Europe's recent action may be to assure a minimum level of interoperability in Europe, the fact remains that the Common Position confers regulatory certainty and therefore a market advantage upon only one type of technology' (FCC, 1999a, 1999b).[173] In July 1999, the US Department of Commerce and the USTR sent a new trade policy letter. They stated that several EU Member States 'apparently continue to adhere to the view that all licenses and radio spectrum to be assigned for 3G wireless telecommunication services be dedicated to a restrictive UMTS standard' (USTR, 1999a). Europe had to standardize other ITU proposals than solely European technology.

---

173 FCC Commissioner Ness echoed the concerns expressed to Bangemann in a speech on telecom policy in January 1999. 'The single standard adopted by ETSI does not enable mobile wireless systems based on other standards a migration path for upgrading their existing infrastructure to be compliant with the standard adopted by ETSI. The ETSI standard also does not require EU 3G systems to be compatible with other types of mobile wireless systems. Thus, consumers using existing mobile wireless systems that are incompatible with the 3G standard adopted by ETSI will not be able to enjoy seamless roaming with their systems in the EU. A second concern is that it may be premature for ETSI to adopt a standard while the ITU standards evolution process is ongoing. To do so risks signaling an unwillingness to join in the multilateral process underway that may recommend more than one standard for 3G systems' (FCC, 1999b).

> Failure to complete these steps would undermine the ITU's global harmonization approach, deny market access to service suppliers using alternative ITU standards, and deny North American and other manufacturers of alternative systems the opportunity for global economies of scale. Where [EU] Member State regulators initiate licensing processes that will provide market access solely for the single standard currently approved by ETSI, the United States will consider all options to address the trade concerns this would raise (USTR, 1999a).

The EC responded firmly to the US allegations and warnings.[174]

> The UMTS Decision requires that Member States grant at least one license to UMTS in order to ensure, in the least trade restrictive manner, the availability of interoperable 3G services in the EU and pan-European roaming for the benefit of consumers. This in no way prevents Member States from granting licenses and assigning frequencies as for the operation of systems other than UMTS. This regulatory approach is notably less restrictive than, for example, the US choice of an exclusive standard for terrestrial digital television, although apparently based on similar justifications. The availability throughout the EU of UMTS as a fully interoperable system is still required to ensure Community-wide interoperability and roaming (EC, 1999j).

In a letter to President Clinton in January 1999, the President's Export Council urged the Administration to promote, in all available domestic and international forums, a multiple standards regime and no single technology proposal (PEC, 1999).[175] Likewise, senior House of Representatives members

---

174 Commissioner van Miert demanded clarification on how the US Administration would grant operators (incumbents as well as new entrants) unrestricted, non-discriminatory and timely access to frequencies and licenses for the provision of services based on any IMT-2000 proposed standard they may wish to implement. 'Regulatory certainty as well as non-discriminatory access and treatment should be available to both new entrants and existing market players to avoid competitive distortions' (EC, 1999j). The US Administration sent a third letter in October 1999 to the newly appointed Commissioner for Enterprise and Information Society Liikanen. They claimed that 'the more flexible US regulatory approach' allowed all operators to respond quickly to evolving marketplace needs and to do so not only through construction of new systems but also through upgrades of existing infrastructure. 'We encourage the EU and Member States to adopt similar flexible and non-discriminatory approaches in their allocation and service rules for the mobile services market' (USTR, 1999b).

175 'Given the fact that US telecom companies have invested billions of dollars in multiple, market-accepted 3G technologies, it would be sharply counter to the interests of US industry as a whole either to advance any particular US technology

(such as Helms and Lott) called for an open, competitive global marketplace and timely provision of 3G services in the US and abroad. The Clinton Administration, they argued, should resist delays in the consideration and approval of multiple wireless standards (Helms et al., 1999). Concerns were still being aired on Capitol Hill, not least from Democratic representatives. Sander Levin, the ranking Democrat (D–Michigan) on the House Ways and Means Trade Subcommittee, sent a letter to the Administration. 'The EU's adoption of an exclusive, mandatory standard for digital wireless technology will exclude American technology from one of the largest markets in the world and is profoundly unfair to American firms, workers and consumers. The adoption of a European standard that is incompatible with American technology will harm the US in numerous ways' (Levin, 1999). The global market for wireless services was expected to grow to around one billion subscribers over the next five years and the opportunity to fulfill this anticipated demand and create thousands of new American jobs would be lost if the EU maintained an exclusionary standard.

> It is imperative that the US insists that any effort to shut out American digital wireless technology from competition in European markets will not be tolerated. In this regard, there is a strong case for including the EU's exclusion of American digital wireless telecommunications equipment and services on any list of priority barriers to free and fair trade (Levin, 1999).[176]

Summary

The European telecommunications sector had historically been highly fragmented. The consolidation of the internal EU market can only be achieved by a certain degree of compliance with a standardized infrastructure to ensure interoperability and by an appropriate regulatory and institutional environment. Interoperability of mobile systems and free circulation of mobile telephones were integral elements of the European policy model. The

---

over another, or to fail to promote a multiple standards regime. It would be to allow our global competitive competitors to assume that multiple standards are not a US Government priority. We urge you to lead this administration in taking an active and aggressive position to promote the interests of US industry broadly and no single US interests specifically' (PEC, 1999).

176 Senator Tom Daschle (majority leader in the Senate from 2001) raised the same concerns to the USTR. 'I urge you to consider the full range of negotiating tools in order to send the necessary message to the EU that the US will not tolerate any protectionist effort to shut out American digital wireless technology from competition in European markets' (Daschle, 1999). The letters from Daschle, Hollings, Levin, and a number of other Representatives were basically copies, which arguably indicate the lobbying pressure from the US CDMA industry.

approach in Europe was based on a three-pronged strategy for mobile cellular communications and wireless Internet. The first action was the regulatory liberalization of services and terminals. A second strategy was to harmonize licensing conditions, in particular with respect to necessary frequency bands. The third approach was the development of common standards where necessary. This chapter has shown the proactive approach of the European Commission (EC) to maximize European welfare with the support of the European industry. The private sector in Europe was interested in promoting a regulatory framework at EU level and a coordinated technology proposal in order to open up for new significant export market opportunities. In addition, an effective EU policy for high technology is conditional on the expected benefits for Member States and thus their political support of the EC and its ambitions to promote competitiveness in high technology industries across Europe. The regulatory and standardization process in Europe, however, was not free from internal and external controversies.

The examination of the EU regulatory strategy for wireless Internet has shown the distinctive role of the EC and EU institutions in structuring the global competition around wireless technology infrastructure. It contributes to our understanding of the differential role of the EU as a more organized supranational political actor and of the strategic use of regulatory policy by multinational corporations in the overall context of global competition in the information and communications technology (ICT) industries. The EU high technology strategy for wireless Internet contrast with the subsidy-based industry support at the national level promulgated by the first wave of strategic trade theorists in the 1980s. As markets and technology change, the character of public support demanded by industry changes. Political institutions try to tweak the rules of the world economy in ways that enhance domestic welfare along a liberal strategic trade perspective.

6. Globalization, strategy and technology

This chapter addresses the issue of reconciliation of established European institutional arrangements and requirements, on the one hand, and challenges brought about by the internationalization of production and corporate strategies, on the other, with regard to Europe's development of the Universal Mobile Telecommunication System (UMTS).

6.1 DEFINING WIRELESS STANDARDIZATION IN EUROPE

The European efforts to standardize critical technology for UMTS would become entangled in a prolonged industry dispute that delayed EU policymaking and triggered concerns within the European Commission (EC) services. The EU R&D activity on mobile communications aimed to compare radio technologies supported by different producer groupings (in the beginning a number of network operators also submitted technology proposals). Those radio technologies were developed in the EU RACE CODIT and ATDMA projects. The major equipment manufacturers in CODIT (UMTS Code Division Testbed), Ericsson Radio Systems AB and Philips Kommunikations Industrie AG, targeted a so-called asynchronous CDMA (code division multiple access) radio technology. The project was promoted both as an alternative approach to the ATDMA project (Siemens and Alcatel) and overseas technologies.

Ericsson and Phillips argued that US radio technology (particularly D-AMPS and CDMA technology) did not meet the global recommendations for IMT-2000 promoted by the ITU organization. Those recommendations included (1) operation in non-synchronized base station subsystems and (2) bit rates for data transmission up to 2 Mbit/s (ITU, 1992). The issue of asynchronous technology (referred to as DS-CDMA) played a role in EU policymaking. Those manufacturers that invested in asynchronous CDMA technology initially referred to the need to ensure European technology independence. The US GPS satellite system used synchronous CDMA (developed by Qualcomm) for wireless base stations, originally a military technology. The CODIT manufacturers claimed that the US Administration could withdraw the availability of services based on that technology. 'A radio interface concept based on multi-rate DS-CDMA offers great flexibility in frequency and radio resource management and service provision for future 3G mobile communication systems. Asynchronous multi-rate DS-CDMA enables a cost-efficient implementation of non-synchronized base station subsystems and makes the mobile network independent of GPS' (Baier and Panzer, 1993, p.259).

In addition, Commissioners Bangemann, Cresson and Kinnock had stepped up efforts in the 1990s to promote European satellite technology. There was a will to reduce the dependence on US technology and the associated economic power and political control. Ericsson, Nokia and Siemens lobbied the EC and national regulators, stressing that US industry may completely dominate satellite markets in the future if Europe did not strengthen European capabilities (ENS, 1996). 'Europe did not want to have synchronized base stations in the GSM specification, because of the reliance on the global positioning system (GPS),' said an Ericsson official (CI, 1998). Ericsson also justified the need to avoid synchronized base station technology on technical performance grounds. 'With synchronized base stations it is necessary to be able to see GPS satellites at all times, which makes siting them indoors very difficult' (CI, 1998). Moreover, the dominant Japanese network operator NTT DoCoMo developed asynchronous technology. For those producers that had a strong market presence in Asia (Ericsson and Nokia) it was important to align with Japan in the international race for future mobile communications. The CODIT manufacturers were eventually able to win support for its 'revolutionary' asynchronous radio technology. Ericsson and the other manufacturers stopped referring to the need to reduce Europe's dependence on CDMA technology and focused on an aggressive globally oriented European strategy for wireless Internet.

The objective of the follow-on EU ACTS project on radio technology called FRAMES (Future Radio Wideband Multiple Access System), a consortium dominated by Ericsson, Nokia and Siemens, with participation from operators and research institutes, was to attempt to narrow down the rival concepts into one unified technology proposal (Mohr et al., 1999). The EC wanted industry to produce an integrated, spectrum-efficient radio technology, referred to in Europe as the UMTS Terrestrial Radio Access (UTRA). 'The efforts to combine these proposals was not least a political decision to harmonize and show a way forward' (Siemens, interview, 28 June 1999).[177] In addition, the results would be communicated to Japanese and US standards bodies in an attempt to have European technology adopted

177 The EU FRAMES project had started out with various proposals that were later narrowed down to three major proposals supported by Ericsson (W-CDMA), Nokia (W-TDMA) and Siemens (TD-CDMA). The remaining proposals were grouped under an umbrella concept dubbed Frames Multiple Access (FMA).

worldwide. The European efforts evolved in conjunction with technology developments outside Europe, particularly in Japan (Anderson et al., 1999).

DG INFSO wanted 'to develop UMTS as a worldwide competitive and versatile mobile multimedia communication standard with capabilities extending beyond a fully evolved GSM' (DaSilva et al., 1998b). Different corporate preferences among Europe's manufacturing powerhouses, however, made it difficult to achieve consensus. Around September/October 1997, Nokia joined the radio technology proposal promoted by Ericsson called 'wideband code division multiple access', or W-CDMA. Ericsson had shifted to this technology proposal for commercial and political reasons, spurred by discussions with Japanese industry and regulators (Ericsson, interview, 17 September 1998). The EU R&D results were presented to the ETSI community in late 1996 and submitted to the Japanese standardization process during 1997 (Anderson, 1998; Berruto et al., 1998).

In 1997 the Japanese standards body ARIB (Association of Radio Industries and Business) had basically decided to select W-CDMA radio technology (Mohr, 1998). Japanese industry showed interest in the European process just as the European industry had been involved in the Japanese process, in particular the FRAMES manufacturers (Ericsson, Nokia and Siemens). In addition, there was a fear factor involved in that European industry wanted to ensure participation in Japanese developments if they would become internationally successful. Japanese industry had failed in the past and wanted to hedge the risks. 'What you have to look at is whether there is a political will for getting to a single global standard as there was a political will for a single European standard' (GSMA, interview, 26 April 1999). European industry wanted to capitalize on the success of GSM and to protect its capital investments, but there were different opinions on UMTS:

> The early efforts toward 3G were too premature because ETSI and European industry were still reaping the benefits from GSM [...] There were collective industrial attempts to slow down the introduction of UMTS. I believe we will see more attempts by certain factions of industry to delay it. They want to capitalize on the GSM market and reap benefits from it before introducing something new [...] (ETSI, interview, 11 May 1999).

GSM had not been formally adopted by the ITU. 'It was a conscious decision not to put the system through the ITU. It was developed by ETSI and has become a global success without any intervention of the ITU' (ETSI, interview, 11 May 1999). The European mobile community intended to submit a formal IMT-2000 candidate technology to the ITU in summer 1998 (Dupuis, 1995; Rapeli, 1995a–c). In 1994–1995, a small team of less than ten people within the GSM Association and its Services Group (SERG) initiated a drive to move toward a 'new UMTS', which endorsed the evolution from GSM networks, Internet capabilities and a clearer international orientation (Samukic and Polese, 1995; ETSI, 1996) (Table 6.1).[178]

Table 6.1 Transformation of the UMTS concept 1995–1996

	Old UMTS	New UMTS
Core ideas	Integration of all existing and new services into a new universal network	Focus on innovative new services Support of GSM services
Partner networks	Broadband ISDN	Intranets and Internet
Introduction	Migration from existing networks	Evolution from GSM networks
Roaming	New development INAP-based	Evolution of GSM roaming MAP-based
Standardization	FPLMTS as a monolithic standard to ITU	IMT-2000 family in ITU ANSI, ETSI, ARIB/TTC

The new UMTS concept included the recognition that close cooperation with standards bodies in the other major regions (ANSI, TI, TIA, ARIB and TTC) would be required in order to promote it beyond Europe. The GSM infrastructure already had the support of hundreds of network operators, manufacturers and equipment vendors worldwide. ETSI had lobbied for a broad international ('family system') concept that acknowledged the ambitions of different regional and national standards bodies and their members. 'This was important because Europe really wanted to build on the GSM footprint and its strength' (EC, interview, 25 June 1999). The other regional standards bodies and the ITU endorsed it: 'It ended the strong confrontation and opened the way to a reasonable successful cooperation between previously hostile parties' (Hillebrand, 2000). Japanese activities accelerated developments in Europe. 'The Japanese were the real pushers for wanting 3G fast. Certainly from our side, the timetable was the Japanese right from the beginning' (ETSI, interview, 1 June 1999). It was felt in Europe that

178 Leading persons in the review group were Hillebrand (later ETSI SMG Chairman), Lilly (Orange and later Lucent), Alan Cox (Vodafone), and Toepfer (Mannesmann). The original model that guided EU/ETSI R&D and standardization work was to integrate all existing and new services into one new universal network, based on B-ISDN. However, this was resisted. 'With all the legacies in 2G it was clear that practically nobody wanted to have a totally new system and since that would be the only way to have a single global standard we have never really believed that there would be a single global 3G standard. An evolution strategy was the only viable way. Evolution meant that we could not have a single standard in the future' (Nokia, interview, 21 September 1998).

more time was needed to match ITU and Japanese work (Figure 6.1). 'Japan and China were for different reasons forced to rely on ITU recommendations. They want to have an international specification because then they do not have to decide whether they should select a European or American standard and you can avoid being accused under WTO obligations and avoid a lot of trade politics' (Ericsson, interview, 17 August 1999).

Figure 6.1 The international process toward IMT-2000

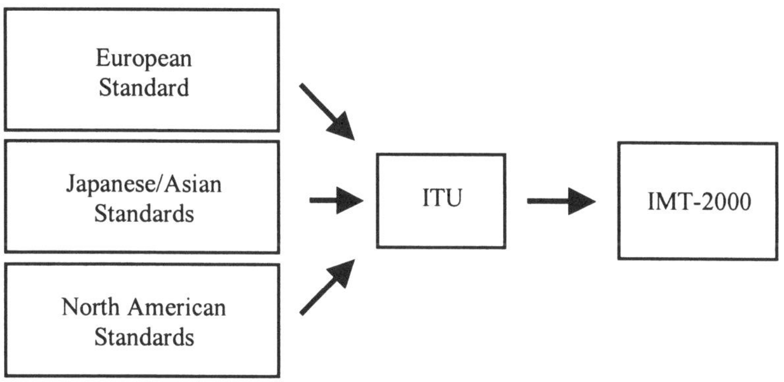

There was a widespread concern that Europe was failing to establish effective influence over and contribution to global standardization work in the ITU. ETSI officials responsible for mobile communications stressed that European industry had to develop a coherent approach to global standardization and the ITU to meet Japanese and US commercial and political activities and to maintain Europe' leadership.

6.2 Rivalry and Alliance Politics

In late 1996, the ETSI mobile community approved the procedure for the definition of the so-called UMTS Terrestrial Radio Access, UTRA (Samukic, 1998). One of the objectives was to narrow down the candidate proposals and submit only one European candidate for IMT-2000 to the ITU. The major rivaling concept groups were the UMTS Alliance, which promoted the TD-CDMA or 'Delta' proposal (which now consisted of Alcatel, Bosch, Italtel, Motorola, Northern Telecom, Siemens, and Sony), and the Nordic Alliance (spearheaded by Ericsson), which promoted W-CDMA ('Alpha' proposal).

Nokia later joined the Alpha group. The Japanese organizations essentially allied with the Nordic Alliance while the North American manufacturers Motorola and Northern Telecom allied with the UMTS Alliance.[179]

Decisions by the Japanese industry and lobbying by NTT DoCoMo across Europe, supported by the Japanese government, would have important consequences for the European standardization process. NTT DoCoMo announced that it would promote the Nordic Alliance and its W-CDMA technology proposal. For those industrial players that were less well connected with Japanese industry and the Asian market, this was a major concern. They thus played on the fear of Asian competition and argued that Japanese pressure could jeopardize the future of GSM. Would Japanese industry attach its radio technology to the evolving GSM core network, or would it opt for either the Japanese PDC core network as it had in the past, or the US IS 41-network technology?

European industry had no interest in using those core networks for commercial and technical-strategic reasons. How, for example, could the integrity of the GSM platform be guaranteed and European investments protected? The problem was made more difficult in that there was a gap in implementation schedules between Europe and Japan. In November 1997, Commissioner Bangemann convened senior officials of the big four European mobile equipment manufacturers: Jorma Olilla (CEO, Nokia), Lars Ramqvist (CEO, Ericsson), Volke Jung (Siemens) and Alain Bravo (Alcatel). 'There were big commercial interests behind all the proposals and solutions' (ETSI, interview, 16 May 1999). Bangemann attempted to exert pressure on industry. 'When the industry is not coordinated and there are conflicts then the EC cannot intervene on one side or the other but they can however stress certain things and push for active and broader industry interests and this was what EC did' (Ericsson, interview, 15 September 1998).

---

179 Alcatel had not participated in the EU FRAMES project but later aligned with Siemens. The latter had made a mistake with GSM and did not start investing significantly in radio technology until after Ericsson and Nokia had already moved ahead. 'The markets were still closed and a deal was made at the national German level to create a consortium (which consisted of Bosch, Philips and Siemens) and to start a system. Siemens did the switching and the other companies the radio and in that way Siemens entered later into radio. They only started developing their base stations after they acquired part of GTE that had had a lot of technology on board. There were also people in the military division and by putting all that together Siemens created a team that was capable of developing Siemens GSM radio base station technology and then the evolving 3G. Alcatel was never very strong in cellular radio, although they were stronger than Siemens in satellite' (Siemens, interview, 27 October 1998).

Bangemann wanted to see a unified radio technology across Europe and to repeat the GSM success. He was against the US approach for 2G mobile cellular, where multiple standards impeded the creation of open standards and interoperability. 'The EC wanted to be assured that industry would be able to reach one standard; that was their basic concern. There have been a number of high-level meetings where the EC wanted to be sure that industry gave a sign of confidence that the standard would finally be in place' (Siemens, interview, 27 October 1998). At the same time, a debate on intellectual property rights (IPR) intensified and the EC was concerned about the implications that IPR issues could have on European standardization. The decision on radio technology would be critical to EU policymaking and DG INFSO made its voice heard. 'We will carefully monitor the decisions related to UMTS that have global impact, and call on parties concerned to contribute in achieving transparency at this juncture. We would consider necessary action if IPR questions threatened to affect the further development of UMTS in Europe and at global level' (EC, interview, 30 April 1998).

The mobile operator community and the GSM Association felt that a rapid implementation schedule was inappropriate from a policy perspective and its members wanted the selection of only one wireless radio technology.[180] It promoted a standardization policy based on the GSM platform in order to not split the global GSM community and regionalize markets, which would jeopardize opportunities for economies of scale in terminal equipment, network economies in services, and other benefits deriving from one global GSM system. Its members opposed the political activity of manufacturers. 'Concern was expressed about the current entrenched and polarized positions within the GSM manufacturing industry sector, often on geographical and regional lines' (GSM Association, interview, 10 August 1999).

Manufacturers argued that their customers would face severe economic burdens in upgrading GSM networks if they did not support manufacturers' technology choice. Moreover, the GSM Association stressed that current institutional and power arrangements had to be reformed to allow operators to influence the outcome.[181] 'ETSI SMG is the *de facto* standardization body

---

180 GSMA called on the major consortia to work out a compromise. They preferred one global radio interface for UMTS that encompassed all regions. In addition, to be successful they believed that the global GSM operator community must wholeheartedly support any ETSI radio interface selection. During the UMTS standardization campaign, GSMA officials (Lilly, Nugter and Nordstrom) went to standards meetings and lobbied Japanese and other East Asian and North American organizations (GSM Association, interview, 10 August 1999).

181 The GSM Association strongly criticized the ETSI General Assembly decision concerning extending voting rights to associate members that failed to reach the required 71 per cent weighted majority, mainly as the result of the actions of two large European countries. This would have increased the political voice of mobile

for the overall GSM community and therefore has to open up to our overall membership to remain credible in its role' (GSMA, interview, 10 August 1999). In particular the Asian Pacific Interest Group (APIG) of the GSM Association expressed its interest in full participation on GSM and UMTS.

Eventually, the GSMA decided to support the Nordic Alliance. This decision was political and based on the signals by Japanese industry that the W-CDMA proposal supported in Japan would be available outside Europe and in the broader Asia Pacific region regardless of the policy outcome in Europe. Though the GSMA supported the W-CDMA radio technology proposal, it stressed the need for a satisfactory IPR situation. Moreover, Japanese industry could align with US industry before a more coordinated Japanese–European approach had been established. 'It was in the interest of operators to choose the same technology as the Japanese in order not to push Japan into the US CDMA Development Group and CDMA2000' (GSM Association, interview, 29 March 1999). The political lobbying and policy preferences of the GSM Association consisted of (1) selecting the technology with the greatest potential on the world market and as a global standard, (2) aligning with Japan to avoid a political alliance between Japanese and US industry that would restrict European options, and (3) ensure a continued strong role worldwide for the GSM industry. 'It was not based on technical reasons but on global commercial reasons. All the political factors made us decide to propose W-CDMA to our membership as the preferred radio interface technology' (GSM Association, interview, 10 August 1999).

When it comes to the European manufacturing community, the Nordic Alliance argued that its proposal offered the best choice for continuous GSM evolution. Ericsson and Nokia stressed that they were in the driver seat in that they had been engaged in the EU projects on W-CDMA radio technology (Figure 6.2). 'Europe has pioneered the development of W-CDMA. The EU RACE and CODIT projects were the first cellular W-CDMA development efforts in the world' (Ericsson, interview, 15 September 1998).

---

non-European operators in crucial decisions such as the European UTRA selection. The GSM Association complained in a letter to the ETSI Director General that although ETSI was currently privileged to be the global standards body to maintain and develop the worldwide GSM standard, the organization had to open up for non-European organizations. 'In the interests of our future cooperation the highest level of the GSM Association required that the GSM global operator community achieved the required amount of status and recognition within ETSI that it undoubtedly deserved' (GSM Association, interview, 28 January 1998).

Figure 6.2 Europe in Wideband CDMA (W-CDMA)

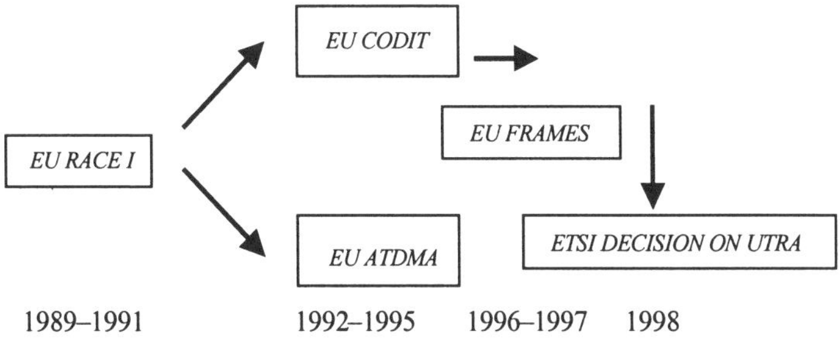

Source: Interview material.

Furthermore, Ericsson and Nokia referred to the success of GSM (as did the UMTS Alliance) in luring governments and EU institutions. They situated the high-level choice of a radio technology within a European context and stressed that W-CDMA had to be within ETSI to allow European companies a competitive position in the battle for a global standard. To select any other technology proposal would isolate Europe from global markets and would be harmful for the European telecom industry. Ericsson and Nokia argued that the success of UMTS globally would mean employment and export markets. The Nordic Alliance stressed the importance of the Asian market. Ericsson and Nokia addressed the Asian audience (to reassure them of their support) as well as operators, national regulators, politicians and EU institutions (to convince them about the long-term strategic implications for the future of Europe).[182] UMTS, they argued, had to be accepted throughout Asia, since

182 The UK Department of Trade and Industry (DTI) had an influential role in coordinating actors. 'This is a rather deep DTI tradition and they have done this all the years. They speak a lot about liberalization and deregulation but if you look at their need, they had quite an intensive national 3G advisory group, Third Generation Advisory Group (TAG), that sought advice from their players but also exerted influence' (ETSI, interview, 16 June 1999). 'I definitely think that the national governments had a role in UMTS and in particular the German, French, Italian and UK governments. I believe we have an old state tradition of coordinating positions in international bodies. The UK members of ETSI would always have a joint meeting before they go to ETSI General Assembly and state a position and they would all vote the same and I believe they have the same types of meeting. The French and Germans had loyalty problems since they felt they had to support Alcatel and Siemens and we could see that government industry policies and employment issues were involved. The UK government supported the Ericsson proposal and the Italian government supported the Siemens-led Delta proposal with Italtel' (GSM Association, interview, 10 August 1999).

investments in Asia would be significantly larger than in Europe and North America together.[183] Any delay would be interpreted as a weakness by non-European players and would be used to disadvantage European industry. 'Manufacturers tried to use their governments to strengthen their positions while the governments thought it would help their local policies. Industry, however, wanted to serve a global marketplace and be global players' (GSM Association, interview, 10 August 1999).

> The European industry actors played on the European card. They say that we need a European solution and that we could not accept a non-European solution in the ETSI process. A European solution had to come from European companies and they were aligning up to get EU support in a document saying that it has been a global process but was initiated in Europe. It had helped the companies to grow, it had given the politicians a big political success with GSM and they allied against the US in mobile communications in Europe (GSM Association, interview, 10 August 1999).

Interviews with senior officials in the Nordic and UMTS Alliance indicated that Ericsson and Nokia were relatively more positive about their chances in world markets and in facing Japanese competition. 'The UTRA selection battle was not so much about technology. The uniting factor among the other companies was a "stop Nokia" and "stop Ericsson" attitude. We think that the only possibility for European industry is that we are able to compete head-to-head with our worst competitors and not to try to establish some artificial barriers' (Nokia, interview, 21 September 1998). The UMTS Alliance, on the other hand, argued that it was far from certain that W-CDMA would imply lower development risks, lower terminal complexity, lower infrastructure costs, and a clearer IPR situation. Rather, Siemens and Alcatel claimed that their technology proposal would better guarantee compatibility with 2G systems and therefore be less risky. Moreover, they argued that Europe lacked experience in the operation of CDMA.

---

183 The Japanese Ministry of Posts and Telecommunications (MPT) supported a similar W-CDMA technology as Ericsson and Nokia did. 'Reflecting the interest that many other countries are showing in the W-CDMA system which has flexibility and expandability, Japan will begin a comprehensive study of the W-CDMA system taking into account the compatibility with the existing narrowband-CDMA systems' (MPT, 1997). Japan was pushing developments in Europe. 'Looking at the timing of the whole 3G development process it was very much dictated by the Japanese needs. No European player needed such a system that early and it was certainly so that the big vendors could have a world market system. It was also in the interest of the operators who have market volumes. Industry policy in Japan was a key driver' (ETSI, interview, 16 June 1999).

6.3 WIRELESS STANDARDS AND INTERNATIONAL COMPETITION

In late January 1998, the ETSI mobile community met to decide on the radio technology standard. The UMTS Alliance played on the fear of IPR claims and argued that the IPR environment strongly favored its proposal. Its supporters pointed out that Qualcomm had not agreed to apply the ETSI IPR policy. The US manufacturer wanted to stall development in Europe and stated that it needed more time to investigate and analyze ETSI IPR policy. At the same time it argued that it had a strong IPR position in all the major proposals. European manufacturers found such statements questionable since it could have become a member much earlier; it had had an office near the ETSI headquarters for many years, and knew what was required under ETSI IPR policy. Representatives from the two major rival industry consortia continued to debate the relative merits of their proposals. The Nordic Alliance received strong support from Asian industry (NTT DoCoMo, NEC, Fujitsu, Matsushita/Panasonic, and Mitsubishi Electric), network operators in the Asia-Pacific region, which aligned with the GSM Association, and British Telecom. In late January 1998, 198 members voted (1523 weighted votes). The W-CDMA proposal received 61.1 per cent of the votes, while the TD-CDMA proposal received support from 38.7 per cent of the cast votes (ETSI, 1998b; TWB, 1998a, 1998b). The result did not meet voting requirements.

The outcome of the voting signaled the start of a consensus-building process. The EC intervened again and stressed the need for a single radio technology as input to the draft EU UMTS strategy document it was about to publish. Moreover, DG INFSO would consider appropriate action to enforce Community law in support of EU policy if abuses occurred with respect to UMTS IPR holders (EC, interview, 25 June 1999). All the leading European manufacturers, with support from major operators (such as NTT DoCoMo and Vodafone), issued a compromise proposal.[184] 'The end game for industry was to get its radio interface accepted at global level. Behind was the industry fight on who is going to get the largest market. I believe Siemens and Ericsson had talked about what rights Siemens would get from Ericsson in

184 The Italian Ministry of Communications, for example, expressed its satisfaction with the consensus proposal, which pointed in the right direction. The UK Department of Trade and Industry (DTI) stressed (like the Austrian Ministry of Post and Telecommunications) that it had concerns about a possible fragmentation and the achievement of a reasonable IPR situation for UTRA (ETSI, interview, 16 June 1999). 'There were at least 40–50 senior executives sitting in different hotels in Paris and who communicated with the senior level at company headquarters. The operator community had an important role in the standstill situation. There were clear and organized initiatives not least from Italian interests to reach a solution that penetrated the standardization work in ETSI' (UMTS Forum, interview, 28 April 1999).

terms of IPR and working agreements, and so on, if they could agree. That would ensure their place in the market' (GSMA, interview, 10 August 1998). On 29 January in Paris, the ETSI mobile community settled the issue 'in the absence of a sustained opposition' (ETSI, 1998b; FT, 1998).[185] The Nordic Alliance claimed that its proposal won with minor amendments (RT, 1998d). 'It was a technology choice conditioned by strong political factors since the reason to select the one or the other was not based that much on convictions about technical performance but rather that it would have a greater potential to generate a global standard' (UMTS Forum, interview, 28 April 1998). The EC and DG INFSO welcomed the agreement:

> The debate on the basic concept of radio access for UMTS has slightly delayed the standardization agenda as originally foreseen. However, we consider that this debate was necessary to achieve a consensus that offers the best conditions for the important work ahead to define the detailed UMTS standard specifications (EC, interview, 31 March 1998).[186]

The ETSI mobile community argued that the agreement could enable European-based industry to maintain a commanding global position. This view was supported by the EC. 'ETSI last week satisfactorily answered a very important question. That is, how should the standard be set for 3G mobile telephones? It is important because this means a mobile European platform,' said Commissioner Bangemann, adding that with this decision Europe increased the opportunity to repeat the success of GSM (RT, 1998e). Commissioner Brittan agreed: 'This is not just a technical matter, it is a huge advance' (RT, 1998e). The EC, and the manufacturers that stood to gain most from alignments with Asia (notably Ericsson and Nokia), argued that delays would endanger Europe's lead. 'European politicians were very positive

---

185 The UMTS solution consisted of 'two standards within one': W-CDMA and TD-CDMA. The plan was for W-CDMA to be used for wide area applications while TD-CDMA would be used primarily for low mobility indoor applications. In a joint statement, nine vendors (Ericsson, Nokia, Siemens, Alcatel, Nortel, Bosch, Italtel, Motorola and Sony) said that the agreement was the first step toward achieving a worldwide standard.

186 The UTRA decision would come to be a point of reference in a looming transatlantic trade dispute in 1998 and beyond. European interests, however, stressed that it had been an open and non-discriminatory process. 'It was important to stress that it was unanimous since Qualcomm later complained very much. They were present with three representatives in the room and their hands were up. It was emphatically asked who was against, who supported it, and who abstained, and it is recorded in the protocol. In other words, the approval was a unanimous decision. In addition, the decision in Kristianstad in 1997 on the scenarios where the US CDMA-based solution was not included was a unanimous decision and endorsed by Qualcomm' (ETSI, interview, 16 June 1999).

about GSM. It was a success story of European policy. The political will for a common standard for 2G was there. The EC and governments got involved at a very high level in different countries as they have in the last few months' (GSM Association, interview, 10 August 1999).

European organizations (ETSI SMG, UMTS Forum, GSM Association's 3G Interest Group, ECTEL and DG INFSO) had embarked on political activities in Japan and elsewhere, acting on a mandate by the ETSI mobile community. These Exploratory Missions, initiated in 1997, aimed to explore opportunities for bilateral or multilateral cooperation. There existed a Japanese willingness to cooperate.[187] Those European manufacturers with significant commercial interests in Japan and Asia and good contacts with their industries pushed for the Exploratory Missions. 'Ericsson contacted the UMTS Forum and said that the Forum should start talking with the Japanese. Then the process was initiated' (UMTS Forum, interview, 29 March 1998).[188]

In late 1997, the ETSI SMG Chairman was granted a further mandate for Exploratory Missions with the aim of identifying a solution for closer and more innovative interregional and international cooperation on GSM and UMTS specifications. The mandate stated that ideally there should be one, and only one, joint working structure among all interested bodies to avoid parallel work, and overhead coordination to save both time and money. The institutional form of such political collaboration would take the form of an ETSI Partnership Project, defined as activity established when there was a need to cooperate with an external body and where such cooperation could not be accommodated with the ETSI organization.[189]

---

187 The Exploratory Missions covered visits to North America, Japan and Korea. They also included China, where GSM had been adopted as a national standard. The Chinese Ministry of Post and Telecommunications (MPT) had been invited to participate in ETSI GSM activities. Its major operators, China Telecom and China Unicom, were members of the GSMA. DG INFSO, not least its international relations and R&D mobile units, lobbied China parallel to the Exploratory Missions (EC, interview, 29 October 1998).

188 There were some tensions within the ETSI mobile community. 'When we were about to set out for the Exploratory Missions Nortel opposed strongly and said that we could not do that while we received support from Alcatel. To expand and foster as large a captive market as possible is the name of the game and Alcatel supported it since it could offer them greater market access. At that period in time, Siemens was against it as well but later came to support it' (Ericsson, interview, 22 September 1998).

189 Some of the targets were the production of an open and complete system standard for UMTS as a member of the IMT-2000 family, the continuous evolution and maintenance of GSM, as well as a productive relationship to other standardization bodies concerning UMTS. To realize those goals, it was stated that the missions should aim for one joint working structure among all interested bodies. One of the results of the first Exploratory Mission was a European–Japanese agreement that would allow European companies without a presence in Japan to participate in the

> There was a clear interest amongst major players in exploring whether we could find a way of closer collaboration. There was an interest, but it was a heterogeneous interest. There was of course an interest on the part of the vendors in finding a solution that could be used in Japan and in the rest of the world. On the part of the operators, there was an interest in finding a final roaming solution with Japan. On the Japanese side, NTT DoCoMo strongly wanted to break the isolation. In sum, there were interests both on the manufacturer side and on the operator side. Several [Exploratory] Missions started before the ETSI radio interface decision that eased the situation (ETSI, interview, 16 June 1999).

The ETSI SMG Chairman (Hillebrand), with wide support from the mobile community, wanted to create a world specifications body to ensure the transition from GSM to UMTS. 'We want to create a model to explore a more efficient way of working around the world. I do not believe we can do this by ourselves' (TWB, 1998c). Firstly, UMTS standardization was aimed at becoming one member of the IMT-2000 family with global acceptance based on (1) the evolving GSM core network and (2) a common radio access standard (UTRA). Secondly, the institutional set-up needed to be stronger than purely cooperation arrangements or cooperation through the ITU, to improve efficiency, timely production, quality and coordination. The GSM operator community pointed out that the objective would be to lay the foundation for common specifications, global interoperability and conflict resolution. The ETSI SMG Chairman pushed for global opening and equal rights for all regions and reported that the concept of one 'global working structure' seemed to be viable.

Ericsson was a leading proponent of a global institutional arrangement. 'Our goal with the new global structure is to come together in this joint grouping instead of conducting the exercise in three parallel places. We wanted to include China and Korea to spread the GSM core network and the W-CDMA radio interface standard and to generate a harmonized platform based' (Ericsson, interview, 22 September 1999). A global joint institutional and working structure would entail significantly more than traditional arms-length bilateral cooperation. 'Globalization has several levels: A first level is

---

meetings of all ARIB committees. In turn, ARIB would be able to make contributions in Europe. 'It eased the exchange of information and constituted one of the channels through which the Japanese and ETSI versions of W-CDMA could be kept together. It was understood in ETSI SMG that there was a risk that there would be separate radio standardization activities in ARIB, in ETSI SMG and probably in ANSI T1P1 in the US (which looked after the GSM evolution). In addition, there would be core network work in ETSI SMG, TTC in Japan and in ANSI T1P1' (ETSI, interview, 16 June 1999).

global acceptance of a standard or specification and a second level is full global participation in the elaboration and decision making. This second level is achieved in the new working structure' (ETSI, interview, 16 June 1999).

The ETSI wireless community reminded ETSI members that the GSM Association had initially been accused of 'neocolonialism' and later avoided references to 'Europeanization' and 'European technology'. They argued that a regional body like ETSI could not own a world standard. Rather, every partner had to be treated equally. 'We thought that 3G was too important to leave to chance so we were looking to ways to globalize our efforts' (ETSI, interview, 11 May 1999). The proponents of a global working structure stressed that it was unimaginable that Japanese and US companies would come to the court of ETSI as 'second-class citizens'.

> With a big and proud nation like Japan and the Japanese institutions you cannot make them second-class citizens in Europe. You can obtain a global leadership not through privileges but through performance – if you perform and contribute then you have global leadership. Power cannot be executed by formal control mechanisms. You would have created another colonial regime. Leading people in the mobile industry had a widely shared view that we would need to have a regime or organization where everybody has the same rights and the same obligations (ETSI, interview, 16 June 1999).

6.4 POLITICAL RESISTANCE TO GLOBALIZATION

The proposal for a multilateral institutional arrangement triggered concerns within the ETSI bureaucracy and fixed network community. Some member delegates suggested that non-European organizations could become 'associate members', and thereby contribute and participate. The fixed operators were particularly vociferous about the globalization of UMTS and institutional multilateralism. 'The fixed network business is far behind the mobile industry because fixed networks are much more geographically focused on national markets and the incumbent operators are very used to working nationally. There was a terrible fight between the European fixed community and the mobile community because the fixed operators were trying to delay the process' (Nokia, interview, 21 September 1998). The proposal for a new global working structure was viewed by some as creating an autonomous entity outside the control of ETSI. 'ETSI cannot afford to have this new forum turn into yet another technical adventure without strong involvement from its members' (ETSI, interview, 13 March 1998).

Representatives of fixed network operators (who were also ETSI Board members) criticized the exploratory missions, which launched a European controversy that would escalate in 1997 and 1998. 'Concerns were aired in some quarters in Europe that ETSI would give up a great deal of its traditional power, not least by some network operators that are less mobile and multinational than the manufacturers' (UMTS Forum, interview, 29 March 1999). In late February 1998, the ETSI Board convened a special executive meeting in Frankfurt, a political signal that globalization was a controversial topic in Europe and ETSI. 'Unrest and concerns had been building up for some time and were visible among the European standards organization's rank and file' (ETSI, interview, 11 May 1999). It had not yet been decided whether ETSI should go it alone, further open up its organization to non-European actors, or embark on a new multilateral institutional working structure. Three political or policy 'communities' can be identified in this globalization-European controversy (Figure 6.3).

Figure 6.3 European roots and globalization

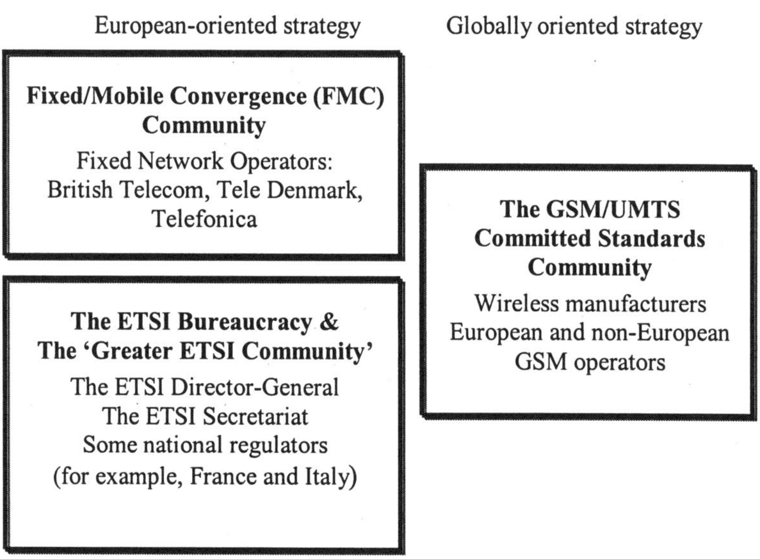

Source: Interview material.

These policy communities reflected different vested interests with respect to the nature and route of the management and globalization of UMTS. Entrenched positions, invested time and money, status and power were at stake. The FMC Group, which had been tasked by the ETSI General Assembly to progress fixed/mobile convergence issues, together with the ETSI standing organization, were critical of globalization. The GSM/UMTS community wanted a joint multilateral working arrangement. 'At that time the Europeans, led by Ericsson and Nokia, wanted to ensure that Europe would be the dominant player in IMT-2000. To globalize from the start was a good insurance policy' (ETSI, interview, 11 May 1999). The ETSI SMG leadership stressed that a clear mandate had been given, approved by more than 200 delegates, including several Board members, the GSM Association, the UMTS Forum, and with the EU Commission as observer.

The FMC community and the ETSI bureaucracy attacked the ETSI SMG Chairman (an 'organized manhunt', according to many standards officials in the ETSI mobile community). Delegates from fixed network operators claimed that the ETSI SMG Chairman was in liaison with Japan at the expense of the interests of the organization's members. Moreover, they argued that his activities jeopardized the international competitiveness of the European telecommunications industry and 'the well being of European citizens' (ETSI, interview, 16 June 1999). The European–Japanese alliance 'threatened' the worldwide deployment and position of GSM. Furthermore, they argued that the Chairman and his Vice Chairmen from Ericsson neglected established rules and working procedures, which undermined the authority of the political representative body, the ETSI General Assembly. Finally, Hillebrand was accused of having accumulated a secret database to which the rest of the ETSI membership did not have access.

In March 1998, the ETSI Board announced that it had launched a new initiative, the UMTS Globalization Group (UGG), to provide a mechanism for discussions on partnerships with relevant organizations around the world. The ETSI Director-General chaired the group (Table 6.2).[190] The UGG objective was to enable specifications 'to be prepared and promoted in a manner which would make them attractive to global partners such that they would be implemented worldwide' (ETSI, interview, 11 May 1999). The UGG should provide strategic management, investigate relations with external partners, and propose an organizational structure that would meet the expectations of ETSI and its partners.

---

190 The ETSI Director-General headed UGG, seconded by Vice Chairmen who had criticized and would continue to criticize globalization. They had support from a Board member (Telefonica) and the ETSI Network Aspects (NA) representative. Among the Board members, there was one camp that supported globalization (Ericsson, Nokia and Siemens delegates), together with SMG members (Vice-Chairman from Ericsson), GSMA and UMTS Forum (Telia Mobile).

Table 6.2 The UGG membership

Chairman:	The ETSI Director-General	UGG Group:	6 ETSI Board delegates 3 ETSI SMG delegates 3 ETSI NA delegates 1 GSM Association 1 UMTS Forum delegate Observer: DG Enterprise
Vice Chairmen:	2 ETSI Executive Board delegates		
Secretary:	ETSI Secretariat		

DG Enterprise, responsible for institutional issues in the standardization domain, had a place on the ETSI Board as a counselor to 'guide ETSI', was regularly invited to its General Assembly, had access to all ETSI Technical Committees, and played a role on trade policy issues. The responsibility of the EC was to interpret legislative activities and to ensure that the trade relations between the EU and US ran smoothly.

> They supported us in our efforts and liked the concept of us working closer with the US because there were already then some strong discussions and trade conflicts between the EC and the US. The EC saw this as another way to demonstrate our openness to work with other regions of the world and to prove to them that we are not anti-competitive (ETSI, interview, 11 May 1999).

Leading ETSI representatives met EC officials to discuss the future structure, scope, direction and implications of UMTS standardization (ETSI, 1998a). The EC argued that the globalization of European technology and industry should improve trading relationships with other regions in the world. However, the EC expressed some concerns. First, it recommended avoiding the creation of a new legal entity for the development of various UMTS specifications. A legal structure could give rise to regulatory problems. Second, the EC expressed the need to take due account of the interests of the EU Member States (EC, interview, 26 May 1999). Leading actors in ETSI and Europe embarked on a number of activities in the late 1990s in close cooperation with standards bodies in East Asia and North America. Apart from the globalization of GSM standards, the UMTS concept and associated specifications and the W-CDMA technology, a major initiative was the UMTS IPR Task Force that was spearheaded by senior officials in Alcatel and other European organizations. It reflected the growing importance of IPR in the globalization of corporate activities and technologies (Figure 6.4).

Figure 6.4 ETSI initiatives in 1998

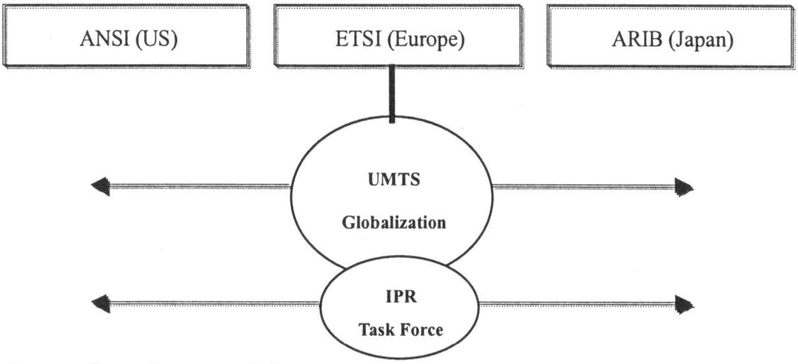

Source: Interview material.

ETSI targeted those countries that had to make a decision on infrastructure and shared the same business interests as ETSI members, such as building on GSM. In 1998, a large number of radio technology proposals were submitted to the ITU (10 terrestrial technology proposals) (Table 6.3).

Table 6.3 Submitted radio transmission technologies to the ITU (in 1998)

Organization	**Technology**
Proposals for satellite systems	
ARIB (Japan)	W-CDMA
CATT (China)	TD-SCDMA
TTA (South Korea)	Global CDMA I and Global CDMA II
DECT Forum (Europe)	DECT
ETSI (Europe)	UTRA
TIA (USA)	UWC-136, CDMA2000, WIMS W-CDMA
T1P1 (USA)	WCDMA-NA
Proposals for satellite systems	
ESA (Europe)	SW-CDMA and SW-CTDMA
ICO	ICO RTT
Inmarsat	Horizons

Source: Trade Press.

6.5 HUFFING AND PUFFING

Japanese and European standards bodies agreed to support jointly the same radio technology platform at international level. 'It was understood that if it was not possible to reach a fully agreed position, then any differences would not be interpreted as an unwillingness to cooperate' (ETSI, interview, 16 June 1999). European manufacturers, and the ETSI mobile community, supported the political and institutional globalization of UMTS activities. Several delegates from the fixed telecom world and the ETSI bureaucracy still resisted globalization and what they perceived as the diffusion of political control. 'The fixed-network players and the ETSI standing organization resisted the development. The ETSI Secretariat was completely against it and certainly the Director-General and the head of the Secretariat. This was clearly an important part of the majority' (ETSI, interview, 16 June 1999). On a broader level, there was opposition, particularly to the Nordic manufacturers, because of their strength in international markets, their ability to push their policy preferences in ETSI, and thus their political power. There was also the looming fear that Asia would take over.[191]

There were two rival institutional solutions with regard to the planned multilateral institutional arrangement (that would be called the Third Generation Partnership Project, 3GPP) associated with the issue of political control: the modular approach and the integrated approach. The European mobile community (ETSI SMG delegates, GSM Association, a number of Board members, and UMTS Forum) supported the integrated approach and criticized the modular approach, which 'ignored market requirements and jeopardized the goal of making GSM-based UMTS (G-UMTS) the most successful IMT-2000 family member' (ETSI, interview, 31 August 1998). The ETSI Director-General and the Vice-Chairmen of the UGG (ETSI Board members) supported the so-called modular approach, which proposed three independent modules and separation of work activities (radio technology, GSM core network, and terminal equipment).

---

191 Up to the mid-1990s, Japanese organizations lobbied to have their cordless technology PHS accepted in Europe. 'The Japanese went through a phase of trying to win the PHS battle and hoping that they would be able to repeat GSM. When the PHS threat came to Europe, before it failed in Japan, the EC was lobbied very hard by some professional lobbyists and became very alarmed. Some people in the EC started to think that DECT had failed due to the handset design and that PHS was going to swamp Europe. It would be a purely imported technology and that frightened the EC' (GSMA, interview, 26 April 1999).

- 3GPP should have transferred to it from ETSI a minimum work program, containing only those work items related to the IMT-2000 family which are essential to progress in a global environment (the modular approach – 'European solution').
- 3GPP should have transferred to it all of the work necessary to define a complete IMT-2000 candidate. The majority of the ETSI SMG work program, including the GSM core network and GSM radio work items, but excluding those items of a strictly European regulatory nature (the integrated approach – 'global solution').

The ETSI wireless community argued that a modular institutional-strategic approach would lead to the fragmentation of specification production, approval procedures, coordination and management processes, with severe consequences for European competitiveness. The bottom line was that the modular approach would separate GSM and G-UMTS and that regional standards bodies would maintain significant parallel structures (ETSI, 1998e). 'The cohesion of GSM and G-UMTS is endangered, since G-UMTS would be separated from GSM work and be distributed between European, US, and Japanese standards organizations and the 3GPP' (ETSI, interview, 12 August 1998).

The supporters of the modular approach argued that the G-UMTS approach would result in a completely self-contained arrangement, which would threaten harmonization with diverging regional requirements across the world. 'ETSI should continue to develop GSM specifications as required by the marketplace and initially move only the 3G generic work required at worldwide level to allow for global interoperability between GSM and 3G mobile systems' (ETSI, interview, 1 September 19988). Furthermore, it was believed among Board members that it would lead to loss of political control. An integrated global working structure could jeopardize European high technology standards policy since European standards would be specified and approved by independent non-recognized interest communities. Therefore, ETSI should not participate in an international coordination activity covering shared specification activities. European work should be transferred directly into ITU. The 3GPP should only allow non-ETSI members to participate in very specific areas in the self-interest of the broader ETSI membership. The fixed telecom world in ETSI set up the so-called ETSI Partnership UMTS (EP UMTS) in a political move to control the UMTS standardization development in the long term.[192]

---

192 This was strongly supported by the ETSI fixed network interests. In their view, 3GPP would be restricted to the development of specifications for the initial phase of a complete 3G mobile system based on 3G radio technology and G-UMTS. It was suggested that the scope of EP UMTS would cover 'the ownership, planning,

Supporters of the EP UMTS argued that the role of the 3GPP was not defined for later phases and that ETSI should retain control in the long term, while the ETSI mobile community stressed that such planning should not be allowed to jeopardize the setting up of the 3GPP. The EC participated in the UGG and closely followed the battle on the globalization of UMTS. 'I think the idea of 3GPP leaves ETSI rather vulnerable because in theory it lessens the control by one organization to outside influences' (EC, interview, 26 May 1999). One of the main issues was the need for a European fall back position at strategic and operational level. This was also a concern for the EC bureaucracy. 'The EC was very keen and had a certain part in this but more in response to a desire to stop the process from falling over. They wanted to make sure that the joint work was successful. Many national administrations had similar feelings' (GSMA, interview, 26 April 1999).

> If the 3GPP presents something to the ITU it will be taken very seriously. It is not a small cowboy forum set up with ten players. This is the strongest grouping in the world setting standards for 3G. Behind the weight of industry you have national governments, so from the European perspective it is the EC sitting behind this work, and you have MPT in Japan. The US State Department is *de facto* sitting behind ANSI, TIA and 3GPP2, since all operations in the US in terms of telecommunications and standards come under the State Department. They are in the background at the highest level. Not only do you have large industrial representation but also quite serious government support that, when you get to the ITU, is a very important influence (ETSI, interview, 11 May 1999).

From a high technology policy and an institutional perspective, the EC was concerned that the 3GPP was not a formally recognized standards body. It was felt that its structure should be subsumed into, or at least annexed on to, the ITU structures. It would then provide an added impetus of respectability and neutrality and achieve harmonization through a clear consensus process at international level. The strategic EC objective was to achieve or facilitate harmonization and to encourage the regions to cooperate early on through an official process. The EC (and in particular DG Enterprise) preferred formal standards bodies and viewed the 3GPP as 'a little bit of an orphan'. It preferred to see industry consortia linked to formal bodies. 'For instance, if one of the regions decides not to play ball in this procedure what do you do? The danger for me is that it could become a destructive environment that falls

---

resource-management, organization and where appropriate the delegation of all the work required to progress UMTS activities within ETSI that go beyond the initial phase' (ETSI, interview, 8 October 1998).

apart. You then go back into a regionalized structure again with each region developing its own, non-harmonized global specifications' (EC, interview, 26 May 1999). The mobile community opposed the idea that the ITU should control detailed specification work. 'There are high-level people in the ITU who see 3GPP as a threatening competitor and this is definitely a political dimension. There are those who think that the 3GPP has damaged the ideal of a single global standard' (UMTS Forum, interview, 26 April 1999).

The EC recognized its limited powers to intervene in a sector with strong international manufacturers. 'If companies go global with a global shareholder base, global R&D, global manufacturing, global standards, what can the EC say. The concern we have is that standards are driven more and more by global industry coalitions [...] The whole weight of this development is shifting' (EC, interview, 25 June 1999). The EC was concerned that the influence of ETSI would be significantly reduced and thus undermine its important role for the EU policymaking environment. In the long term, industrial influence could move across the ocean. Therefore it was important to stimulate developments in Europe and to try to safeguard European capabilities. 'We risk that the US and the companies over there start to completely dominate the future and the market in the fundamental shift toward the complete "Internet Protocol network paradigm". The industry may eventually disappear, not necessarily the production capability but the intellectual capability' (EC, interview, 25 June 1999). The EC was concerned that progress evolved outside the formal European trade groups and standards bodies and that there was a lack of strategic direction in European industry.

> It makes me raise the question whether we should not have a Community competence in the whole development of information and communications technology (ICT). All the small initiatives in Europe may not be able to weigh up to the major drive and thrust that you find in the US. It is our role, a major role for Europe. I think we may need to go to a supra-regional government. At the end, it is all about employment (EC, interview, 25 June 1999).

The ETSI mobile community and related organizations viewed a stronger collaboration with Asian organizations as imperative to promote and protect European interests and investments.[193] The Japanese could go ahead alone and develop a 3G system that by chance could be the dominant system. 'Japan needed a system yesterday. They have huge subscription growth and not enough radio spectrum. The Japanese government has assigned radio

193 Japanese organizations had initially tried to challenge their interests through the ITU but their contributions met delaying tactics from European and US actors. 'I believe that out of frustration they became open to suggestions to work directly with ETSI' (GSMA, interview, 26 April 1999).

spectrum for IMT-2000 but it could not be used until 3G standards were ready. The Japanese were therefore strongly committed to have results ready by the end of 1999, to wave those results at their government and get new licenses issued for a new system. The pressures in the Far East are quite different' (ETSI, interview, 11 May 1999).[194] The Koreans had been made aware of the ETSI 3GPP proposal directly through their Japanese counterparts. According to ETSI, it appeared as if there was some reluctance on the part of the Korean government but eventually Korea joined the 3GPP. 'The Koreans and the Japanese were falling into the same trap and decided that they really wanted to go for having some competition and alternative technology for the next generation. Qualcomm, which had a total monopoly on their market, was pushing them into a corner over licensing. By joining us on the 3GPP they were assured that when people go to Japan from Europe their systems would work' (ETSI, interview, 11 May 1999).

Wireless operators in Europe stressed the need to move ahead and fully open up ETSI to non-European organizations. Rather than focusing on the balance of resources, power and responsibility within ETSI, they argued, ETSI had to provide an internationally oriented standardization process within ETSI in order to become marginalized. Eventually, the UGG settled for an open recommendation: 'To transfer to the 3GPP those elements of the 3G mobile system where special attention is agreed as necessary to encourage rapid global specification production' (ETSI, interview, 16 June 1999). It was further agreed that the priority in the negotiations with Asian delegations would be to spur a harmonized radio interface within 3GPP. It was decided that an Extraordinary General Assembly should be convened to enable the whole ETSI membership to be consulted and vote on the remaining questions and controversies. At the final ETSI vote in fall 1998, 94.5 per cent voted in favor (1394 weighted votes in favor and 81 against) of creating an ETSI Partnership Project (3GPP) that would develop specifications for the initial phase of a complete 3G Mobile system (ETSI, 1998f).

In late 1998, the European, Japanese, Korean and US standards bodies formally agreed a framework document on the new interregional entity (the 3GPP) as a sound basis for further elaboration. It had also been presented to China. 'The 3GPP will provide globally applicable technical specifications for 3G mobile systems based on evolved GSM core network and UTRA, to

---

194 Furthermore, European delegations (ETSI, UMTS Forum, and DG INFSO) continued to strengthen European–Chinese bilateral relations. 'It was agreed that the objective of the visit to China was to ensure that they selected UTRA as the access technology for their 3G mobile networks. To achieve this, ETSI was offering an open hand to work with them. The ETSI future plans were not rigidly fixed and comments should be invited from China on any issues that were of concern to them' (ETSI, interview, 16 June 1999).

be transposed by relevant standardization bodies (Organizational Partners) into appropriate deliverables (for example, standards). The technical specifications will form the technical basis of an IMT-2000 family member' (ETSI, 1998d). ETSI was the dominant player, although the partners were formally equal. It had the GSM weight and experience. 'But we could only guarantee the use of the system in Europe. We could not guarantee that it would be adopted worldwide' (ETSI, interview, 11 May 1999). The external organization of 3GPP illustrated in Figure 6.5 shows the intermediating role of the proposed multilateral structure between European regulatory structures, on the one hand, and the global IMT-2000 activities, on the other.

Figure 6.5 The external dimension of 3GPP

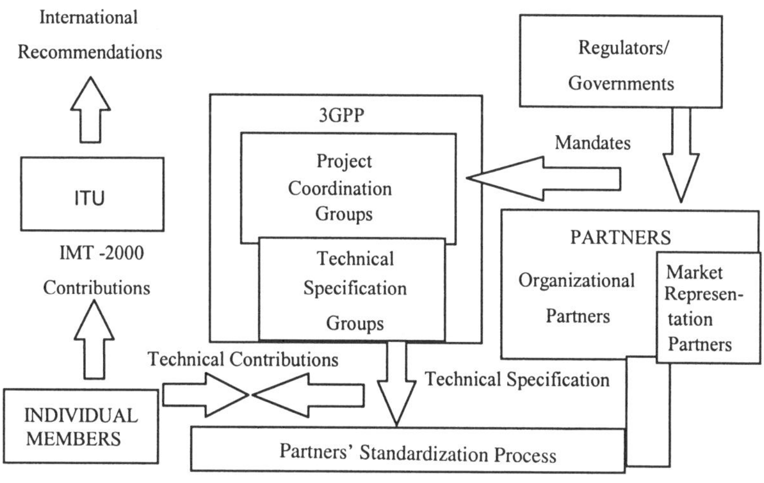

Source: ETSI (1998c, 1998d).[195]

195 The standards bodies constituted the organizational partners. Trade associations consortia were eligible as market representation partners and three bodies had been accepted into the 3GPP by 2001: the GSMA; the Global Mobile Suppliers Association (initiated by the marketing units of Ericsson, Nokia and Motorola); and the Multimedia Wireless Forum (an international industrial consortium).

6.6 Global Consolidation and EU Policy

In summer 2000, it was formally decided to extend the scope of 3GPP to cover GSM activities in line with the recommendation of the ETSI mobile community, which was accepted by standards development bodies from China, Japan, Korea and the US. ETSI SMG had already transferred all its work related to UMTS in early 1999 and all common GSM/UMTS work in the third quarter of 1999 (Hillebrand, 2000). It had its last meeting in June 2000 and was formally closed down shortly thereafter, putting an end to the group that had actively worked to ensure a continuation of the GSM success story. 'Participants in both ETSI and 3GPP judged that the time was now right to import the ongoing GSM standardization work into the Partnership Project. This will ensure that cohesion between GSM and the 3G specifications may be managed efficiently' (ETSI, 2000c). The ETSI SMG Chairman, who now stepped down, took the opportunity to repeat the value of an integrated international institutional structure. 'The new arrangements concentrate all GSM and 3GPP specification work into one body, which is open on equal terms to interested companies worldwide. It will ensure the integrity of the GSM/3GPP platform, eliminating the risk of incompatibility and inefficiency that might have arisen had the work remained distributed among independently acting groups' (ETSI, 2000c).

Global cooperation had been significantly restructured under the guidance of the ETSI mobile community (ETSI, 2000a). Around 1996–1997, a long-term strategy emerged that a wide global acceptance of GSM required a global structure and international orientation to maintain the compatibility and integrity of GSM beyond Europe. This strategic reorientation process can be divided into a number of milestones (Figure 6.6). First, the *European–North American Cooperation Phase* emerged around 1996–1997. It was the first stepping-stone from a situation with separate regional institutional structures to a multilateral order. The underlying rationale was to minimize the risk of incompatibilities between the scope, structure and service features of the various GSM standards. In 1997, a transatlantic cooperative agreement was agreed and implemented in gradual steps. It improved the integrity of GSM between Europe and North America and the rest of the world. In the *European–Chinese Cooperation Phase*, the European mobile standards community targeted the Chinese GSM community to integrate all of Chinese requirements into the GSM specifications.[196] ETSI SMG had persuaded the

[196] Chinese network operators had implemented GSM network infrastructures based on ETSI specifications. As in North America, however, Chinese organizations had modified the European specifications through unique additional requirements by independently working on the key GSM switching protocol.

Chinese authorities and industry to align their specific requirements with European specifications to strengthen the international integrity of GSM, and to open up to participation in ETSI.

Figure 6.6 The evolution of global cooperation

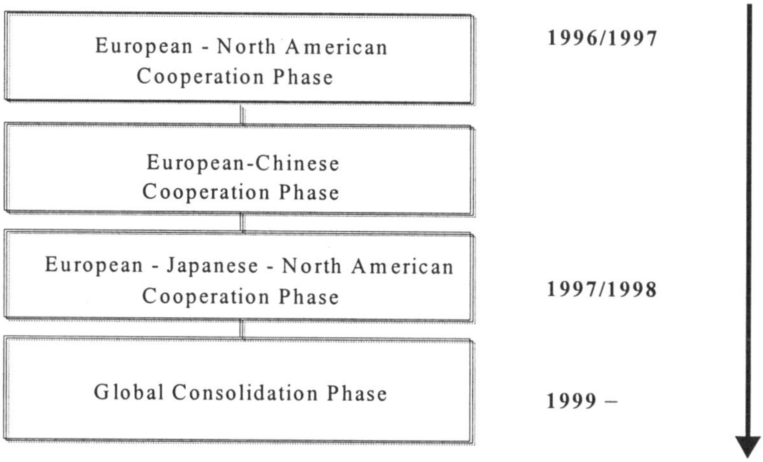

In the *European–Japanese–North American Cooperation Phase* (1997–1998), the European mobile community accelerated efforts to foster stronger ties with non-European standards bodies. They concluded that there was an opportunity for strategic cooperation. The European strategic objective was to base the cooperation on an integrated approach to G-UMTS (the evolving GSM core network infrastructure, the UMTS services concept, and the selected UMTS radio technology, UTRA) in order to ensure evolution from the worldwide installed base of GSM. Finally, the European wireless community was increasingly concerned that a globally decentralized, trilateral cooperative management and political structure would not provide any overall decision making body for conflict resolution and integrated coordination. As part of a *Global Consolidation Phase*, they concluded that a global strategic agreement required an innovative global solution. 'Having relinquished some of the control of UMTS standardization in order to ensure its positioning as a global standard, the challenge for ETSI today, is to continue to be a driver in the development of what is hoped to be the world dominant technology. The contribution from ETSI to the 3GPP has been to find a balance that ensures global acceptance of UMTS and minimization of loss of control' (ETSI, interview, 11 May 1999).

The US manufacturer Qualcomm (with background support from Lucent) initiated a political campaign against UMTS activities in Europe. In late 1997, the ETSI Director-General asked Qualcomm whether it claimed any IPR essential to the UMTS radio access technology. Would the company be prepared to grant licenses for the patent rights if it claimed to own essential IPR in accordance with the terms and conditions set out in ETSI IPR policy? Qualcomm responded that it owned IPR that were essential to the proposed radio technology, including the leading proposals supported by the Nordic and UMTS Alliance. Qualcomm lobbied intensively on Capitol Hill and presented some of its activities to ETSI SMG, among which was a letter to US Senator John McGain calling for convergence between its own technology proposal (CDMA2000) and W-CDMA.[197] Moreover, Qualcomm argued that the ETSI standardization process was 'unfair', which triggered concerns by European industry.[198] The projected strong growth in the wireless sector with the introduction of 2.5G and 3G services turned standardization into a major trade policy issue.

ETSI responded to the US Congress that the ETSI standardization process was marked by openness, balance of interests, an appeals process and consensus, and stressed that 80 per cent of its funding came from the private sector (ETSI, 1998g). 'The GSM standard was built by finding a balance of interests leading to mutual benefit for the greatest number of market actors through a process of collaborative competition' (ETSI, 1998g). ETSI stressed that the single European market project required harmonization and standards constituted a useful means to achieve it. 'Just as the US government recognizes the importance to encourage long-term growth for enterprises and

---

197 Qualcomm consistently argued that it was of critical importance to achieve convergence between the ETSI proposed specification for W-CDMA and Qualcomm's CDMA2000 before it could commit to the ETSI IPR policy. The company stated that if W-CDMA were to remain as it stood, it would not agree to grant licenses under its essential intellectual property for the W-CDMA standard according to ETSI IPR policy.

198 The President of Alcatel, on the one hand, and the Presidents of Ericsson, Nokia and Siemens, on the other, aired concerns to the ETSI Director-General. Alcatel was of the opinion that, given the current situation of a fundamental IPR blocking situation, the existing UTRA standard should be revisited by either modifying those elements of the standard blocked by IPR or by exploring a negotiated settlement with the patent holder. The other manufacturers were more aggressive. Qualcomm, they argued, was seeking to change the 3G W-CDMA standard toward its own technology for reasons of sole self-interest. It represented a blatant attempt to enrich itself at the expense of the industry and the manufacturers resisted the proposed technical amendments sought by Qualcomm. The company had not contributed to the UTRA decision process and had not submitted its technical contribution in time. The European mobile operator community was concerned that the conflicts spilled into standards work in Europe and globally.

promote efficiency and economic competition through harmonization of standards, so does Europe' (ETSI, 1998g).[199] The EC supported ETSI and the European standardization model. 'The US pressure was considered as highly negative and contrary to the spirit of WTO and of technology neutrality. A "double language" characterizes US foreign policy' (EC, interview, 25 June 1999). ETSI went to the US to 'set the record straight' and met a complex lobbying environment. 'You have a terribly complex situation in Washington D.C. with intricate links between the Administration and Congress. Officials in the former, it appears, even stimulate people in Congress to pressure the Administration and USTR, etc.' (ETSI, interview, 1 June 1999).

In 2000, ETSI established a Mobile Standards Group (ETSI MSG), which would be the entity with responsibility for the production of harmonized European Standards as needed for regulatory purposes related to 2G and 3G mobile systems (ETSI, 2000b). The EC had urged ETSI to produce European Harmonized Standards (in accordance with the EU R&TTE Directive, prepared by DG Enterprise) due to the need to avoid barriers to international trade.[200] 'We were going from national to European and now we are going from European to global as the next phase and it is very much linked with trade issues. The Americans seem to be far more insular than ever the "Fortress Europe" has been viewed' (ETSI, interview, 11 May 1999). The EC

---

199 ETSI also responded to criticism presented in July 1998 by US CDMA proponents at a hearing on 'Trade Relations with Europe and the New Transatlantic Economic Partnership', the US House Subcommittee on Trade/the Committee on Ways and Means. First, 'ETSI's adoption of a standard that lacks technical or economic advantages over competing standards and, unlike the alternative, was incompatible with most existing standards, was an action that creates an unnecessary barrier to trade in violation of the Technical barriers of Trade (TBT).' Second, 'The EU had closed its markets for years to all but one wireless technology, one that happened to be manufactured by large European concerns. The EU had embarked on a policy of denying GSM's competitors entry into the European market.' Third, 'Once Europe had its common cellular standard, the game changed from devising legitimate technical standards to create an exclusionary industrial policy that would enable European manufacturers to market GSM around the world from their protected home market.' Finally, 'Faced with a protectionist European industrial policy and ETSI's non-objective standard setting process, Qualcomm had been forced to fall back on its IPR to protect its current customers and its position in the next generation of wireless technology' (ETSI, 1998h).

200 The EU R&TTE Directive (1999/5/EC) was first published on 7 April 1999 and came fully into force in April 2000. It was intended to complete the European internal market for R&TTE, introduce a much lighter regime, and replace the current regulatory system, which involved third party testing, or the obligation for a manufacturer to have an accredited quality assurance system.

called on ETSI to ensure the proper existence of the different members of the IMT-2000 family/non-European standards (Figure 6.7).[201]

Figure 6.7 EU objectives for GSM/UMTS standardization

Summary: Standardization, Globalization and Political Control

Proposals from the ETSI wireless community had met with political resistance on the part of the ETSI standing organization and a number of ETSI members, notably in the fixed network operator community. In their view, it threatened to reduce the control of full ETSI members over standardization activities. The ETSI wireless community, in particular its chairman and colleagues together with those manufacturers with strong positions in Asian and US markets, supported the globalization initiative as a way to foster multilateral coordination. The EC was concerned about this development in that it could jeopardize the basis for EU policymaking and a central source for political support. European industry, however, had strong commercial and political interests and wanted globalization.

The resistance to a globally oriented institutional structure had been based on political ideas about the appropriate balance between globalization (the partial relinquishing of control that it entails) and regionalism (the protection of established identity and control). One the one hand, there was the principle of globalization and the need for cooperation with other regional entities and interests to promote 'European' solutions to the global marketplace (export

201 A small delegation within ETSI produced a 'strategic framework', identifying key regulatory requirements and targets for the standardization work. It concluded that ETSI needed to produce the proposed harmonized standards for all IMT-2000 systems, which led to EU Mandate M.284.

sales, market share and the build-up of sunk costs and therefore long-term dependence). The ETSI membership was interested in capitalizing on its engineering and management prowess and reliability. One the other hand, there was the need to preserve the long-established political model of European integration with the view of further consolidating the European market. There was the principle of retaining political control and influence over important projects and protecting the identity of ETSI as an officially recognized European standards body with European roots. There is a looming tension between retaining the regional anchoring of ETSI in Europe (and its role with respect to EU policymaking), and the challenges to its identity and its role in an increasingly liberalized and competitive standardization and technology environment.

This chapter has examined the new realities of global competition in information and communications technology (ICT) markets. It has indicated the distinctive role of EU institutions in structuring the global competition around wireless standards along a more liberal strategic approach to high technology policy. EU public support for 'domestic' multinational firms' preferred international forum for standardization might be the new economy's equivalent of direct or indirect export subsidies. The EC is more concerned with job creation and Europe's overall competitiveness in high technology industries than with safeguarding access to critical military technologies. However, the EC will be more willing and able to support an effective high technology policy when it has the support of those manufacturers with the strongest position in the international structure of competititon and that have the resources that the EC needs to formulate and implements policy. This chapter has also shown the changing role of intellectual property rights (IPRs), which matter far more in the case of the third-generation mobile wireless systems than for IPRs in the GSM development. Furthermore, a critical driving force behind efforts to prevent market fragmentation in future mobile communication markets were network operators. European standardization policy was based on efforts both to promote scale economies in equipment markets and to support network economies in services with regard to mobile phone systems.

7. Digital radio in Europe

Radio is a Latin word and means 'ray'. Its roots go deep into the nineteenth century and its origins are associated with European scientists such as Faraday, Hertz and Marconi. In the late 1990s, the world of radio was on the brink of a transition toward a digital-based delivery system, which would enable the transfer of images and text, as well as digital sound of a higher quality. Europe has developed a digital radio system, the Eureka 147 or terrestrial Digital Audio Broadcasting (DAB-T) system.

7.1 BACKGROUND TO DIGITAL RADIO

Much of the early technology selection and effectively the standardization of DAB – terrestrial and satellite – were done under the European R&D program Eureka.[202] The objective was to finalize the work on the Eureka 147 DAB system and develop a standard with the aim of having it adopted by a formal international standards organization.

> After the agreement of this standard, the path will be opened for the European entertainment electronics industry to stimulate a virtually saturated market with new products for car and domestic audio broadcasting units. In turn, this will encourage considerable innovation from European microelectronics manufacturers. The drawing up of a new digital audio broadcasting standard will therefore provide a long-term counterbalance to the increasing dominance of the countries of the Far East in the consumer electronics sector. Transmitter products will provide a new impetus for investment goods industries (EURP, 1987).

---

202 Eureka is a Europe-wide research network with the aim to promote market-driven R&D in advanced technology as a complement to R&D programs organized by national, EU and other European organizations. The Eureka 147 project commenced in 1986, with manufacturers, radio stations and research institutes, and ran over four years. DAB had also been developed in Eureka JESSI projects.

The implementation of digital radio in Europe has been debated throughout the 1990s among broadcasters, manufacturers and policymakers. The story officially started in the 1980s when the first digital (satellite-based) radio broadcasting service, Digital Satellite Radio (DSR), which could be received throughout Europe via a small but costly satellite dish, entered operation in Germany. In 2001, Eureka 147 (DAB) was still the only system available in the world to meet all the requirements drawn up by the International Telecommunications Union (ITU) Recommendations 774 and 789. In 1992, the Inter-Union Technical Committee of the so-called World Conference of Broadcasting Unions recommended the system worldwide. In December 1994, the ITU formally recognized DAB as a world standard for sound terrestrial and satellite audio broadcasting, respectively, to vehicular, portable and fixed receivers in the VHF/UHF radio frequency range (ITU-R Recommendations BS 1114 and BO 1130). In March 1997, ETSI adopted Eureka 147 DAB as an agreed European Standard (ETS 300401).

The first introductory terrestrial-based digital radio services started in the mid-1990s. In 1998, the UK Radio Authority awarded the first commercial digital (DAB) license (local digital multiplex license) in the world. It was followed in November 1999 by the launch of the UK's first national digital radio (multiplex) service through Digital One (MW, 2000a), which had an obligation under its license to promote digital radio.[203] One year later, there were 38 digital radio stations in the London area. 'The radio industry as a whole will continue to work to make digital happen, but we need help, both from the government and retailers, to ensure its success,' said Oldham, director of strategy of Capital Radio, UK (MW, 2000b). In the late 1990s, DAB services were available from terrestrial transmitters in several European countries, Australia, Canada, China, India and other parts of the world. Hong Kong, India, Israel, Malaysia, Singapore, South Africa and South Korea were all engaged in testing and evaluating DAB. In Japan there was competition between Eureka 147 and homegrown radio technology. The Japanese industry was developing a system called ISDB-T, which in the late 1990s, however, was not yet seen as a major competitor to the European DAB system (WDABF, 1999b). According to a Bosch official, there were both opportunities and serious threats:

---

203 The initial focus in Europe was the car radio market, which would turn out to be a less than fortunate decision. It required extensive infrastructure, while margins would be low; it was difficult to perceive clear profits coming back to industry.

> The market share of the European consumer electronics industry has unfortunately diminished continuously in the last ten years. DAB is a European technology. European institutions are leading [...] After the digitization of communications, digital radio is probably, after digital TV, the last chance for Europe to enhance its competitiveness in the consumer electronics sector [...] Europeans, who developed the system and invested most in DAB, have to put all their efforts to participate in the exploitation of the system. With such a joint European efforts DAB can and will repeat the success story of GSM (see EC, 1998m, p.16).

The WorldDAB Forum argued that while DAB was a great opportunity for Europe, competitors would penetrate the European market if Europe failed to coordinate its activities (EC, 1998m, p.31). There was a perception, however, that DAB had significant competitive advantages if harmonization of national regulatory regimes and policies could be achieved across Europe. 'We would emphasize the fact that the European industry has already established a world lead in digital radio technology. DAB is an international standard. European companies are expected to win a significant share of the audio market over time, creating and securing many jobs' (WDABF, 1998b).

> Regulation needs to ensure a critical mass of good quality content across Europe as quickly as possible as a crucial driver of the take-up of digital radio. WorldDAB therefore urges the EC to encourage national regulatory authorities to respond positively to broadcasters' needs in managing the early stages of this radio revolution. To do so will, we believe, reduce the risk of piecemeal and uncoordinated implementation of the new technology, which would be to the considerable disadvantage of consumers and the audio-manufacturing base in Europe (WDABF, 1998b).

In Europe, a major challenge to DAB has been the political and industrial support for Digital Video Broadcasting (DVB), a system developed for digital television, which in principle can use this platform to deliver radio services. European industry associations such as the European Broadcasting Union (EBU) and the European Association of Consumer Electronics Manufacturers (EACEM) wanted the European systems to develop independently (WDABF, 1998a). This is an interesting case since there is a transatlantic competition dimension involved (DVB in Europe and elsewhere versus ATSC in the US). 'What the politicians want is European standards that they can bring to their colleagues in the US, so they can tell European voters that they have something better than America, and that they have helped to develop it' (EACEM, interview, 1 March 2000).[204]

The US industry and political bodies have been reluctant to consider the European DAB system, preferring to investigate so-called IBOC DAB (in-band on-channel) technology.[205] The approach toward digital radio in the US is different from that in Europe, Canada and elsewhere, which may impede the interworking of digital radio systems. '[A] few broadcasting companies recognized early in the 1990s that digital technology could drastically alter US radio's competitive landscape. To preserve their considerable investment in stations, several broadcasters began an effort to provide technology that could forestall US deployment of a system that would compete with their stations. One such system was Eureka 147, which was already under development in Europe' (EDN, 1999). The result was the creation of USA Digital Radio (USADR) in 1991 by Westinghouse Electric Corporation, CBS Inc and Gannett Corporation to develop a viable DAB system (BC, 2000).

---

204 According to DG INFSO, the ATSC Grand Alliance based in the US (the body that has promoted high definition television) and Europe's DVB Alliance had not competed in each other's territories (NAFTA and EU) but they were competing for third countries, notably Argentina, Brazil, China and Australia. One central political issue has been the allocation and management of frequencies. 'Frequency allocation for terrestrial television is a political process. There is a strong allegation that the US is misinforming on the potential of Europe's standard that seems to be better. In Asia, Japan was developing a Digital TV Broadcasting System very similar to Europe's DVB. If the EC will act to prevent the spread of alleged misinformation from the US, there may be common interest to dialog with various Asian countries to avoid the spread of the US's allegedly inferior standard' (EC, interview, 12 April 1999). For a description of EU political intervention in the late 1980s and early 1990s, see Hart and Thomas (1995) and Kaitatzi-Whitlock (1998). In early 2001, the US, South Korea, Canada and Argentina were using the ATSC standard for terrestrial broadcasting services. Japan has committed to its own development of technologies used in DVB, the ISDB standards. In 2001, the DVB consortium had been signed by 290 member organizations in more than 35 countries and its members were examining links with the Internet and mobile communications. The DVB organization has its industry-member-driven body based in the European Broadcasting Union (EBU) offices in Geneva.

205 This technological platform is expected to make amplitude-modulation (AM) sound like the current frequency modulation (FM) band, and the current FM to broadcast with CD quality, with an improved audio capacity and a capability to offer data services. It will surround an existing analog signal in the same band and on the same channel, hence its name (IBOC). The idea of allocating and using new spectrum to carry digital transmissions, opposed by the FCC, was dismissed. A transition period was expected to last some 10 to 15 years, during which both analog and digital signals would be broadcast (so-called 'simulcasting').

These activities signaled the beginning of IBOC DAB technology in the US as a relatively inward-looking industrial venture.[206] In short, US broadcasters hoped to avoid systems being developed or deployed in Europe and Canada. In 1990, the Federal Communications Commission (FCC) opened proceedings to consider authorizing digital broadcasting both on the ground and by satellite. The DAB subcommittee of the National Radios Systems Committee, co-sponsored by the so-called National Association of Broadcasters (NAB) together with the Consumer Electronics Manufacturers Association (CEMA), has evaluated digital radio and IBOC since 1993. The first attempts to design an IBOC technology failed. The FCC ruled that terrestrial digital radio technology was not mature enough and the activities were suspended (BC, 1999b).

In December 1997, the DAB subcommittee relaunched investigations on digital radio, not least since Lucent had proclaimed that a completely tested IBOC digital system would be available in the near future. In 1997 and 1998, Lucent and USADR cooperated through a joint venture but ended it and started working on different technological paths toward implementation. The problem was that there were different proposals for digital radio systems and the US government and the FCC might have to intervene to select a standard if there were several viable options.[207] On 1 November 1999, the FCC formally launched the rule-making process to consider a technological

---

206 The attempt to generate a specific technology for terrestrial digital radio in the US is interesting in that it was not consistent with claims by the US government and legislators in the field of wireless cellular communications that multiple standards are good for competition. In the looming transatlantic trade war on wireless equipment, the US government demanded that technologies other than homegrown European ones would have to be considered. In other words, while the myth of a consistent US 'free-market approach' is evident for political economy scholarship, digital radio illustrates again that regulatory policy is often a central component, and that the nature of political intervention varies according to circumstances and industry pressure. Nor was its approach to digital radio consistent with the stress on interoperability and open standardization in the field of satellite navigation.

207 Satellite and Internet broadcasting can also distribute digital audio. In the light of expected competition from satellite digital audio radio services (SDARS), broadcasters called on the FCC in late 1998 to elaborate rules for technical standardization and the transmission of terrestrial digital signals. In 1997, the FCC had issued licenses to the satellite radio broadcasters CD Radio and XM Satellite Radio in the so-called S-band (approximately 2.4 GHz), which intended to launch subscription-only systems online by 2000, and thus had a different business model from the planned free over-the-air terrestrial digital radio services (BC, 1999a). One of the terrestrial radio companies behind the appeal to the FCC, Cumulus Media, linked the issue with public interest and fair treatment. 'If terrestrial radio is unable to compete effectively with satellite and with other media such as Internet broadcasting, there is a possibility that radio stations will fail, thus depriving the public of free, locally originated programming' (BC, 1999a).

standard for terrestrial digital radio (DAB-T). There was industry pressure for a unified technology. 'A single standard is necessary to promote rapid consumer and industry [acceptance] of the system and to ensure interoperability of radio receivers,' argued Digital Express (BC, 1999a).

7.2 THE EUROPEAN RADIO INDUSTRY

The amount of deregulation in the radio industry varies between countries. By 1998, in France (whose private stations had 74 per cent of listening), Greece (90 per cent) and Spain (77 per cent) the new private broadcasters had formed national networks and competed on equal terms with the public broadcasters. On the other hand, in Denmark, Finland and Sweden, the public broadcasters retained over 65 per cent of the radio audience. National private stations had only been awarded licenses recently (WDABF, 1998a). Most of the radio companies derive most of their revenues from their home markets. In value terms, Western Europe is the world's largest market for audio equipment. Germany has consistently been the largest and most valuable market for all audio products, followed by the UK and France. In 1996, it was estimated that the European market for audio products was worth Euro 14.8 billion. Regarding total employment in the European radio industry, it has been estimated at 295,000 jobs (WDABF, 1998a) (Table 7.1).

Table 7.1 Employment in the European radio industry

Direct employment by public broadcasters	35,000
Direct employment by private broadcasters	20,000
Freelance employment	25,000
Jobs in related industries/companies	115,000
Employment by audio manufacturers	100,000

Source: WDABF (1998a).

The radio and DAB sector has operated in what can be seen as a 'horizontal' structure in that program providers, operators, and receiver manufacturers are separate from one another. They make few contractual arrangements with each other. Rather, they rely, as they did with terrestrial analog TV and radio, on the combined effects of independent actions to start and popularize the new system. The introduction of DAB would thus require a complex co-ordination process to encourage its introduction. 'Radio in Europe has traditionally been a mass of individual companies, under different ownership, meeting local needs. There are no multinationals with large shareholder finance to risk capital in new technology or subsidize receivers' (WDABF,

1999a). Radio consisted of micro markets, with no dominant players (BC, 1999c) (Tables 7.2–7.4). In addition, since radio consumers did not pay for reception, digital radio could not follow the television or mobile subscription models, which subsidized the hardware. Radio was independent from consumer devices and entirely disconnected from manufacturers.

Table 7.2 Selection of broadcasters committed to DAB, 1998

Public Broadcasters/Service	Commercial Broadcasters
BBC (UK)	Antenne Bayern Hornfunkanbieter
Radio France	Bayerische Rundfunk
Radio France International	Deutsche Welle Radio
Sveriges Radio	GWR (UK)
	NRJ (FR)

Source: WDABF (1998a).

Table 7.3 Selection of digital transmission companies, 1998

Digital Transmission Companies	Headquarters
CTI	Warwick, UK
Deutsche Telekom AG	Bonn, Germany
NTL	Farnborough, UK
TDF (Télédiffusion de France)	Paris, France
Teracom	Stockholm, Sweden
Sogetec	Paris, France

Source: WDABF (1998a).

Table 7.4 Digital radio/DAB-related receiver manufacturers, 1998

European	European	Non-European
Alpine, France	IDEO, UK	Delphi Delco, US
Arcam, UK	Meridian, UK	Hitachi, Japan
Bosch/Blaupunkt, Germany	Philips, Netherlands	JVC, Japan
Clarion, France	Psion (UK)	Matsushita/Panasonic, Japan
Cymbol, UK	Terratec, Germany	Pioneer Kenwood, Japan
Grundig, Germany	VideoLogic, Germany	Sony, Japan

Source: WDABF (1998a).

The European radio industry was not well organized in the first part of the 1990s. A firmer structure for political activity emerged after 1995. 'This had been the situation in Europe for some time. Nothing seemed to be happening whereas the Digital Video Broadcasting (DVB) project had political support. The latter had started later but seemed to be going places. Certainly there is a difference in money that we are talking about. There is a lot of money to be made on television and not so much to be made on radio. The industry was not run efficiently' (EC, interview, 23 November 2000). In 1997, EuroDAB changed its name into WorldDAB (Figure 7.1). 'The WorldDAB Forum's particular focus is to assist the smooth introduction into the market place of a highly successful European invention, Eureka 147 – which is the DAB system. Implementation of this new digital technology will ensure a vibrant and dynamic radio sector in a world of convergence and mark a worldwide European success' (WDABF, 1998b).

Figure 7.1 Organization of WorldDAB Forum, 2001

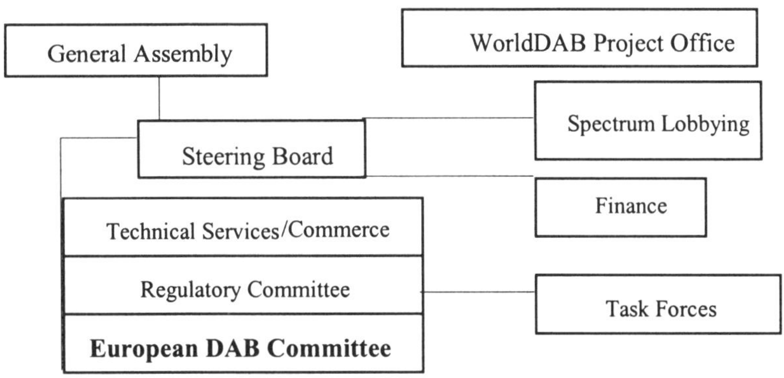

Source: The WorldDAB Project Office.

The EC referred to the major organizations that pushed politically for EU intervention as the 'EuroDAB consortium', which handled general political activities toward the EU institutions. This consortium consisted of AER, the Association of European Radios, EBU, the European Broadcasting Union, EACEM, the European Association of Consumers Electronics Manufacturers (EACEM), and WorldDAB. 'Around 1999 people were a lot better organized and to a certain extent we helped them to get better organized in that they would know how to speak to the EC internally, etc. We requested them to do a number of studies, in particular on market failure scenarios' (EC, interview,

23 November 2000). The EACEM joined the EuroDAB committee in February 2000. In addition, the EC became an observer. 'The EC is now an observer in WorldDAB, which is perceived as much less sexy than DVB for politicians. If you look at the volumes and the money involved in the radio market it is much less. The impact of telecom is much more important than radio' (EACEM, interview, 6 March 2000).

7.3 INDUSTRY ARGUMENTS FOR PUBLIC INTERVENTION

The EuroDAB consortium lobbied the EU institutions and the EC with arguments that stretched from technology and employment to regulation. Neither DAB, nor digital radio more broadly, had been covered in the EU political-regulatory framework. 'What measures are needed at European level? First of all, the recognition of the importance of the Radio sector and its inclusion in the Community's definition of audio-visual, which for some reason only includes the television and the cinema, and lately with the new technologies also the multimedia. This is not just a detail since all political decisions are taken based on this definition' (AER, 2000).

According to its European promoters, DAB was the best technology around, the standard was designed for mobile communications, and it even allowed the transmission of data, pictures, and so on. 'DAB is a broadcasting system for high power sound transmissions, and responds to the needs of sound broadcasting in the future. DAB-T is presently the only platform that guarantees the transition from analog to digital technologies' (CDAB, 2000). Moreover, it was strongly associated with Europe. 'It is a European standard that has been created through a European research program, which is a very strong argument' (EACEM, interview, 6 March 2000). Terrestrial DAB (DAB-T) supporters voiced political concerns about the relationship with satellite DAB (DAB-S). 'DAB-S is an interesting possibility, representing a complement to DAB-T. Its development may enable the emergence of pan-European or regional channels, which underline European cultural and linguistic diversity. There is nonetheless a short-term concern that DAB-S may slow the transition from analogue to digital technology' (CDAB, 2000). Moreover, WorldDAB repeatedly argued that DAB and DVB were not substitutable and served different purposes.[208] Tyssen Krupp Multimedia, a German conglomerate, argued strongly for the potential of DAB:

---

208 According to the WorldDAB Forum, the distinction between DAB and DVB is the following. 'There are two entirely new digital broadcasting systems that have been developed to replace existing analogue services in the years to come: DAB and DVB (Digital Video Broadcasting). Both systems offer significant advantages: better technical quality, greater ruggedness of signal, greater spectrum efficiency

> As far as digital radio is concerned, there is only one system available which is suitable, and that is the Eureka System DAB (Eureka 147): a system which has been developed by the European institutions, is standardized and has been introduced in many countries in Europe and beyond. It is also internationally recognized as the only system that meets all the requirements of the whole radio value chain. Digital radio and digital broadcasting in general in Europe have a leading position, and the consumer electronics industry has a golden opportunity – and perhaps the last one – to win back lost ground, with the obvious impact on the European economy. Several hundred thousand jobs and billions of turnover in Europe depend directly on the successful introduction of DAB in Europe and beyond (Tyssen Krupp Multimedia, 2000).

Industry argued that DAB was internationally competitive and had great potential. 'Digital radio is a stimulating opportunity for the development of the European electronics industry and for the development of the new economy. DAB will be a global success if it succeeds in overcoming the challenges now faced by Europe' (CDAB, 2000). There was, however, a need for concerted European action to ensure the large-scale introduction of digital radio and DAB and to secure its economic potential. A number of preconditions had to be met: (1) a stable regulatory framework for the introduction and sustainability of digital radio; (2) the acceleration of licensing processes; and (3) Europe-wide management, coordination and sufficient allocation of frequencies. If met, 'the European success story of GSM, in the field of communication technology, could be mirrored in the broadcasting domain by a success story for digital radio' (Tyssen Krupp Multimedia, 2000). WorldDAB also stressed the importance of ensuring an open, non-discriminatory technology platform (WDABF, 1999a):

---

and huge potential for additional program and data services, as well as lower energy costs. However, the two systems have been developed for different purposes. Both have merits and advantages to meet specific objectives. The DAB system design has been optimized to provide reliable mobile and portable reception in relatively narrowband channels. It addresses the way radio is used by the vast majority of consumers. DVB, on the other hand, has been designed to provide larger channel capacities (normally required for high-quality television signals) primarily to stationary receivers and, in the case of DVB-T, also to portable receivers. In DAB, the same modulation/coding system works for the different delivery media (terrestrial, satellite, and cable), which means that a common receiver circuit can be used for different broadcast channels. For DVB, channel coding has been optimized for each delivery system, and thus different circuits may be required for terrestrial, satellite and cable delivery. Since the size of DAB frequency blocks is smaller than for DVB, DAB is also more suitable for efficient frequency planning' (WDABF, 1998b).

> Standardization work in Europe is very important, with broad and if possible global acceptance. This philosophy has underpinned the Eureka 147 program since its inception and now drives the development of DAB applications. If, going forward, consensus cannot be reached the EC should consider regulation of certain key technology areas in order to ensure an open and fair market throughout Europe. A divergent market is in direct conflict with the interests of consumers (WDABF, 1998b).

The importance of radio, it was argued, called for public support. 'In the past, well-meaning European initiatives to help new media have not always been successful, and community fingers have been burned. Shouldn't we leave everything to market forces? These are all reasonable questions. But the answers lead to a clear conclusion. Radio does need European action coordinated by the Community' (EBU/WorldDAB, 2000). WorldDAB argued that radio was the most trusted medium and promoted democracy. In spring 2000, there were an estimated 10,000 radio stations across Europe. According to a 1997 survey in EU countries on attitudes to the media (the press, TV and radio), radio was the most 'trusted medium', scoring 63 per cent overall (highest in Finland and lowest in Greece), with TV at 56 per cent and the press at 40 per cent. 'Despite the explosion of other services, millions of Europeans continue to cherish radio as something special in their lives, an essential part of their culture and democracy. Radio cannot be squeezed into the margins of the audio-visual debate as an afterthought' (WDABF, 1998b).

Industry had identified a number of problems. First, there had been a lack of incentive and commitment on the part of the manufacturing community to increase the production volume of receivers. The EBU/WorldDAB Forum argued that free-to-air digital radio would not be able to take off by itself according to market economic principles since the manufacturing industry was too weak. 'One thing is crystal clear. Europe's national receivers market, on its own, is too small to make digital radio succeed. There has to be pan-European confidence building, and pan-European coordination, if DAB is to succeed. This cannot be done without the Community' (EBU/WorldDAB, 2000). The EuroDAB consortium stressed the role of audio manufacturers:

> The benefits to the Community as a whole of digital radio will be massive, and clearly justify a large investment, but the return the set maker gets is going to be only a small part of the total returns. The most gain of digital radio will be for Europe as a whole, because of the continuing social and cultural benefit that radio will bring. The largest investor (at least initially) – the set maker – is making the smallest gain, the profit of the receiver sale. Initially the set maker is carrying any broadcasters' gain, and the social and cultural gain, on

> his shoulder. He will not mass-produce digital radios simply because it is in the long-term interests of the people of Europe [...] There are some products, which have a high social or cultural gain, which market forces alone cannot make successful – 'Merit goods' (EBU/WorldDAB, 2000).

Furthermore, the broadcaster community believed that they had invested enough as reflected in the accumulated investments since around 1996–1997. The WorldDAB Forum referred to the classical chicken-and-egg situation, that is, without support from the manufacturers and EU institutions the marketplace would have difficulty taking off. 'The transition for free-to-air digital broadcasting of any kind must have formidable barriers to success. When set makers think of investing in the mass production of digital free-to-air radios, where these barriers seem highest, they must worry. To have the courage to surmount the obstacles, the European Community has to build confidence in the value of radio, and its place in the information society' (EBU/World DAB, 2000).

> When European Community action on media is suggested today, there are those who reasonably argue that the lessons of the past should be learned. Efforts to help television through MAC and HDMAC by the Community years ago did not succeed. Though the MAC television world was a different planet to digital radio, the argument goes that since these efforts were unsuccessful, the future of the media must be left to market forces alone. The Commission and the Community should stand back, and let nature take its course. Why make the same mistakes again? The answer is that we do not have to. This time we can do it with the benefit of hindsight [...] True, if we do nothing at all, we cannot make a mistake, but neither can we contribute to society. Doing nothing is a luxury that industry can never afford, and the Community must have similar courage. Doing nothing is not the right thing to do – it is too easy. The European Community must take this matter seriously. Radio has what economists call a major 'externality', and its preservation into the digital world is a matter of pan-European importance (EBU/WorldDAB, 2000).

Industry wanted a stable framework for the radio sector and its technology, which it pointed out to the European Parliament (EP). 'A substantial concerted effort is now required from all stakeholders – broadcasters, manufacturers and legislators – to carry the transition forward. The situation is not helped by the lack of a public policy lead from Europe' (BBC, 2000). Furthermore, lack of interest and coordination among national and international regulatory authorities further complicated implementation.

Frequency scarcity, and the absence of long-term planning in this regard, constituted 'a major barrier to the growth of DAB and could stifle it altogether' (WDABF, 1999b). WorldDAB argued that European spectrum planning had to consider the problem of 'simulcasting', that is, the transition phase to digital where analog and digital would coexist, which would require additional spectrum.[209]

> Because so far the sector has run smoothly in the analog world, no measures have ever been foreseen to support this sector at European level. Today, however, Radio is going through its digital transition, as other media do, and the process is not easy because of the lack of a concerted action among the different players but also because of the lack of the necessary regulatory frameworks in the different Member States. This situation implies a clear risk of market failure. Based on this, the European Institutions should play a relevant and necessary role in the implementation of the Digital Radio in Europe, which ultimately will benefit radio companies but also the citizen, who will continue enjoying Radio but with new services and programs. Despite the operators' requests, the EC continues in the view that the Radio sector does not need any action at European level since the market failure case has not been demonstrated. European Radios hope that the European Institutions will not wait too long to support the sector because then the market failure will be a reality (AER, 2000).

The EuroDAB consortium pointed out that DAB should be judged according to its potential to spur the creation of new services and employment, technological innovation, European competitiveness, and economic growth. 'Radio broadcasting and its development are a highly regarded source of information and entertainment and as such must be taken into account in the regulatory framework for the Information Society' (WDABF, 2000a). It should be assessed in line with the economic value of spectrum, which would clarify that DAB would provide greater welfare. The EU institutions, they argued, should officially publish the DAB standards with references to standards and/or specifications, technical interfaces and/or network functions. The inclusion of DAB would be 'a suitable response to the need to create

209 'WorldDAB believes it is now essential that spectrum is made available for a broad implementation of DAB. Radio services currently broadcast in analog need also to be available in DAB. These include existing public and commercial AM and FM broadcasters, operating national, regional, local and community services. There must also be appropriate allocations for new audio, data and multimedia services. These are essential to stimulate the sale of new receivers and create opportunities for new revenue streams for broadcasters' (WDABF, 1999b).

certainty and, at the same time, not hinder technical innovation or limit consumer's rights' (WDABF, 2000b). While recognizing obligations under the World Trade Organization (WTO), they stressed that the EC should acknowledge politically the importance of those systems that were driving the take-up of digital services, were suitable for open, non-discriminatory, efficient access, interoperability and service interconnection, had significant potential as world standards, and had been recommended by formal standards bodies. 'A rigid application of technological neutrality may well impede the progress of digital radio and threaten its market failure' (WDABF, 2000a).[210]

7.4 The EU Audiovisual Policy and Digital Radio

The EU audiovisual policy has been heavily focused on the cinema, television and video sectors with the aim of promoting the diversity of cultures in Western Europe and enhancing the competitiveness of the European content and programming industry in the new economy. It was recognized that the audiovisual sector not only had a cultural dimension but also a growing economic significance as a facilitator of technological development. The EU, in coordination with national authorities, thus had the task of strengthening the cultural dimension, protecting the 'audiovisual heritage', promoting industrial activity and supporting the role of the audiovisual sector at the international level.

In April 1998, a large European Audiovisual Conference was held in Birmingham entitled 'Challenges and Opportunities of the Digital Age'. A major conclusion was that there was a need for structural change in the European audiovisual sector in order to foster, through national and European policies, its overall competitiveness at the international level. The EC's Directorate-General for Education and Culture launched a communication ('Audiovisual Policy: The Next Steps') that recognized the need for public financial and promotional support in order to strengthen the competitiveness of the European audiovisual industry (EC 1998t). Another result was the initiation of consultation on a public support program for the audiovisual industry by encouraging and facilitating the development, distribution and

---

210 The WorldDAB Forum demanded that it should be allowed political access to the High Level Communications Group (HLCG), which included the EC and national regulatory authorities. Together with the so-called Radio Spectrum Policy Expert Group (RSPEG) these bodies could initiate EC recommendations or mandates (for example, regarding spectrum allocation and standards). 'It is essential that the procedures and work of these groups are transparent and fully publicly accountable [...] WorldDAB should be entitled to play a role in the preparation of any relevant measures to be submitted to the HLCG for endorsement' (WDABF, 2000a).

promotion of European audiovisual works (Council, 2001). The High-Level Group on Audiovisual Policy delivered a report in late October 1998 ('The Digital Age: European Audiovisual Policy'), which focused on the cinematographic and audiovisual industry and on public intervention to support these fields through the DG EAC MEDIA program. Commissioner Marcelino Oreja chaired the group, assisted by Gregory Paulger, Directorate-General for Information, Communication, Culture and Audiovisual Media, who would play a central role in the politics of digital radio (EC, 1998o).

A Forum on Audiovisual Policy ('European Content for the Digital Millennium'), organized by the Finnish EU Presidency, followed in October 1999. In December 1999, DG EAC issued a communication that defined the priorities in the audiovisual sector for the period 2001–2005. In this communication, DG EAC emphasized that the content industry carried the promise of generating highly skilled jobs but that it was important to ensure that 'this growth benefits the European audiovisual industry and does not simply result in higher levels of imports' (EC, 1998n, p.2). DG EAC pointed out that the EU's deficit with the US in audiovisual trade amounted to around Euro 7 billion annually; it played on the threat of 'content imperialism' in order to legitimize public support. In December 2000, its political activities resulted in a Council Decision on financial and other forms of support through the MEDIA program. 'The competitiveness of the audiovisual program industry depends on the use of new technologies at the program development, production and distribution stages [...] Public access to the European audiovisual heritage should be improved, in particular through its digitization and networking at European level' (Council, 2001, p.36).

The EU again defined the audiovisual sector and the new economy in the digital age by focusing on television broadcasting and cinema distribution; there were no references to the role of digital radio as a new technology in this context. DG EAC and the Council stressed new technologies, digitization, distribution of audiovisual works and the need to protect the cultural and linguistic diversity in Europe, while acknowledging the importance of the economic dimension (exactly those arguments that the European digital radio sector used when lobbying the EU). DG EAC had touched upon radio as a budget line around 1994. It was not until later, however, that DG EAC from a 'contents point of view' got more involved in defining what digital radio really meant (rather than from the 'transmission point of view' associated with DG INFSO). 'Governments will be less prepared to look into something if it is not going to bring results quickly. It has to be balanced against what it provides the public. If the EC has an approach that is more citizen-oriented as we are told, then I would say that the most listened to medium or even most believed medium is radio' (EC, interview, 23 November 2000). Digital radio, however, was excluded in the political definition of the 'new economy'. DG EAC's internal annual publication *Le Magazine* illustrated DG EAC's orientation toward television broadcasting (satellite and terrestrial), cinema and the e-Europe initiative.

As a result of political lobbying on the part of the EuroDAB consortium, some people in DG EAC started considering the formulation of a definition of (digital) radio. One of the defining features of radio was its local element. 'Radio had by necessity to be local, not by satellite or cable where it would lose its localization feature. It has a value particularly for those who want to listen to, for example, the news in their own language. The only system that existed that was local and digital at the same time was DAB. It was from there we progressed into DAB' (EC, interview, 23 November 2000). DG EAC was responsible for EU audiovisual policy and the only instrument that existed to look into broadcasting, the Television without Frontiers Directive. It argued that there was a similar need for digital radio as for television. 'It was considered that since we were doing this for film and television, then we should do it for radio given that from now to the day radio is in place we are talking about 500 million radio sets. It is a market of large proportions' (EC, interview, 23 November 2000). There was, however, a dilemma. 'If indeed the world wants to adopt a European system, you could not claim that this system ought to be adopted if Europe itself could not find a way to adopt it. On the other hand, the effect of the high definition television debacle, which EC made efforts to introduce in the early 1990s and the failure that ensued, had made us shy, if you like. We did not want to try any longer, which meant that the principle of technology neutrality was adopted, which stated that we could not support standards' (EC, interview, 23 November 2000). DG INFSO had argued that high definition television was a telecom issue. It should therefore be within its bureaucratic department.

DG EAC administrators recognized that the principle of technology neutrality was less applicable for television and mobile wireless than for radio. '[T]hat is exactly the same as with DAB so why is it that DAB is not now supported?' (EC, interview, 23 November 2000). In March 1998, DG EAC organized a conference on radio in the digital era. This conference was followed by a second conference in Spring 1999 (EC, 1999n). DG EAC Director-General Pappas supported digital radio politically. Pappas stressed that the EU audiovisual policy had focused on television and the assumption that the image was the most effective means of communication. He claimed that the emphasis on television threatened radio. 'It is my assumption that DG EAC, responsible for information and communication, should make the maximum use of radio' (EC, 1998m).[211] Some EP Committee members (for

---

211 Pappas stressed the cultural effects of radio and its importance for minorities. 'We have already examples like in Sweden where there is broadcasting in the Finnish

example, the Committee on Legal Affairs and Citizens' Rights and the Committee on Foreign Affairs, Security and Defence Policy) agreed that there was a disproportionate focus on television. Pappas firmly believed that the future of broadcasting was digital. 'And saying that the future is digital I should not neglect to remind you that this is a European invention. Eureka 147 is a European system, which has been recognized as the standard for digital radio broadcasting not only in Europe but also in many countries across the world. The system demonstrates its potential to become a great success like GSM' (EC, 1998m).

Following the EU workshop on digital radio in 1998, DG EAC launched an initiative and action plan to promote the status of digital radio in the EU, dubbed *European Radio Industry and its Transition to Digital*. It stressed that the cost of a possible market failure in the implementation of digital radio would be high for the sector and European society. The EU could play a considerable role in helping industry to overcome barriers and guarantee radio its place in the new digital environment (EC, 1999m). 'WorldDAB has warmly welcomed this initiative, establishing as it does for the first time, a commitment to take full account of radio in the audio-visual debate' (WDABF, 1998a).

DG EAC believed that there was a 'window of opportunity' to support digital radio and DAB, which should not be missed by Europe since the leading system for digital radio was European. It envisaged initiatives along three dimensions: regulatory, political and financial support. It argued that radio had to be an integral part of all EU policies and that it needed a concerted promotional action across national frontiers and industries. Political and regulatory support could be decisive in setting the transition to radio in motion and the European institutions could function as catalysts for bringing that about. Financial support could prove necessary 'to promote industry investment and avoid a market failure in the implementation of digital radio' (EC, 1999m). The shortage of frequencies and lack of a coordinated EU regulatory framework (market entry, licensing and operating conditions) reduced the regulatory certainty and therefore the incentives to invest.[212] 'If, in the absence of a concerted implementation strategy, the

language and we understand how important it is for the minorities, and how important it is for the diversity of our cultures to make use of the radio in the digital era' (EC, 1998m).

212 In July 1995, a frequency plan was agreed for digital radio. It did not, in the view of industry, provide sufficient frequencies. The 'pre-Commissioner Reding' DG EAC stressed the need for a supranational dimension. 'If the current Wiesbaden frequency allocation might be sufficient for an initial market implementation of digital radio, there is a clear barrier to further growth. This barrier is a significant discrimination in favor of early adopters against those operators that, for financial or structural reasons, might be endangered by such competition distortion caused

European radio industry were to fail and digital radio were not to take off, there might be a very high price to pay for the European society as a whole. Not only in economical and financial terms but also socially and culturally' (EC, 1999m). The combination of the lack of a commonly adopted and definitive implementation strategy and schedule, a clear business model and uncertainty among industry leaders resulted in a deadlock situation. 'The horizontal problem of the absence of an implementation strategy is considered by the sector as the most important obstacle to market implementation and development of digital radio' (EC, 1999m).

The action plan agreed with the EuroDAB consortium that the DAB system had the potential to become a worldwide success just like GSM (EC, 1999m). However, a large-scale rollout of digital radio was not in sight. A failure to implement digital radio in a timely manner would result in loss of investment, business opportunities and turnover. In addition, it would constitute 'an important loss of prestige at international level' (EC, 1999m). The action plan stated that the investments that had gone into the development and implementation of digital radio amounted to around Euro 450 million. Implementation failure could result in stagnation of radio broadcasting turnover and radio receiver sales, no innovation in the program market, and loss of patents, trademark and royalties for European industry.

> A worldwide implementation would enforce Europe's role in the consumer electronic market with obvious positive consequences for employment in this sector. It would also strengthen Europe's prestige in R&D and result in revenue streams through royalties. Losing or substantially diminishing the business opportunities created by European invention would result in a significant defeat. Radio broadcasting would possibly be marginalized in a multimedia environment losing its vital role and its independence as a medium. This is a risk that Europe cannot afford to take (EC, 1999m).

The major portion (Euro 230 million) represented investments at national level to establish the transmission infrastructure. At the manufacturing level, around Euro 180 million had been invested in R&D, product design, development of prototypes, and so on. 'Most of these investments, approximately between 85 and 93 per cent, would be irretrievably lost if a market failure scenario came about. Very little could be reused for different purposes, even if another non-European system prevails, or if other media technological platforms (that is, DVB) take over digital radio' (EC, 1999m).

---

by technical access barriers to frequencies. In this case, it might prove extremely difficult, if not impossible, to reach the critical mass of digital radio receivers, in absence of which digital radio would be condemned to failure' (EC, 1999m).

7.5 The European Commission and the Radio Industry

The proposed DG EAC communication would never see the light of day outside the EC bureaucracy and would 'die' with the Santer Commission. First, the timing was not the best. In March 1999, the EC announced its collective decision to resign. The EC services were reputedly instructed not to issue any new policy initiatives for about nine months.

> We waited for a moment when there would be a political will to continue but unfortunately this has not materialized. The previous Director-General [Pappas] was more interested and definitely more favorable. He identified an opportunity that we should go ahead. This is politics I can tell you. The issue of resources became a topic. It was decided that we have no option, presumably, that there were not enough human resources in DG EAC to deal with digital radio. At the political level it was passed over to DG INFSO around October 2000, which meant by definition that they had the resources (EC, interview, 23 November 2000).

It was recognized, however, that it was not only the restructuring of the political and administrative levels of the EC that impeded the support of digital radio in Europe. 'The new Commission literally decided to throw away the communication that had been prepared because they felt it was intervention into the market and the market should decide' (EACEM, interview, 6 March 20000). The resistance was clearest from DG INFSO. 'The communication died due to resistance from DG INFSO that argued that the radio sector had to elaborate another strategy that was integrated with the existing framework. This DG is more liberal than DG EAC and stresses the need of market forces to promote digital radio' (WorldDAB, interview, 20 November 2000). The DG EAC report on digital radio was in some quarters tactically defined as a new policy initiative. 'The planned communication was the end of the DG EAC support to DAB' (WorldDAB, interview, 8 March 2001). The EuroDAB consortium and the WorldDAB Forum acknowledged that lobbying strength was a major factor in EU policymaking:

> Stronger lobbying and interest help to explain why certain technologies receive more support. We cannot really understand the lack of support. The radio industry is more fragmented and has a very horizontal structure with a lot of actors involved. There is less money involved. There is no direct return of investment. Basically, we do not have a business model, which makes it hard to talk politically. For us that should mean exactly the contrary – we need the support all the more (WorldDAB, interview, 1 March 2001).

The radio industry could not muster the same political resources as, for example, the television industry. 'When we talk about political support there is organized political support and political support properly. Organized industrial benefits, and the political support that it brings forward, is something totally different from support from politicians (who I believe are interested in radio since it gives them direct access to the voting public). It is about big industry. Television has big muscles and that is all there is to it' (EC, interview, 23 November 2000).

Moreover, there was some controversy as to how to perceive radio in a broader context and the benefits it could generate. The EuroDAB consortium clashed with the top level in the new DG EAC: 'We claim that DAB is a world standard, although the US has not adopted it. DG EAC is telling us that DAB does not have the potential to be a world standard. That was the last thing we heard and we were of course appalled. DAB is a world standard and has been recommended as such. If you look at how many countries have adopted DAB, there are quite a few. There is great potential. The DG EAC argument is not a good argument and we are trying to oppose it' (WorldDAB, interview, 1 March 2001). Yet another issue was the relationship between culture and industry with regard to radio 'There is always this argument in Europe whether radio is culture or industry. If it is culture they say it is going to lose money and if it is industry it will make money. I do not particularly think so. If you take cinema it is also culture. I do not think radio is any more or any less culture than cinema, but it is culture that touches everybody everyday' (EC, interview, 23 November 2000). According to industry officials, the problem was not the sub-cabinet level of DG EAC, which by and large was open to supporting digital radio. Rather the resistance was at higher level, that is, in the Cabinet, and in particular the Head of Cabinet, Paulger. In a keynote address to industry, the political and cabinet level in DG EAC made its position on digital radio clear:

> The sector has in fact repeatedly been asked, within the constraints of the principle of technological neutrality, to justify why action should be taken in order to support it. Even though the sector has responded vigorously to the EC's request, it has, so far, not been able to demonstrate the market failure, which would justify Community action [...] The EC is in no way attempting to discriminate against radio, particularly by comparison to other media. Simply, the radio sector has been able to operate without creating the economic deficits prevalent in other parts of the audio-visual. This is due to a number of factors inherent to the sector (for example, less exposure to 'foreign competition') – but whatever the reasons: it is good news. As the saying goes: if it is not broken, don't mend it (CDAB, 2000).

Understandably, the EuroDAB consortium and WorldDAB members did not agree with such arguments and conclusions.

> The question should be why the present DG EAC, advisors and Commissioner Reding are not sensitive? The person [Paulger] is against digital radio for political interest. He has always been against and not supportive of radio, and is not willing to spend time on radio. It is not his 'baby'. Therefore Commissioner Reding is against radio. Reding simply follows what her Cabinet says. Moreover, Reding comes from Luxembourg and has a lot of pressure from, and is pushing hard for, satellite aspects because they have a big satellite company. Obviously, there are satellite people lobbying against DAB because it will take part of the radio spectrum that they need to do their own things with satellite, including radio (WorldDAB, interview, 2 March 2001).

EuroDAB wrote to Commissioners Liikanen (DG Enterprise and DG INFSO) and Reding (DG EAC) asking to see them. In February 2000, the EuroDAB committee had a meeting with DG Enterprise and the EuroDAB chairman Michael Green (BBC) explained what had happened with the DG EAC communication. EuroDAB members asked what DG Enterprise could do to support this European technology. DG Enterprise admitted that radio was important, that it would experience a transition to digital radio and that European industry had to support this transition at the same time as the market should decide the outcome. However, DG Enterprise admitted that Europe needed a coherent strategy to move to digital and the radio sector should help the EC to design a rationale to convince people to replace their existing investment by digital products. According to EuroDAB members, DG Enterprise said that the EC tend to respond only when people start to shout and that it had not yet heard a strong voice from the manufacturers. It had thus concluded that industry did not have a problem.

> Whatever argument DG EAC can muster against radio they use it, such as that they are technologically neutral, cannot push for a single standard, or that radio is doing very well and does not need any support. The EU only intervenes when it is needed and there is no deficit. They always claim with television and cinema that there is a big deficit with the Member States in terms of production of programs or films [...] It is true that analog radio is doing very well. But because of finite capacity of spectrum there is no further growth opportunity in the analog world, so it either goes digital or it gets marginalized (WorldDAB, interview, 1 March 2001).

The EuroDAB members realized that it was important for the EC that DAB was a formal European standard, that is, it was necessary to make sure that it became one of the core technologies in ETSI. The EC indicated to EuroDAB that the EC could help to promote digital radio indirectly through funding ETSI activities, which would ultimately promote the DAB standard. Essentially, the EuroDAB was told to talk to industry to push for DAB as an ETSI core technology in order to rally support from politicians and the EC bureaucracy. 'EC is a very complex machine. They have to build consensus among themselves. If you are serious about lobbying you need a plan not only to lobby the EC but also governments in the Member States because they will vote on a Directive and later implement it. Lobbying is not a small task and you have to build consensus among all these layers and players' (EACEM, interview, 6 March 2000). Moreover, the radio sector recognized that the EC and governments were focused on other technologies. 'The politicians are looking to UMTS as the next battle. This is what is going to make them successful and the message is simple: extrapolation of GSM. The next thing is that UMTS will replace everything (telecom and radio) so all frequencies will be given to UMTS' (EACEM, interview, 6 March 2000).

> Before the WTO agreement the EC could openly support a single standard. Now it must be technologically neutral. They cannot do it but they still do. Galileo and UMTS are cases where they do it. That is exactly the question we ask, that is, why they do not do it with DAB. Industry needs to help them, like DVB. They can help us but they claim we can do it on our own. They claim, for example, that they are not helping the DVB platform, which is not true at all. They are helping DVB. The reason why they support Galileo openly despite being technologically neutral should apply to this as well (WorldDAB, interview, 1 March 2001).

7.6 Digital Radio and Market Imperfection

In April 2000, the Portuguese EU Presidency organized a conference dedicated to DAB in Lisbon with the participation of administrators in DG EAC and a large number of industry people (CDAB, 2000). In May 2000, European broadcasters and manufacturers called for joint action to promote DAB. The political message was the need for 'concerted moves by public and private broadcasters calling on the EU to, at least in principle, encourage national regulators to make available adequate spectrum space for digital radio, and to create the regulatory framework needed for its success' (EBU, 2000). In addition, the radio industry lobbied the new EP and managed to include radio in the media program financed by DG EAC. The Chair (Gargani) for the EP Committee for Culture, Youth, Education, the Media and Sport was open to support digital radio and organized a public hearing in May 2000 (EP CCYEMS, 2000a). 'With all that, DG EAC is sort of obliged to come back and support DAB as well. Reding will not be politically able to deny all this' (WorldDAB, interview, 1 March 2001).[213]

In October 2000, the EuroDAB consortium met with representatives from DG EAC, DG Enterprise and DG INFSO. Its objective was to make sure that those EC officials made a positive recommendation to their cabinets about the request for a high-level meeting with Commissioners Liikanen (information society and enterprise) and Reding (education and culture). The aim was to generate a stronger political relationship with the EC. The EC delegation, represented among others by the DG INFSO sector head dealing with media and TV, and administrators from the DG EAC Audiovisual Policy Unit, responded that the radio industry was successful. There was no immediate threat in terms of 'content imperialism from the US'. It was thus unclear what would justify intervention politically.

In addition, the EC delegation argued that convergence among technologies, which was something that the EC looked upon favorably since it was assumed to have 'federalizing implications', implied that all technology platforms would deliver the services that citizens would need. Therefore, it was not possible to single out radio specifically for special treatment and as a target for intervention. DG INFSO argued that the convergence of technologies made it difficult to promote specific technologies (delivery mechanisms) without distorting competition or discriminating against other platforms. Its policy was to allow the market to work alone and intervene with *ex ante* regulation only under certain conditions, that is, when 'gate-keeping difficulties' arose. DG INFSO argued that the 'digital dividend for radio' was not as strong and efficient as that of television. In addition, if the competitive advantage of digital radio was significant, why was there a need for political intervention? Intervention in the digital radio market would also include regulators at national level, which could create political problems since the EC was reluctant to put pressure on the EU Member States.

---

213 Industry stressed that radio should be included in industrial policy initiatives. 'The Parliament can play an important role to change this situation by asking the Commission to integrate Radio in the Media Training program, since the 200,000 European professionals that work in this sector should also have the opportunity to get training to be ready to enter the digital world. Furthermore, and because Radio is content and content is "King", the Media Plus program should also foresee a pilot project to create a favorable environment to the production of new content for Digital Radio' (AER, 2000).

The EuroDAB consortium presented a number of justifications for supporting digital radio (for example, analog radio characterized by finite spectrum, no further growth opportunities, and consumer problems). They argued that radio could not remain analog in 'a digital world' and that the other major technological platforms (DVB-T, UMTS, satellite carriers and the Internet) did not meet radio's needs and could (or perhaps should) not therefore replace the DAB system. Rather, DAB and DVB-T together constituted promising, competitive answers to non-European solutions and developments. The EuroDAB consortium further repeated the case that the current situation was characterized by an uncertain market and slow rollout due to a number of factors, the most significant of which was that manufacturers required a single European marketplace to drive down the cost of receivers and create a mass market. Manufacturers identified a patchwork of licensing regimes, migration policies, spectrum allocation and different timetables for the introduction of DAB services (Table 7.5). The political message delivered to the EC was clear. The EuroDAB consortium stressed that the market alone would not guarantee the success of digital radio. It was politically imperative to intervene to foster a truly European dimension.[214]

Table 7.5 National regulatory regimes

	Licensing	Frequency allocation	Network deployment	Private/Public broadcasters	Consumer awareness
UK	Available	Available	Available	High	Low
Belgium	Low	Low	Available	Low	No
France	Low	Low	Low	High	No
Germany	Low	Low	Low	Low	No
Italy	Not available	Low	Low	Low	No
Spain	Available	Low	Low	High	No
Sweden	Low	Low	Available	Low	No

Source: Trade press and interviews with European radio officials.

Industry stressed that there was a clear need for different mechanisms of a political nature to be instituted. 'Without successful digital transition, radio will atrophy. Industry needs new opportunities, the marketplace needs new players to reinvent it, new creativity, more choice for consumers, delivered

214 The business models applied to drive digital TV and telephony (pay-per-use, subscriptions and hardware subsidy), it was argued, would not be appropriate for radio. Moreover, political definitions of spectrum criteria and allocations discriminated against digital radio in the absence of a coherent digital radio policy. Finally, no consumer technology had recently succeeded at the national level only.

by a common proven Europe-wide standard' (AER et al., 2000). The creation of a truly European market had to be supported where licensing regimes would be in place for both public and private broadcasters. Regulators had to allocate sufficient spectrum to satisfy the needs of digital radio. National migration policies toward digital from analog radio had to be harmonized. Finally, the advantages of an agreed common standard such as the DAB system had to be recognized politically. Failure to do so would strengthen the 'market failure for a European world-beating technology' and result in the 'encroachment of non-European technologies' (AER et al., 2000). Industry argued that digital radio deserved at least the same support as television from the European examples (WorldDAB, interview, 1 March 2001).

In December 2000, EuroDAB representatives were invited to an EP meeting on digital TV standards. 'This was a fruitful occasion to voice radio's perspective about standards and the need to promote interoperability. Although the meeting had mainly focused on digital TV standards, it had given a good opportunity to show that a case by case approach was necessary, and that DAB was showing similarities with the GSM case' (WorldDAB, interview, 2 March 2001). DG INFSO had likewise participated in the meeting on digital TV standards. The EP Committee on Industry, External Trade, Research and Energy produced two reports in 2000 that had a bearing on broadcasting (EP CIETRE, 2000a, 2000b). Following these reports (by the EP Members Thors and MEP van Velzen), and with support from German EP member Ruth Hieronymi, the EC proposed to establish a sub-group on digital TV standards issues called the Open Network Provision Digital Broadcasting Expert Group (ONP DBEG). 'DG INFSO managed to have DAB included on the agenda of this group. The EC asks us regularly for input and information to be used in this debate and we are helping. They hope that Member States will put in place similar license regimes as for example in the UK. Commissioner Liikanen gave the go-ahead to all this' (WorldDAB, interview, 1 March 2001).[215] The DBEG would work alongside the existing Licensing Committee and ONP Committee (for incumbent operators in the telecom field) and would include technical experts. It would enter operation in early 2001. In addition, it would provide an arena and opportunity for industry representatives to influence the process. DG INFSO,

215 Representatives from the radio sector stated that the highest level of UK Radio Authority was a key driver. The latter had announced its preparedness to assist the WorldDAB Forum and proactively push for the implementation of DAB in Europe by opening a dialog with colleagues from other EU Member States and by trying to get the UK Government to support these activities. In particular, it would suggest that the newly established ONP DBEG should include radio as a priority item on the agenda. The radio industry had clearly stated that DAB was facing regulatory and implementation problems. Around the same time, Ford UK announced a commitment to undertake to line-fit DAB receivers in cars in 2004.

which now in principle had the main responsibility for radio, stated that the mandate of the group the EC announced was broad enough to encompass digital radio and that the EC would use existing channels with Member State experts in the ONP committee.

Taking into account the political arguments expressed by industry, DG INFSO proposed two lines of action. Firstly, the EC would approach the EU Member States and inform them about the situation of digital radio in Europe. 'There is a different situation in each country. What we try to convince DG INFSO to do is a benchmarking exercise, that is, to look at best practices that are in place, such as in the UK. We need to convince them that bad practices may stall the whole market' (WorldDAB, interview, 1 March 2001). If a sufficient level of consensus resulted there could be an opportunity for benchmarking activities with the aim of examining implementation methodologies on a country-by-country basis, which could then be followed by political action. Secondly, the EC proposed that digital radio could be included in any relevant current or planned events, agendas, statements, and other political activities. Though WorldDAB welcomed indications that the EC would discuss general issues, it recognized that was far from a coherent EU high technology policy for digital radio in Europe. 'We have told them that this is not about money but about political leverage. It is about a European accent because a single European market is not enough' (WorldDAB, interview, 1 March 2001). In late 2000, these lines of action were forwarded to DG INFSO.

In November 2000, Commissioner Reding outlined the contours of a new EU audiovisual policy. The central message was that DG EAC intended to move from 'Television without frontiers' to the broader principle of 'Content without frontiers'. In addition, the DG EAC services would launch a number of studies by the end of 2001.[216] The ultimate objective was to launch a consultation process based on the results of the studies, which would be followed by a communication that would include any necessary proposals for amending the Television without Frontiers Directive. DG EAC proposed that the EuroDAB committee should have a separate meeting with the administrative level in DG EAC. Moreover, venues for collaboration would be explored between DG EAC and the WorldDAB organization. Within the EC bureaucracy, DG EAC and DG INFSO would work toward a more coherent division of labor within the context of the new EU audiovisual

216 Two of those studies could be relevant to digital radio: one on the impact of technological and market developments for the future audiovisual sector, and another on the development of new advertising techniques. A third study would examine the impact of measures to promote the distribution and production of European television programs.

policy envisaged by DG EAC.[217] The political activities of industry have the potential of changing the resistance within the EC toward digital radio.

In order to proceed, the EC stated that it would need assistance from the radio industry in the form of detailed information and a common position from manufacturers, including views on actual problems and barriers to the implementation of DAB. It would also need contact information on national officials (both from government and regulatory authorities) responsible for the implementation of DAB in Member States. However, EC representatives stressed that this political activity would depend to a large extent on the opinion and reactions from Member States on those issues that remained essentially within the competencies of Member States and where the principle of subsidiarity applied.

Furthermore, the EC stressed that the objective would be to create a dynamic for digital radio more broadly. In mid-January 2001, the EuroDAB committee organized a meeting of 10 of the biggest manufacturers to discuss the problems encountered in the implementation of DAB. 'Manufacturers need a pan-European market and volume sales to cut down prices. What many manufacturers see is a patchwork of license regimes. By putting together all the manufacturers and then doing a position we managed to convince DG INFSO that an action plan is required at that level' (WorldDAB, interview, 2 March 2001).[218] Proposals to combine DAB with GSM handsets, and to have DAB operated with a GSM service, had emerged (WDABF, 2000c). The digital radio industry may take advantage of the volumes and the components within the DAB receiver to keep the costs of a combined product down. '[To survive] DAB either has to take advantage of a world that has changed and adopt a niche or be adapted into a multimedia [technology like] UMTS,' said an EC official (CWI, 2000).

---

217 DG INFSO announced that it might include the DAB transmission standard in the list of standards that EC would publish as part of implementation of the broader telecom package. 'Little by little, with the lobbying we have been doing, we have managed to have radio more included in the picture of the framework directives, which is an important step forward. For example, the EC has agreed that they will list DAB among standards in the *Official Journal of the European Communities*. This is an important recognition. It is, however, not clear when that would happen. It is something that is envisaged to be done in general – not DAB in particular' (WorldDAB, interview, 1 March 2001).

218 The support of not only specialist manufacturers but also major industry actors would be central to achieve affordability and availability. 'What you really need is a big player like Sony to throw its weight behind the sets, but is has not even got around to thinking about it yet. It will eventually, because digital is the future', said an EMPA radio chief executive (MW, 1999).

In 2001, the European digital radio industry honored the 100th anniversary of Marconi's successful wireless transmission across the Atlantic, from Poldu (Cornwall, England) to St John's (Newfoundland). The EuroDAB consortium and its members continued calling on Member States and the EU institutions to encourage the development of DAB radio in the information society to allow European citizens to enjoy the associated cultural and entertainment services of digital radio. In early 2002, a conference hosted by Digital One in London gathered representatives from around 25 consumer electronics producers, car manufacturers and retailers. There were also increasing attempts from the satellite DAB (S-DAB) providers to gain much-wanted frequencies. There were few signs from the EU audiovisual authorities, though, that they would specifically point to the importance of digital radio and DAB in legislative documents beyond 2001.

Summary

The lack of a concerted European implementation strategy and a clear business model, shortage of frequencies for digital radio and the absence of a pan-European regulatory framework contributed to problems in the transition to digital radio in Europe. Parts of the EC were reluctant to support digital radio since there was no visible import competition threat ('content imperialism') from the US, a lack of a strong and unified political front on the part of manufacturers, a presence of what DG INFSO regarded as alternative technologies that they already supported (for example, DVB, Galileo and UMTS), and no clear institutional opportunities for the EC to intervene actively. In addition, the EC did not expect EU governments to support digital radio in the absence of visible economic and political benefits. The driving force behind the development of the Eureka 147 (DAB) system has been the public broadcasting sector.

The radio broadcasters lacked strong support from manufacturers and were not able to persuade the EC that there was any imminent import-competition challenge. The investments that have been poured into R&D, infrastructure building and coordination reduce the risk that a non-European competing technology will substantially penetrate the European market in the near future. Supporters of digital radio industry in Europe argued that the problem was not technological performance but lack of regulatory certainty and political success at the European level. This argument, however, was evidently not sufficient in order to generate positive distributional consequences for the radio sector in competition with the *grand projets* (DVB, Galileo and UMTS). The WorldDAB Forum lobbied to persuade national and regional administrations and EU institutions to provide support in order to ensure the viability of the DAB technology infrastructure. It stressed that the relative lack of political commitment on the part of the authorities had impeded a coordinated introduction strategy for DAB (the absence of radio policy issues on the agendas of the EU institutions and of harmonized frequency allocations for DAB across Europe, which its proponents stressed was needed for the transition from FM to DAB).

The electronics manufacturing industry and the EC (and possibly national administrations) were cautious with regard to investments in high-technology products and administrative support in the wake of the endorsement of the failed MAC and HD–MAC standards for high-definition television in the early 1990s. The WorldDAB Forum/EuroDAB consortium did not perceive the EC to be a neutral political actor (for example, its 'technological neutrality' argument towards DAB) but a negative one. They argued that other technology infrastructures, such as GSM, would probably not have succeeded without the direct intervention of the EC to harmonize frequency allocations in the late 1980s. Digital radio was content-oriented and suffered from a lack of significant investments in and supply of programming, which resulted in a weaker European dimension due to the lack of a common language. Moreover, the European radio industry had a relatively weak position in the structure of international competition and consumer electronics manufacturers were reluctant to invest in hardware. Altogether, this situation reduced the expected institutional and domestic returns on the part of EU institutions and governments and thus their willingness to lend ideological, legal and political support to the DAB technology infrastructure and its proponents.

8. Concluding evaluation

In Lisbon in March 2000, the EU heads of state and governments made a long-term economic-strategic commitment to transform the EU into the most competitive and dynamic knowledge-based economy in the world. The development of advanced technology infrastructures is an integral part of this ambition. This study set out to specify conditions under which the EU, and particularly the European Commission (EC), makes policy choices with respect to large-scale industrial projects, and to explain variation in the outcome, that is, in EU high technology policy. This concluding chapter summarizes the development of satellite navigation, wireless Internet, and terrestrial digital audio broadcasting. It will assess the observed variation in EU high technology policy. Furthermore, a comparative political-economic analysis is performed guided by the analytical framework. Finally, the last section of the chapter provides an interpretation of the empirical results and concluding remarks.

8.1 RECAPITULATION OF THE BOOK

A general proposition was derived from the basic demand-supply model of industrial policy creation. When industrial policy is provided, this proposition is complementary to the propositions that are derived from the more refined model. The basic model of high technology policy states that if there is an economic incentive for producers and interest groups to demand and for the public bureaucracy to supply such policy, it will be provided.

The technology-intensive infrastructure projects that were examined in this study, however, enjoyed different levels of support through EU political-economic intervention despite sharing a number of seemingly similar characteristics. First, the EU transport ministers have supported, and granted a central role to the EC in organizing, satellite navigation and the Galileo project. It benefited from significant public financial support, although some ministers criticized aspects of the overall project. In late March 2002, the EU transport ministers released significant public funding for the development of the Galileo infrastructure and reached an agreement on the regulation of the management structure for the program. Second, in the case of UMTS, the EC together with the Member States elaborated and supported a comprehensive regulatory framework, dedicated to mobile communications and wireless Internet. Finally, the transition to terrestrial digital audio broadcasting in Europe (DAB-T) has not enjoyed any similar political and public financial support, even though its supporters have pointed to a similar potential to that of current mobile communications. In the audiovisual broadcasting sector the EC and the EU supported the attempts to introduce analog high definition television in the past (which failed), and, interestingly, they have provided political support for the digital video broadcasting (DVB) consortium and project for a number of years.

The European Satellite Navigation Program: Galileo

In the early 1980s an interest for Global Navigation Satellite Systems (GNSS) gradually emerged in the international civil aviation community. Its members were searching for ways to increase airspace capacity, provide substantive economies in flight operations and further enhance the regularity of air transport. It became clear at an early stage that a worldwide and centralized operating structure for GNSS would not be feasible due to entrenched positions and vested interests. Instead, the major actors promoted regional operating structures and interregional coordination through bilateral agreements, thus opting for 'globalization via regionalism'. In particular the US, which had a *de facto* world monopoly, did not want to relinquish political and institutional control over its GPS constellation as a way to create an international, civilian-controlled public utility with a coordinating role of the International Civil Aviation Organization (ICAO). Since the US government has world hegemony in the field of satellite navigation and positioning it has promoted GPS as an international public good.

In the 1990s, Europe was still essentially a non-entity in satellite navigation and was under pressure from the US GPS infrastructure, which could achieve a stronger entrenched position and institutional deadlock in world and European markets. It would be difficult for European industry to capture those markets in the future unless swift actions were taken. The development work of the European Tripartite Group (the EC, ESA and Eurocontrol) resulted in more focused efforts to promote an independent European satellite navigation capability for communications, navigation, surveillance and transport applications. In mid-1994, DG Transport and Energy (DG TREN) launched the first official dedicated report on satellite navigation in Europe with the view of rallying political support in Europe behind the first-generation of GNSS (GNSS-1). It strategically committed to satellite navigation at an early stage and thereby signaled its determination to actively support the development of a European capability in this field.

The major EU countries and their domestic industrial interests in the space and air traffic control industries invested in the so-called Egnos program, which technically was spearheaded by ESA. European policymakers, industrialists and national experts considered that it would be crucial to engage in a build-up of satellite navigation as early as possible to keep up with developments in the US and elsewhere, and to gain experience and access to technology. During the period between 1994 and 1997–1998 European organizations were busy raising funds for this European GNSS-1 project. All the resources that went into this development implied that those who invested wanted to protect their investments. Sunk costs and capital entrenchment reduced the incentives to abandon the project. The distribution of contracts throughout Europe implied that both governments and economic actors defended the project, with the exception of Denmark and Sweden, which did not participate in Egnos for economic and technical-strategic reasons, respectively. This is a good illustration of how infrastructure projects generate various types of coalitions.

Around 1998, the political and industrial debate on the creation of European GNSS-2 capabilities intensified. While European Air Traffic Service (ATS) providers and governments invested in GNSS-1/Egnos, a consensus emerged that Europe had to allocate resources to GNSS-2 and start developing an independent high-capability satellite navigation system to participate in world markets. During the Gulf War and Kosovo crisis, some European countries were displeased with the total reliance on the US, which, although sharing intelligence gathered by satellites, did not grant direct access to imagery and other classified data. Furthermore, the push for satellite navigation was placed in the context of the weakness of European electronics and components industries. The EC considered that there was a window of opportunity for a market-pull development of its space industries, which were perceived to be in something like a crisis due to lack of large projects. European industry could gain a foothold in the high-end market segments by a timely entry into the satellite navigation market and a large-scale project could protect and foster the employment of highly skilled personnel and create new jobs in a high-capability industry.

Eventually, a group of political and industrial patrons rallied behind the Galileo project and the plans to promote European GNSS-2 capabilities started to gain momentum in the period 1998–2000. According to the EC, it was clear from the outset that the US would not be interested in a 'real transatlantic partnership'. EU high technology policy to support satellite navigation came to follow an infant-industry argument in that policymakers and industrialists considered there was a need to foster a new high-capability industry by temporary subsidies and political backing. EU high technology policy provision to support the development and implementation of satellite navigation in Europe has similarities with European developments in the space launch and commercial aviation markets, which were also promoted on commercial and infant-industry grounds. The Arianespace consortium succeeded to grab more than half of the world launch business after the US declined to offer Europe access to its technology, and Airbus Industries achieved parity with Boeing in certain civilian aircraft markets.

The majority of transport, industry and research ministers in the EU, together with the EC, ESA and Eurocontrol standing organizations and the space industry, supported the initiative. The Galileo project, however, ran into political controversies and delays starting in late 2000. Some major EU countries such as France, Italy, Spain and Portugal were prepared to proceed with the project based on the need for a European-controlled system without any major conditions attached. They supported the idea that Galileo would be a public service to ensure commercial and strategic independence from the US GPS constellation. Germany, the Netherlands, the UK and Denmark (and to some extent Sweden), on the other hand, preferred a firm commitment from industry to help fund Galileo early on. They stressed that Galileo should be a civilian-controlled system and run on a purely commercial basis. In particular the UK and the Netherlands expressed similar arguments as the US government, that is, the Galileo system would have difficulty competing with the GPS infrastructure that already was operational and that provided basic services free of charge. In April 2001, a consensus was reached among the contending EU governments that Galileo should be a civilian-controlled system under private sector control. The Netherlands continued to call for a clearer official wording that would ensure that industry would be in charge of the development and commercialization of Galileo.

In June 2001, the EU Galileo program received further support by EU transport ministers who called on DG TREN to elaborate an interim institutional structure. In 2001, the EU's Galileo program had gained such a momentum that it was difficult to envision the EU backing away from the project. The cost of the Galileo program was estimated at around Euro 3.25–3.6 billion, depending on the source. Prior to the agreements between EU governments on Galileo in April 2001, the EC succeeded in convincing the European satellite industry the need to announce that they would commit significant funding, although industry officials stated in private that this was not a binding commitment. At the end of 2001, national ministers to ESA approved significant funding for the development phase of the EU Galileo program (Euro 550 billion). The corresponding commitment by the EU transport ministers was delayed. As part of one of the 'priority actions' ('Connecting European Economies') identified by the European Council meeting in Barcelona in mid-March 2002, the European Council stressed the importance of the Galileo program in the field of transport. It welcomed the progress on Galileo and asked the EU Transport Council to take the necessary decisions in late March 2002 with regard to both the funding and launching of the program and the setting-up of the Joint Undertaking in cooperation with the European Space Agency.

By this time the German government had already reached a cabinet decision to commit public funding for the further development of the Galileo program. Shortly, thereafter, the UK government announced that it would reduce its opposition to the required public funding, whereas it reiterated its position that it had to be developed as a purely civilian program, that economic actors that would be considered for contracts had to be excluded from management structures, and that the Galileo program should not conflict with NATO capabilities. The Dutch government, in particular, shared these viewpoints, and to some extent Germany and some smaller EU Member States, such as Denmark and Sweden, although Sweden was interested in an integrated security policy framework for Galileo. Arguably, the main reasons for the variation in government positions and preferences were (1) their relative Atlanticist orientation combined with their relative association with the British versus a more continental model of European integration and of political economy, and (2) the relative expected benefits for major domestic economic groups combined with past investments in GPS infrastructure (including defense infrastructure) and preferential commercial contracts with the US GPS government and industry. In the case of Finland and Sweden (which had weaker domestic industries), the expected strategic-technical benefits deriving from better coverage of a high-accuracy Galileo satellite navigation and positioning system resulted in a favorable general position despite concerns about the public cost and management structures.

Wireless Internet: UMTS

The European wireless industry is one of the few high technology fields where Europe can boast of being a world leader. The growth of mobile cellular communications has been spectacular. Compared to the planned satellite navigation system where Europe was a non-entity in the 1990s, European industry has positioned itself strongly in mobile communications worldwide through the GSM infrastructure. In the 1980s, representatives of the European telecom industry and regulatory community started to explore the idea of an integrated solution for a second-generation (2G) mobile cellular system. Back then, the EU backed mobile communications to promote a pan-European solution through increased regulatory certainty (for example, support for pan-European standardization, administrative support and consultation, and a coordinated frequency strategy). The main objective was to replace existing analog systems and thereby harmonize the European market and help industry to create economies of scale in terminal equipment and network economies in services.

A milestone in the evolution of EU high technology policy and strategy was the establishment of the UMTS Task Force in 1995, which was convened by DG INFSO in order to generate a more coherent approach to mobile communications in Europe. This exercise gathered the major actors in the field of wireless standardization, R&D, regulatory policy and corporate management. A major strategic question was how Europe could ensure that its expertise and emerging leadership could be maintained and incorporated into the planned IMT-2000 umbrella concept promoted by the ITU. Industry called for an appropriate regulatory framework that would ensure investment confidence in mobile Internet. In the UMTS Task Force report produced in 1996 Industry officials argued that high technology policy for wireless Internet would be a central element in order to further strengthen European technology, industrial competitiveness and job creation.

The strategic goal, which was supported by the EU institutions and the private sector, was to extend the competitive advantage and the international footprint that the European industry and operators had gained with the GSM family of standards. In 1996, the UMTS Forum was established as the central body charged with (1) the elaboration of European policy toward market implementation as well as with (2) the coordination of different views in Europe on regulation (licensing and frequency management). This platform provided the EC with political support functions required to justify the efforts, to gain information, to increase its regulatory reach and bargaining strength with governments. In December 1998, the EU ministers responsible for telecom matters eventually approved a European strategy for wireless communications and wireless Internet, which formally was enforced in early 1999. The auctioning of UMTS licenses in Europe generated substantial revenues for some governments. It likely represents the largest financial reallocation ever from the private to the public sector. By March 2001, DG INFSO estimated that European industry had spent around Euro 130 billion ($117 billion) for 3G wireless communications licenses and that operators had spent at least as much to prepare the rollout of new services. There were substantial economic and political issues at stake.

Concerns started to emerge in the European wireless sector around 1995–1997 that Europe was failing to establish effective influence over and contribution into global standardization work. A central strategic issue here was whether the Japanese industrial activities and regulatory push by the Japanese government would result in a successful technology, whether such a technology would be compatible with the European GSM infrastructure and whether Japan would align with US industry. The Japanese government pushed for an earlier market implementation than Europe. In early 1998, the

European mobile communications industry and non-European actors in the European Telecommunications Standards Institute (ETSI) opted for a technology proposal called UMTS Terrestrial Radio Access (UTRA). Japanese industry backed this technology choice, which was different from the CDMA (code division multiple access) technology developed by Qualcomm. It would require major investments in infrastructure and mobile handsets, which was supported by the wireless manufacturers. Moreover, the European technology would operate at around 2.1 GHz, while the current GSM technology worked at 900 MHz and 1800 MHz, imposing additional challenges on infrastructure, design and planning. In sum, Europe's choice to focus on access to the lucrative Asian markets was a high-risk game that could either help to maintain its leading international position, or lead to shifts in the international division of labor in that Japanese and US industry would gain market shares with alternative technologies.

Terrestrial Digital Audio Broadcasting (DAB-T) in Europe

The European radio market is undergoing a transition toward digital radio, which in the long term is set to replace the existing AM and FM audio broadcast services. The development of the DAB-T system started in 1986–1987 with the objective of gradually replacing analog FM radio and developing radio broadcast services in CD quality. Moreover, the goal was to develop a standard that would be promoted to formal international standards organizations. Digital radio is moving beyond merely digital radio transmission into a multimedia system, which is able to transmit texts, data and moving images simultaneously (for example, listening to audio services while looking at weather or road maps), and to receive them interference-free in moving vehicles. In 1994, the ITU formally recognized DAB as a world standard for sound terrestrial and satellite audio broadcasting, respectively, to vehicular, portable and fixed receivers. Its supporters have pointed out that the Eureka 147 DAB system is a European standard, which has been partly funded by taxpayer money, and that it represents a case of European technological excellence. Several European countries, Australia, Canada, China, India and other parts of the world have tested the DAB technology and proceeded with deploying the system. At the same time, market implementation has been slow in Europe. Interest groups in the radio industry argued that Europe had to demonstrate the potential of the DAB system to the world, which, according to those interest groups, requires public support at national and European levels in order to foster a coordinated regulatory framework that, in turn, would further stimulate investment, production, implementation and user acceptance.

8.2 Variation in EU High Technology Policy

The categorization of EU high technology policy (1995–2001) divides it into three categories: 'EU political support', 'EU institutional coordination' and 'EU financial support'. Table 8.1 summarizes information on EU political support for the infrastructure projects examined in this study.

Table 8.1 EU political support

Technology Infrastructure	Galileo	UMTS	DAB
EU Political Support	1998–2001 4 EC COM documents 2001 Council Resolution* 1999–2001 EU Council Conclusions: Cologne (1999), Feira (2000), Nice (2000), Stockholm (2001), Laeken (2001) (1999 EP Resolution) Logistical support to user groups*	1997–1998 3 EC COM documents 1999 Council Decision 2001 EU Council Conclusion: (Reference) Stockholm (2001) (Meetings: EC President, Commissioner Bangemann with industry officials) 1997 Presentation to the Council	1999 1 EC action plan blocked 2001 Digital radio included in the Telecom Package (2000 EP Hearing) (Request to meet Commissioners Liikanen and Reding declined) (2001 Digital Broadcasting Experts Group?)
EU Political Support	High	High	Low

Note: *To this should be added the Resolution that ESA adopted in Fall 1999 on its contribution to Galileo, endorsed by the ESA Member States. Also, the EU prepared a Draft Regulation on a Galileo Joint Undertaking in Fall 2001 and DG Research produced a communication on space policy in December 2001 that supported Galileo.

The category of 'political support', the measurement of which is dichotomous (low = no endorsement, high = endorsement), refers to whether the technology infrastructure project in question enjoyed formal political endorsement through dedicated official EU communications with collegiate support from the EC services, with subsequent adoption at the EU Council of Ministers level, at least in the form of a Council Resolution. (Other events are included within parentheses only in order to enrich the categorization and allow for some further comparison.) Both Galileo and UMTS received substantial political endorsement from the EU (scoring high). No EU policy document dedicated to DAB, however, was endorsed at the collegiate level within the EC services (scoring low). Secondly, the concept of 'institutional coordination' refers to the extent to which the EU institutions, and in particular the EC, were granted a centralized administrative and management role in decision making arrangements (Table 8.2).

Table 8.2 EU institutional coordination

Technology Infrastructure	Galileo	UMTS	DAB
EU Management	Steering Committee, Program Management Board (PMB), Galileo Interim Support Structure, Joint Undertaking		
EU-Private Sector Interaction	1998 GNSS-2 Forum	1995–1996 UMTS Task Force 1998 UMTS Globalization Group	(2000 DAB Conference 2 days)
EU Institutional Coordination	High	Medium	Low

Comment: *One example of logistical EC support is that provided to the European GNSS Maritime Advisory Forum (EGMAF)/Maritime User Forum from 1998, which in 1999 changed its name to the European Maritime Radionavigation Forum (EMFR).

This category includes a qualitative assessment of the nature of interaction between the EU institutions (and the EC in particular) and the private sector. The information is ranked along a qualitative ordinal scale (low, medium, high). When it comes to the Galileo project, the EU Transport Council delegated a central administrative role to the EC in the definition and development of the project. In addition, the EC had influence over the policymaking and technical decision making within ESA and through the interaction structure that has evolved between the EC services and ESA within the context of the elaboration of a EU strategy for the space sector (scoring high). Regarding mobile communications (UMTS), the EC is a formal observer in ETSI but was not granted a central administrative role aside from its regulatory powers (scoring medium). The EC has no formal administrative role in the digital radio sector (scoring low).

Table 8.3 EU financial support

Technology Infrastructure	Galileo	UMTS	DAB
EU Funding	RTD Projects Framework Programs 4 and 5 31.67ME (68.67ME) (1) + 59.45ME (2) =**91.12ME** (3)	RTD Projects Framework Programs 4 and 5 **94.76ME** (4)	RTD Projects Framework Programs 4 and 5 **11.42ME** (5)
EU Financial Support	High >50 ME	High >50 ME	Low <25 ME

Comment: (1) This data excludes financial contributions from ESA, which, however, should be regarded as public funding. Moreover, the EU funding for Egnos up to the end of 2001 is excluded, which amounts to Euro 68 million. (2) The financial commitment of the EC in feasibility studies and related activities for Galileo up to the end of 2001 is included. (3) This total amount excludes GNSS-1 and GNSS-2 complementary Trans-European Networks (TEN) studies, to which the EC contributed around Euro 6.9 million up to end of 2001. (4) This amount excludes the financial support of the EC to standardization activities in the European Telecommunications Standards Institute (ETSI). It also excludes investments by the European Investment Bank (EIB) in wireless projects, which may amount to US$3.9 billion. (5) No data has been found on funding of DAB in the DG EAC's MEDIA (audiovisual sectoral funding) programs in the 1990s or beyond.

Finally, EU 'financial support', also ranked along an ordinal scale, refers to the relative financial support the project enjoyed out of the EU budget and focuses on EU funding for research and technology development (RTD) projects and related financial commitments (Table 8.3). (See Lembke, 2002d, for more details on EU financial support for each project.). Both the Galileo and UMTS infrastructure projects benefited from high levels of EU funding.

Substantial funding has also been channeled into satellite navigation and Galileo. Again, the real amount of EU funding for Galileo is bound to be considerably higher than the data available on EU RTD funding. A number of pilot projects were initiated at the end of the time frame of this study; however, the level of funding was not available at the time of concluding this study. The data included in Table 8.3 incorporates announcements from the EC that Euro 37 million would be available from the EU GROWTH program and its subprogram 'Sustainable Mobility and Intermodality' under the 5th Framework Program for investment in Galileo. Together with the 'Intelligent Infrastructure and Mobility Management' subprogram (part of EU funding for the 'Transport and Tourism' program under the 4th Framework Program), EU funding for satellite navigation amounts to at least Euro 91 million up to the end of 2001. Moreover, significant amounts have been invested in the Galileo project through ESA. When it comes to DAB-T and digital radio, this project has received considerably less EU RTD funding. In addition, the so-called EU MEMO project, which represents Euro 5.59 million of EU funding, has been included although it is also directed at GSM. Finally, in March 2002 EU transport ministers authorized the release of an additional Euro 450 million, which would be complemented by funding commitments on the part of the ESA and European ministers for research and space affairs.

Overall industrial competitiveness, strategic ownership, employment, rationalization of ground-based air navigation infrastructure and more efficient and integrated transport systems in Europe provided major arguments for high technology policy. A European satellite navigation system would allow industry to provide a number of services, including satellite-based tracing and tracking services, communications, sophisticated route planning for aircraft to improve the efficiency and safety of air traffic control, highway traffic management and ocean shipping navigation. A European system would complement, as well as compete with, the US GPS *de facto* monopoly in the market for satellite-based services. In the 1990s, GPS technology and services already penetrated the European and overseas telecommunications, maritime and automobile industries and markets. The European satellite hardware industry stressed the importance of satellite navigation, which would have socio-economic and industrial benefits of such a magnitude that EU high technology policy was required and desirable. European firms pointed out that the US industry received substantial funding from military sources and that the Japanese industry also benefited from government-supported space programs. Table 8.4 summarizes variance in EU high technology policy through a qualitative assessment of the categories of the dependent variable. In this scheme, high technology policy is ultimately either high or low. It is ranked 'high' for UMTS since it has no low values but two high-value scores.

Table 8.4 A qualitative assessment of EU high technology policy

Policy Sectors	Transport	Telecom	Audiovisual
		European Projects	
Categories of High Technology Policy	GALILEO Satellite Navigation	UMTS Wireless Internet	DAB Digital Radio
EU Political Support	High	High	Low
EU Institutional Coordination	High	Medium	Low
EU Financial Support	High	High	Low
EU High Technology Policy	High	High	Low

The argument that European industry should benefit from high technology policy to obtain a more competitive balance or level playing field has been central throughout the development of satellite navigation in Europe. Industry forwarded concerns to the EU and government agencies at national level that US companies would most likely be given privileged access to technical specifications and planned modifications, thus ensuring US industries a commercial advantage over foreign competitors. In short, GPS enhancements were likely to be disclosed first and foremost to US companies, putting European firms at a competitive disadvantage. In this context, industry argued that public support to develop a European satellite navigation capability was not only an option but also a strategic necessity. Moreover, the creation of new industries would retain highly skilled employees in European industry and organizations would allow them to upgrade their skills and would generate new jobs in Europe. Participation in the world satellite navigation market would permit European industry to develop technological capabilities and long-term commercial and industrial leverage.

Moreover, DG TREN and the space manufacturing industry argued that Europe could not allow its transport systems to depend on satellite technology controlled by non-European, governments and industries. The US controlled the GPS system (for example, possible jamming and encryption) and the EU (and countries such as Japan, and more recently China) has an interest in gaining control of the data source. A major drawback of the GPS system, according to the EC and industry, was that the US Administration did not provide legally binding guarantees of its accuracy, whereas a future Galileo operator would provide such guarantees. User communities such as commercial aviation, rail and marine traffic communities, insurance companies, and so on, would need such guarantees before they would heavily invest in and depend on satellite navigation. The EC thus strongly pushed for a European-led component to a global satellite navigation system to balance the commercial and strategic US monopoly.

The US tried to dissuade foreign governments from investing in their own systems to retain its monopoly as a means to promote its domestic industry in world markets. DG TREN reckoned that a main added value of Galileo would be as an insurance policy in the event US military authorities would decide to charge for access or make it unavailable to users and commercial interests. A global navigation system financed and operated by one nation's military and other organizations, it argued, was by definition unstable. Europe could not rely on foreign entities to provide open access to GPS satellites. Finally, support of satellite navigation would promote European transport infrastructure and the consolidation of the European markets.

Regarding the wireless sector, the political activity behind the promotion of wireless Internet in Europe in the early phase included (although to a minor extent) arguments about technological independence. The US wireless manufacturer Qualcomm sponsored CDMA technology, which also specified the use of the GPS system to synchronize base station timing technology. At times in the early phase, manufacturers argued that European mobile communications infrastructure should not depend on CDMA that had been developed for military purposes in the 1960s. Instead, Ericsson AB and Philips Kommunikations Industrie AG targeted non-synchronous CDMA radio technology for base station subsystems. The idea that Europe would roll-out infrastructure that could be disabled by a command from the US Department of Defense (DoD) gave them a political reason to pursue the slightly different W-CDMA technology. This argument was complemented by the commercial logic to differentiate their technology from alternative solutions promoted overseas through a homegrown solution.

However, making the case for the importance of wireless technology by reference to military technology and thus couching the analysis in realist terms distracts more than it helps. The neo-mercantilist literature made arguments about the importance of certain key civilian technologies in times of tight military budgets and potential 'spin-on' effects. It seems more plausible, though, to assume that the EU's and the EC's support for 'their' wireless manufacturers and technologies has been driven by welfare maximization interests. In the case of wireless, the EC has probably been more concerned with job creation and Europe's overall competitiveness in high technology industries than with safeguarding access to various critical military technologies. If the 'spin-on' logic drove behavior in this domain, for example, one would perhaps expect the US government to play a much more prominent role. In fact, one of the instances of direct US government involvement in this area was its pressure on China to build a 2G CDMA network based on Qualcomm's technology. A realist 'spin-on' logic suggests that states should be wary of exporting dual-use technologies, especially to potential future competitors.

DG INFSO viewed mobile communications as one of the critical technologies that could support economic self-determination, job creation and the overall competitiveness in high technology industries in Europe throughout the value chain, from components and equipment manufacturers to network operators, service providers, software designers, and so on. The European mobile communications industry clearly favored EU high technology policy in order to reduce regulatory uncertainty and to facilitate economies of scale in equipment and network economies in services ('network effects'). A major strategic objective was to maintain the European leadership in mobile communications by promoting an export-oriented technology that was compatible with the current GSM infrastructure deployed across the world. Therefore, it was important to try to ensure that Japanese or US industries did not succeed in gaining the upper hand in the development of future mobile technology.

The European economy and industry depended on GSM. A loss of significant market shares, or the penetration of the European market by non-European technologies, would constitute a real industrial-economic problem. This situation helps to explain the strong reactions by industrialists and policymakers in Europe when US industrial interests used the US government and regulatory bodies to raise criticism against the European way of doing things. DG INFSO argued officially that Europe had only a 'narrow window of opportunity' in light of emerging international competition from Japan and the US. A strong home market would help European companies compete in world markets and allow them to consolidate the leading position that they enjoyed with GSM. Wireless technology was situated within the context of industrial competitiveness and the 'ideological' promotion of a 'wireless information society'. Industry wanted to repeat the success of GSM and hoped that UMTS would be the preferred global standard.

In 1997, DG INFSO estimated that the European mobile phone market would amount to around Euro 100 billion of annual revenues by 2005. The GSM operator community wanted Europe to select the technology with the greatest potential on the world market (and a global standard), align with Japan to avoid a political alliance between Japanese and US industry that would restrict European options, and ensure a continued strong role worldwide for the GSM industry. The leading European manufacturers argued that employment and exports would depend on whether UMTS would become a global success. A timely, consistent and harmonized approach in Europe could provide industry with the opportunity to get first to the market, build up a strong home base and gain market shares in terms of infrastructure and the expected Euro multi-billion markets for various applications.

The EU institutions regarded digital communications technologies as key contributors to the consolidation of the internal European market, industrial competitiveness, a balanced economic development, and the emergence of an information society and a 'new economy'. European DAB proponents argued that terrestrial digital audio technology (DAB-T) was the best digital broadcasting technology and stressed that it allowed for advanced services, mobile communications, the transmission of data, and so on. In short, DAB-T was envisaged as the only broadcasting technology available to permit the transition from the analog to the digital world. As was the case in satellite navigation and wireless Internet, industry officials argued that several hundred thousand jobs and substantial turnover would depend directly on the success of digital radio and the implementation of the Eureka 147 DAB infrastructure in Europe and beyond. Again, industry argued that Europe had a window of opportunity to promote the consumer electronics industry and the European economy. Therefore, there was a need for concerted European action to ensure a large-scale introduction of digital radio in order to secure its economic potential. The radio industry argued that there was a need to maintain the 'technology leadership' of the DAB system. Initially, DG EAC argued that ensuring that radio developments remained at the cutting edge of technology required a long-term EU and industry strategy.

The demand for a coordinated regulatory policy and framework (licensing and frequencies) was essentially the same as the demand expressed by the economic actors in the European wireless and space sectors. The EuroDAB consortium argued that the Eureka 147 DAB system would help the EU in its attempts to promote the digitization of the European economy and therefore was a natural target for EU high technology policy in light of market failure. Free-to-air digital radio would not take off by itself along market economic principles in the presence of market imperfections. It was a mobile technology, which would fit into the EU goal to support a 'wireless information society'. Moreover, the consortium claimed that radio was a relatively trusted medium and a key contributor to cultural diversity, to a citizen-oriented European society ('an enabler of local values'), and ultimately to democracy. Radio, the EuroDAB consortium argued, had positive externalities and its preservation into the digital world was a matter of pan-European importance. The EuroDAB consortium stressed that radio provided important services to European users.

In 1998, DG EAC acknowledged the importance of Eureka 147 DAB as a European system, which had been recognized as the standard for digital radio broadcasting, not only in Europe but also in many countries across the world. It compared the potential of DAB to the success of GSM. According to the EuroDAB consortium, the radio sector had an annual turnover in 1998 of around Euro 7 billion (split equally between license fees and advertising revenues). The audio retail market, with a turnover of a further Euro 6 billion, complemented this market. While employment in the mainstream radio sector was estimated at around 80,000, the total employment in the European industry was estimated at around 300,000 jobs in 1996.

8.3 COMPARATIVE ANALYSIS: INDUSTRY CHARACTERISTICS

The intention of a more refined analytical framework has been to allow a comparative political-economic analysis of the provision of and variation in high technology policy, thus integrating the different empirical stories into a more coherent whole through a theoretically guided analysis. The model presupposed that the position of economic actors in the structure of international competition (the exposure to import competition and relative domestic or international orientation and strength) has significant bearing on their organizability and preference structures (voice) and, in extension, the EC's ability and willingness to provide high technology policy. The EC need the political support functions of those actors who can provide the largest impact and benefits in line with the EC's search for institutional voids that it could fill in order to promote broader European preferences. When economic actors support the ambitions of the EC, the latter will benefit from their support and will use it in its strategic interactions and bargaining with EU governments and other relevant actors. However, when those economic actors on whom the EC depends have different preferences, the EC will be less willing and capable of counteracting or aligning their activities and preferences. Under such conditions, the EC will have more difficulty in moving ahead swiftly, although it may have identified an area over which it may be able to claim policy 'property rights' and thus may have a rationally based interest in formulating and providing high technology policy.

The import-competition argument was a salient feature in the satellite navigation sector. The satellite manufacturing industry and the European 'space technology community' more broadly actively played on the import-competition threat in European domestic markets. According to the EC, European industry's market share in the late 1990s was only around 15 per cent of the European market and 5 per cent of the global market. From the early 1990s, the satellite manufacturing industry argued that without an active European participation in the future GNSS infrastructures, and the ensuing dominance of non-European industries, the European space industry could be eliminated as a supplier of future international system(s). Lack of action would further strengthen the 'institutional deadlock' of the use of GPS in Europe. Europe needed to leapfrog in order to reach and potentially surpass the US dominance in satellite navigation.

The US GPS industry had already started to expand their activities and market share in Europe. Public authorities, European manufacturers argued, had to face this challenge by supporting the satellite manufacturing industry. Since Europe was largely a non-entity in the world satellite navigation market, European industry thus had marginal market shares and international operations in this field compared to US firms. However, the European space technology community (in particular ESA, ministers responsible for space affairs and satellite manufacturers) expressed concerns about the risks of being marginalized in overseas markets where European firms had established, or hoped to establish, some presence. The close relationship with ESA provided political support for the EC services to move ahead. The strong global position of the US GPS industry, and the relatively limited expected impact of retaliatory measures by other countries on European private sector actors if the EC/EU would actively provide high technology policy, offered European industry an opportunity to express strategic demands and call for substantial political support and high technology policy.

The European satellite manufacturing industry provided political support functions to the EC that the latter would find useful in its strategic-distributional bargaining with the EU governments and with the major user communities. The European space industry further facilitated a momentum behind Galileo in that the major companies were able to organize themselves at the industrial and political level through the establishment of the Galileo Industries SA consortium. The stronger and more internationally successful satellite manufacturers, such as Astrium, officially welcomed participation by non-European industry in the development of the Galileo infrastructure and associated technology. Other companies, such as Racal (Thales), who would not build the satellite infrastructure and would depend on commercial revenues from the services that Galileo will generate, stressed that there was a strong need for public procurement and public regulated services to create a captive home market. When Galileo ran into political controversies, the space industry announced that they would provide funds to the development costs of Galileo, which was used by the EC in their contacts with governments. However, manufacturers were, as were the major user communities, reluctant to provide significant private sector funding at an early stage, which was not helped by political delays in Europe.

European Air Traffic Service (ATS) providers (with support from 'their' transport ministers) supported the replacement of conventional ground-based infrastructure with the aim of reducing the costs of the operation of navigation and guidance systems, which represented considerable annual expenses. ATS providers, in particular, drove the first-generation satellite navigation capability in Europe, the Egnos project. Their support was important for the EC to bring about a political momentum behind a European satellite navigation capability. The more commercially and internationally oriented airspace user community, however, would intensify their criticism against the European Egnos project in the late 1990s. The civil aviation and maritime communities are the most internationally oriented user groups and are well organized at the international level. The Association of European Airlines (AEA) complained to the highest political levels in the EC bureaucracy and national governments that they should not pay unduly for the costs of the Egnos infrastructure since they would not gain from it and since other user communities would be able to use it. The AEA was less critical about the development of Galileo, which was viewed as a back-up system to current navigation guidance technology, but was unwilling to accept an 'unbalanced' distributional burden.

The AEA was concerned that the space technology industry appeared to gain preferential treatment at the expense of the European civil aviation user community. Likewise, the European maritime user community was concerned about the strong role of the space technology community in defining requirements for Galileo. Moreover, it was concerned about what they experienced as ignorance on the part of the EC regarding the importance of complying with established international regulatory arrangements. The maritime industry operates on a global basis, with many ships arriving in Europe and European ships trading with countries throughout the world. Recognition of Galileo by the International Maritime Organization (IMO) would pave the ground for and offer Galileo a unique opportunity to secure a competitive advantage over existing GNSS constellations. GPS and Glonass were recognized only for maritime use in open water areas (high sea and oceans), but not for safety critical navigation in coastal waters or harbor approaches. In sum, IMO recognition of Galileo would give it competitive added value; it would potentially gain a widespread acceptance in the maritime world and further strengthen industry and EU policymaking.

The European civil aviation and maritime communities supported the Galileo project but as internationally oriented user communities they remained cautious. They were both interested in multiple sources of navigation and did not want to see Galileo be introduced as a sole means of navigation, which would threaten their already made investments (capital entrenchment). DG TREN couched the political support for satellite navigation as an effort to develop a more efficient and integrated multi-modal transport infrastructure in Europe, that is, satellite technology should integrate many different modes of transport. It was not pleased that the aviation community continued to hold a dominant position in the debate on navigation; it tried to extend the scope of the project in an effort to mobilize other private and public sector actors.

In the wireless sector, economic interests were concerned about the aggressive Japanese push to obtain a 'first-mover advantage'. They also strategically played on the Japanese competitive threat that Japanese industry had been spurred by a regulatory push at home, and that it was at the forefront in the development and implementation of infrastructure and services. Leading European wireless manufacturers used such arguments to strategically align EU high technology policy with their own preferences. The European market had been relatively closed to Asian and US competition, but Japanese firms expanded their market and production presence in Europe. In the 1990s, the fixed network industry was still far behind the mobile industry and worried about the Japanese efforts.

Even the wireless operators had still a long way to go until they could call themselves truly pan-European. In many cases, they were still more focused on national markets. Manufacturers and operators agreed on the need for a coordinated regulatory policy. They opted for a European strategy for mobile Internet that capitalized on current GSM mobile infrastructure rather than going for a completely new technology infrastructure, or one that was more oriented toward the interests of the fixed telecommunications operators. The largest European manufacturers (Alcatel, Ericsson, Nokia and Siemens) had established good relations in matters such as trade policy and regulatory issues. They often produced common position papers through their corporate affairs offices in Brussels (except Alcatel), Paris, Stockholm, Helsinki and Munich. Industry and DG INFSO acknowledged that a common industry front made it easier for the EC to launch initiatives and gain political support.

The battle in Europe over the selection of radio technology was strongly influenced by the policy preferences of the Nordic Alliance (Ericsson and Nokia) to align with Japan despite concerns that Europe could lose out in competition with Japanese industry. They managed to pressure the EC to stress the importance of the Asian market. The selection of European technology was partly based on a strategy to differentiate technology and business models from major competitors on world markets. For those producers that had strong market presence in Asia, however, it was important to align with Japan. Japanese industry, which participated in EU R&D programs on mobile communications, had shown interest in the European process just as the European industry had been involved in the Japanese process, in particular Ericsson, Nokia and Siemens. The Japanese standards body ARIB (Association of Radio Industries and Businesses) and the dominant operator NTT DoCoMo announced that they would select a version of W-CDMA radio technology. There was a certain fear factor involved. European industry wanted to ensure participation in Japanese developments if Japanese firms would be internationally successful. The stronger players in international markets (Ericsson and Nokia) were the actors with the largest potential in supplying information and political support to DG INFSO and accordingly carried significant weight in the development of the EU strategy.

Equipment manufacturers are today powerful players due to their technical competence and international orientation. The EC need their support to move ahead with a coordinated provision of high technology policy. The EC was not able to cajole the producer industry to deliver an integrated radio technology solution in a timely manner. Though DG INFSO and various Commissioners put considerable political resources into trying to align European industry around a common technology proposal, and accepted the compromise that was reached in early 1998, it could not determine the behavior of major industrial actors. The 'globalization' of European manufacturers, technology and institutional arrangements provides an additional point of observation. The outcome in this regard reinforces the argument that those manufacturers with a presence in Asian markets dominated this process and that they were confident enough to take on competition from Japanese and Korean industries. The EC was concerned about losing administrative control over one of its major governance instruments (ETSI), which was given the task of supporting EU regulatory policy and formal standardization recognized under the EU legal framework. Without the support of the dominant manufacturers, the EC would not have been as able to elaborate a European strategy and it had to meet their corporate strategic demands.

In the trade policy arena, Ericsson and Nokia asked the EC to adopt a low political profile since they played a two-level diplomatic game. In other words, they played the European card to gain benefits from EU policymaking, while lobbying non-European governments and private sector organizations. The strong political push by the major internationally oriented manufacturers jeopardized the entrenched positions of the fixed operator industry and threatened to destabilize their business models. In the battle over 'globalization' within and outside ETSI in the late 1990s, it was clear that there was a conflict between more European-oriented and internationally oriented players. Entrenched positions, invested time and money, status and power were at stake. The more international players and the mobile community gained the upper hand. This story indicates how the contest between rival technologies in the context of European UMTS deliberations became an international issue that went far beyond Europe. In addition, it shows the strategic use of formal international standardization bodies by multinational corporations in the context of global competition in the information and communications technology (ICT) industry.

The supporters of globalization in the European wireless sector (both wireless manufacturers and wireless operators) argued that the fragmentation of specification and standardization work into several national or regional standards organizations and complex bilateral arrangements resulted in inconsistencies and risks to global roaming, to scale economies in terminal equipment markets and to network economies in services. The European and international landscape was different from when the GSM wireless infrastructure was introduced in Europe in the 1990s. There was political resistance among fixed telecom operators and the ETSI bureaucracy to an integrated global institutional arrangement and a strong European–Asian alliance with equal rights of the participating actors. However, the leading European approach was eventually characterized by an internationally oriented strategy aimed at ensuring the support of Asian industry and standards bodies. Similar European attempts such as the support of a European standard for high-definition television (HDTV) failed because the latter was a defensive standard intended to protect Europe against Japanese competition, whereas the introduction and implementation of GSM was promoted as a means to integrate European markets. To globalize from the start was a good insurance policy for internationally oriented wireless manufacturers and operators in order to gain a first-mover advantage and access to Asian and other markets, to seek to have other regions adopt their preferred infrastructure and technology, and to support economies of scale in terminal equipment markets and network economies in services.

The efforts and willingness of particularly Ericsson and Nokia to overcome resistance in Europe to increasing Japanese (and Chinese and South Korean) influence over wireless standardization suggests that head-to-head competition does not necessarily exclude significant international cooperation in a globally competitive environment. The prospect of vast Asian markets certainly prompted European manufacturers to seek a deal with Asian players and to prevent independent technology developments in Japan, China and Korea. GSM was the first viable digital wireless standard and is well entrenched. European manufacturers did not significantly fear that rival Asian 3G technologies could take over a large portion of the European

3G market. If anything, European operators and the EC may have been concerned about market fragmentation due to incompatible systems. The idea that European activities and ambitions in the wireless sector was a defensive approach would be consistent with the old idea of using standards as non-tariff barriers for protectionist purposes. Such an interpretation, however, do not pay sufficient tribute to the aggressive and internationally oriented approach of private sector actors in the wireless domain or to the specific nature of the support they received from the EU institutions. The European strategy for UMTS/W-CDMA should not be regarded as a story of defensive standardization to keep Asian competitors out of the European market.

Finally, the European consumer electronics industry and producers in the radio sector are relatively weak in the structure of international competition. The broadcasters in Europe still generate the bulk of their revenues from domestic markets and are relatively less European than even the fixed telecom industry. The radio sector actors could not muster the same resources and provide the same degree of private information, and thus political support functions, to the EC as the leading players in the mobile communications and space sectors. The digital audio broadcasting infrastructure has been adopted in Europe for terrestrial broadcasting as the replacement for analog FM technology. It had been included as an ETSI Standard but not yet as an ETSI 'core technology', which would give it higher status. Though the WorldDAB Forum agreed that DAB was a great opportunity for Europe, it argued that competitors could penetrate the European market if Europe would fail to coordinate its activities. The EuroDAB consortium, which thus consisted mainly of domestically oriented broadcasters and producers that were relatively weak internationally, called for strong political support. It had pointed to Asian competition as a serious challenge to the European consumer electronics industry. Other DAB systems had been developed in other parts of the world, notably the so-called ISDB-T in Japan and IBOC DAB in the US.

The defensive import-competition argument, however, was not played out often in the lobbying efforts of the European digital radio actors. The proposed ISDB-T (Japan) and IBOC DAB (US) did not constitute major import-competition threats in the 1990s. An additional factor was arguably that the major competition came from within the European marketplace, for example, in the form of terrestrial digital video broadcasting (DVB-T), digital broadcasting by satellite (DBS) and possibly UMTS. Deployment of broadcasting satellites that would allow adequate mobile reception competed for radio channels in the frequency bands between 1 and 3 GHz, which could threaten the terrestrial DAB system. The WorldDAB Forum had produced technical reports indicating that a satellite system using current technology would not be able to provide for mobile and portable reception. Rather, it was suggested that a combined satellite-terrestrial approach would have distinct advantages. DAB supporters argued that satellite broadcasters had delayed the frequency planning on DAB since 1999 and those broadcasters hoped to obtain the same frequencies that the DAB-T industry was interested in.

A main industrial-political and industrial-economic argument, expressed by the proponents of DAB, was that digital radio represented the last chance for Europe to enhance its competitiveness in the consumer electronics sector, which had a relatively weak international record in terms of competitiveness. Indeed, if the EU, national governments and national and European regulators would provide a stable regulatory and investment climate (and avoid piecemeal and uncoordinated implementation) the DAB story could well be as successful as the European GSM story, which was viewed as the European flagship in high technology industries. This was certainly quite an optimistic claim. The question, thus, was whether European public authorities would listen. However, the EuroDAB consortium was neither able to persuade the EC that there was any imminent Asian or US import-competition threat, nor that DAB and its supporters could be as successful a story as GSM. However, the EuroDAB consortium was neither able to persuade the EC that there was any imminent import-competition threat, nor that DAB-T and its supporters could be as successful a story as GSM.

The European radio industry was organized politically through the EuroDAB consortium. The WorldDAB Forum, its umbrella organization, provided a political platform for broadcasters and manufacturers in their relationships with the EU institutions and the EC. However, the organizability and political strength of the EuroDAB consortium was influenced by the unwillingness of the manufacturing community to commit substantial resources into the production of digital radio receivers. In other words, it had difficulties in generating a coherent industrial lobbying effort with political leverage due to the lack of an organized push by the manufacturing sector. The radio sector was relatively horizontal, with a large number of smaller companies and radio stations. The manufacturing community agreed on a 'common position' in early 2001. It could give a political impetus for the EC to look into digital radio in a more focused manner, although digital radio would not receive the same political, institutional or financial backing as Galileo or UMTS.

8.4 Comparative Analysis: Political Entrepreneurship

The 'political entrepreneurship' condition is divided into two components: bureaucratic cohesion and external support. The provision of high technology policy is more likely when the EC expects benefits in line with its institutional self-interests (a precondition to promote broader interests such as European welfare, employment and overall competitiveness in European high technology industries) and when it perceives that EU governments expect domestic returns. In other words, political entrepreneurship suggests that the EC will be more likely to support and provide high technology policy when it identifies opportunities to expand its institutional and political influence. In addition, the provision of high technology policy is more likely when the EC expects that EU Member States identify domestic returns (for example, support of projects that may increase government revenues, that may respond to national security considerations, or that may strengthen European capabilities from which governments and domestic economic actors could benefit, in particular those upon which governments rely for political support functions). If the EC expects that governments have the backing from important domestic economic interests, its resolve would be greater. Even in cases when a number of EU Member States oppose major parts of the overall EC proposal, the EC can continue promoting the project in a proactive manner if it expects that it can rally a majority of Member States behind a project in the future.

Regarding satellite navigation, Galileo represented a window of opportunity for the EC to expand its administrative control and regulatory reach through institutional and policy coordination. It argued that the EU institutional framework would enable a robust structure for the making of strategic decisions. The EU institutional structures were suitable considering the broad range of political, international, economic, industrial and security issues involved. The responsibility for satellite navigation was initially allocated to DG TREN Directorate for Air Transport. However, this portfolio was subsequently transferred to DG TREN Directorate for Trans-European Networks for Transport. (In 2001, the EC's Galileo Unit, which was incorporated under DG TREN B5 in the past, was transferred to DG TREN Directorate E: Inland Transport.)

DG TREN wanted to mobilize broad support for satellite navigation and not restrict the efforts to the aviation market and the associated economic interests. The aim was thus to interest other modes of transport to enhance momentum and partly to counteract the political power of airspace users. It acknowledged the opportunity to significantly rationalize the European infrastructure as well as make it more efficient. EU high technology policy to support satellite navigation would have distributional-strategic implications.

For example, the promotion of replacement of conventional ground-based navigation infrastructure would threaten the business models of suppliers of conventional technology. Furthermore, DG TREN promoted the idea of satellite navigation as the sole means of navigation, which was subsequently opposed by the civil aviation community and it thus modified this approach.

The Galileo technology infrastructure project thus constituted a critical technology that had the potential to significantly strengthen the European transport infrastructures. A European second, independent and interoperable satellite navigation system, DG TREN argued, would promote and protect European investments in important future markets. This represented a major window of opportunity for the EC. DG TREN expected that a large number of future EU regulations and decisions would rely on the availability of a certified Galileo technology infrastructure over which the EU would have control. In addition, such an infrastructure would function as a tool for the implementation of various European policies in different sectors, such as agriculture, environment, fisheries and regional development. It could also represent an important policy instrument for providing a legal framework for law enforcement in Europe. From 2001, DG TREN was also increasingly willing to debate the relationship between Galileo and different aspects of European defense and security policy, something that perhaps was facilitated by political pressure from US military authorities. DG TREN assured that the Galileo infrastructure and its satellite navigation signals would not threaten GPS military signals. At the same time, it also stressed that a large number of European defense systems (land, air and sea) would increasingly incorporate and rely on satellite navigation and positioning. It thus boiled down to the argument that Europe needed to reduce its dependence and increase its autonomy also in defense areas and applications.

The Galileo project would provide a major industrial opportunity for the space technology community, which airspace users criticized due to its unfair competitive implications for other industries. Commissioners Bangemann (industrial policy and telecom), Cresson (research) and Kinnock (transport) gave unconditional support to the European satellite navigation ambitions. This political support within the EC bureaucracy weakened attempts to resist DG TREN's policy proposals and demands for budget resources. Moreover, DG Research viewed Galileo as a central European federated project that would help strengthen the institutional and political linkages with ESA and, according to the former, provide grounds for a gradual integration of ESA within the EU institutional structure, much like the case of NASA in the US. The EC and ESA were involved in high-level efforts to develop a closer cooperation on and coordinated investments in space programs; Galileo was viewed as the main flagship in this context.

Apart from reducing Europe's dependence in transport and defense areas, and supporting the implementation of various European policies, a political agreement to provide the remaining Euro 550 million of the Galileo development phase in late March 2002 would thus stimulate European space and aeronautics policy. (The total cost of the Galileo program had been estimated at Euro 3.25–3.6 billion.) This objective was strongly supported by EU research ministers, ESA, the aeronautics and space industries, and a large number of Members of the European Parliament. DG INFSO also had interests in promoting satellite navigation. The Galileo project, and in particular the applications markets, could be defined as telecom-related applications and the cooperation between DG TREN and DG INFSO intensified in 2001, notably in the area of intelligent transport systems and the creation of a global wireless infrastructure including both UMTS and Galileo. The Galileo project largely sailed through the EC bureaucracy with strong collegial support.

Moreover, the examination of EU high technology policy in support of wireless Internet indicated that the EC (and DG INFSO that is responsible for mobile communications) used a number of governance instruments. Firstly, it facilitated the creation of an ideological platform for the new technology by arguing for its importance for Europe and the creation of a 'wireless information society'. Secondly, it helped to support coordination at European level by mobilizing industry and providing an institutional arena where relevant public and private actors could jointly discuss issues that needed to be addressed. Finally, it provided flanking support for 'European technology' overseas. EU high technology policy was officially justified on industrial (competitiveness and global market shares), technology (leadership) and macroeconomic reasons (employment and the provision of new services).

DG INFSO wanted to repeat the international success of past policy in support of mobile communications. This success provided DG INFSO with an expanded regulatory reach (although the coordination of licensing regimes and procedures remain in the hand of national governments). In the area of formal standardization, the EC has exerted pressure to align industry with the aim of generating European solutions that will enhance its administrative control and regulatory scope. In short, no evidence has been found that the ambitions of DG INFSO to support mobile communications met with any significant resistance. There have been hints at some minor tensions with DG Enterprise that stressed the need to retain some political control of mobile standardization to ensure that industry players provided solutions that met with EU requirements associated with the consolidation of the European market. DG INFSO was a relatively liberal-oriented bureaucratic actor in this regard but aired some concerns about the 'globalization' of the sector.

DG Enterprise, in particular, was concerned that the EC's institutional influence over ETSI, which is challenged by the internationalization of economic actors and markets, would further weaken and thus jeopardize opportunities to coordinate policymaking. ETSI has been important for the EC in providing formal standards that are important for European harmonization and the integration of the European market. The European wireless industry and policymakers supported a coordinated, outward-oriented approach in order to avoid fragmentation of the European and world markets, to promote the opportunity for lucrative market shares in the Asia-Pacific region, and to boost the European economy, employment and the overall competitiveness in advanced ICT industries.

The European UMTS/W-CDMA initiative represented a global ambition to capitalize on the success and footprint of GSM, which had become the international standard for second-generation mobile cellular systems. The EU policy for the wireless sector, and the role of the EC, arguably represents a new form of critical industry support and illustrates the distinctive role of the EU as a more organized political actor in structuring the context of global competition. The development of and policy toward UMTS represents primarily an offensive story of co-opting Asian developments in light of vast Asian markets, not a story of defensive standardization to keep Asian competitors out of the European market. Japanese manufacturers will export handsets to European markets since they will use the same technology as European firms; however, increased competition at home on the same technological platform is arguably a small price to pay for European manufacturers in light of the promise of huge markets in Asia and elsewhere. This outlook stimulated the EC to act since there were expected institutional returns both in terms of maintained regulatory powers as well as in terms of reputation (the association with a European high technology success story). In addition, most EU governments stood to gain from industry and regulatory support in the form of domestic financial and industrial returns and this further strengthened the EC's opportunities and resolve.

Finally, when it comes to digital radio the situation becomes more complex. With proactive support from DG EAC Director-General Pappas and administrative staff, an action plan was elaborated and distributed within the EC services in early 1999. While the proposed action plan in support of digital radio in Europe ran into difficulties due to the collective resignation of the EU commissioners in 1999, it was clear that the proposal was not well received by DG INFSO, which is among the most powerful actors in the bureaucracy. Moreover, the new political leadership of DG EAC, headed by Commissioner Reding, was not willing to support the transition to digital radio. In fact, Reding argued that the radio sector was doing well and that there was no risk of market failure, despite the arguments to the contrary by interest groups representing the sector. The responsibility of radio would largely be transferred to DG INFSO, but the latter bureaucratic actor did not seem to be willing to intervene politically. Even though the radio sector could gain more support it will be significantly lower and no way near the support that Galileo or UMTS received.

The external support condition for the provision of EU high technology policy suggests that in the face of expected political opposition from EU Member States, the EC will be less willing to risk confrontation and thus promote such policy. The capacity and willingness of the EC bureaucracy to intervene through high technology policy will be conditioned by whether EU Member States expect domestic political and economic returns. The mandate for high technology policy ultimately rests with the EU Member States. The support among EU transport and research ministers for satellite navigation was significant. Even the UK government, which had come to reduce its support for space programs within ESA in the late 1980s, considered Egnos and Galileo as a major opportunity for Europe. Denmark and Sweden remained outside the Egnos project for economic and technical-strategic reasons, respectively.

The ECAC (European Civil Aviation Conference) and EU transport ministers called on DG TREN to launch proposals for the development of a European satellite navigation capability in the early 1990s. They stressed that Europe should develop a component of GNSS-1, and more importantly to put Europe in a position to make a contribution to GNSS-2. Controversies would emerge about public funding of the Galileo project, what services it should provide and how it should be managed. In March 2002, the EU Council of transport ministers would meet in Brussels to decide on funding for the Galileo development phase (2002–2005) on 26 March 2002 and they agreed to commit additional funding to the Galileo project. In general, though, the transport ministers agreed that Galileo was of strategic importance. DG TREN and Commissioner Loyola de Palacio stressed that this would be 'the last chance' for the European transport ministers to show political willingness and to support their continuous statements that they recognized the strategic importance of the Galileo infrastructure.

Regarding wireless Internet, EU governments supported high technology policy to support a European strategy and regulatory approach. EU governments began to realize how important it was to get the UMTS wireless in operation also for their own self-interests. There is a lot of money in the form of taxes and user fees that would be coming into their coffers once it was in place and being used, not to mention the profits and income taxes derived from the companies manufacturing and operating the systems. There was some reluctance when it came to the allocation of frequencies. Italy, for example, had already allocated large portions of the assigned spectrum to military authorities. In addition, a number of countries that had experienced less growth in mobile communications up to the late 1990s were reluctant to agree to a speedy licensing and implementation process. In late 1998, however, a common position was formally agreed upon, which went into effect in early 1999. The government support in Europe was thus strong.

Finally, regarding the government support for digital radio, it has been weak in many EU countries. Also a country such as Sweden, which has achieved extensive coverage for digital radio, has been reluctant to open up for preferential treatment of public broadcasters and has thus reduced the incentives for the private sector to commit to digital radio. The UK has taken the most progressive approach toward the implementation of digital radio and has experienced relatively strong growth in revenues and the setting-up of new radio stations. In 2000, the UK regulatory authority appeared to become increasingly interested in digital radio, which can be seen within this context. DG INFSO admitted that it was reluctant to intervene since this would result in conflicts with national regulatory authorities, and would entail the risk of losing political capital in an area where DG INFSO did not perceive the benefits to outweigh the expected battles with national authorities. The arguments by the interest groups for the radio sector – that it was precisely the weak position of radio in Europe that triggered the need for high technology policy – did not convince the EC bureaucracy.

In summer 2001, European satellite digital broadcasters were invited to a meeting with the Digital Experts Broadcasting Group (DBEG), which consisted of representatives from the EU Member States. According to the EuroDAB consortium, the supporters of the Global Radio and WorldSpace systems claimed that they needed the planned frequency spectrum for DAB services more than the DAB-T industry. Their lobbying activity had, according to the EuroDAB consortium, delayed frequency planning since 1999. Moreover, it claimed that the political leadership and the cabinet of DG EAC had strong interests in satellite broadcasting, notably Global Radio that was supported by the Luxembourg government. Though it is difficult to ascertain the connection between DG EAC and satellite broadcasting interests, it is not perhaps too implausible to suggest that the support of major domestic economic interests in Luxembourg (and perhaps of Alcatel and WorldSpace that appeared to politically join Global Radio in Brussels) impacted on the prospects of the lobbying efforts of the DAB-T industry.

8.5 INTERPRETATION AND CONCLUDING REMARKS

This study set out to specify conditions under which the EU, and particularly the EC, makes policy choices with respect to large-scale technology infrastructure projects, and to explain the variation in EU policy for similar projects. It examined the pre-implementation and policy formulation stages.

A basic model of high technology provision was presented as a template built around three cornerstones: an incentive, lobbying and acceptance of lobbying. This framework stated that there must be an economic incentive for industry to demand and for a public bureaucracy to opt for high technology policy. While the cases that have been studied all carried economic potential according to their supporters, variance in the outcome of EU policy was observed empirically. This observation triggered a need to take a further step beyond the basic framework and search for variables that cast further light on the mechanisms under which EU high technology policy occurs. The main argument is that industry characteristics (derived from import competition pressure and relative international competitiveness) and the nature of political entrepreneurship (the EC's bureaucratic cohesion and expected opportunities to expand its institutional and political influence, and its expectation that EU governments perceive domestic returns and thus to be more inclined to support its initiatives) influence the provision of EU high technology policy.

The empirical results indicate that the import-competition condition was stronger in both the satellite navigation and mobile communications sectors, which provided a stimulus for economic actors to ask for high technology policy support and for the EC to use it as justification. Secondly, the relatively greater domestic orientation of the space manufacturing and radio sectors led to strong calls for European-oriented intervention. The EuroDAB consortium, though, was not able to politically gain preferential treatment. Economic actors in the mobile communications also called for EU TIP provision. However, it was different in character compared to the aforementioned sectors in that they called for a more internationally oriented EU strategy. In addition, the manufacturers with strong presence in overseas markets engaged in activities to promote international cooperation with Asian and North American actors in an attempt to further strengthen the lead they had achieved. This process reflected the potential strategic and distributional conflicts between European regional ambitions on the part of the EC, on the one hand, and those of internationally oriented economic actors on which the former relied to provide political support functions, on the other.

Regarding political entrepreneurship, the EC particularly identified opportunities to act upon an institutional void in satellite navigation to expand its prerogatives in a shared power regime at the EU level. The EC used satellite navigation strategically as a means to assert more influence beyond its original assignment, for example, by linking it to an emerging European defense and security policy and to the transformation of ESA into a *de facto* EU implementation agency, which was strongly supported by DG Research and eventually the ESA Executive. When it comes to mobile communications, the institutional returns to the EC are more problematic to assess and categorize. DG TREN couched the need to support the Galileo project in terms of various forms of dependence: socio-economic dependence; transport policy and European regulation of safety applications and law enforcement; time reference and synchronization of digital communications networks; political dependence; technology and industrial dependence; scientific dependence; and security dependence.

The deployment of a second, independent and interoperable (with GPS) satellite navigation system would not only provide a better platform for European investments in major markets but also open up for a strong European (and EC) institutional, political and regulatory control of satellite navigation. DG TREN stressed that the European transport and defense infrastructures would become increasingly dependent on satellite navigation. The US government had recognized that the GPS infrastructure was vulnerable to unintentional and intentional disruptions, which further increased the need for a back-up European infrastructure to ensure proper regulation and efficiency in terms of the implementation of various other European policies (agriculture, fisheries, environment, regional development and so on). The planned Galileo Public Regulated Service (PRS) would use large bandwidths and be encrypted in order to reduce the vulnerability of the infrastructure and thus provide a more secure system to governmental and specific civil users.

Apart from the issue of dependence on GPS, DG TREN also stressed that Galileo represented an indication of Europe's ambition to accept a fair share of the burden of protecting a European and global critical service infrastructure. This would benefit not only European but also US and other governmental and civil users. While the GPS alone was expected to generate a Euro 17 billion market annually by 2020, this market would be worth around Euro 28 billion annually by 2020 with a complementary Galileo infrastructure. In addition, Galileo would (apart from improved service performance, signal integrity and new search and rescue services) include 'guarantees of service' and greater 'liability' commitments in order to strengthen reliance and customer acceptance. It would also improve the precision and accuracy of time referencing and synchronization that are vital for high-speed data transmission.

Politically, DG TREN pointed out that the US controlled and publicly funded the GPS infrastructure and operations as well as the users. Rather than pouring European investments into GPS, DG TREN argued that Europe could better spend the money by investing it into a European-controlled system. The problem was that, when the benefits of satellite navigation would become even more visible, there were no insurances for users and political institutions that they would not face governmental levies or have to meet and adopt requirements that were not in line with their preferences. Europe could thus be held to ransom. Finally, European members of NATO relied on the GPS Precision Position Service (PPS) for military applications. The modernization of GPS in the US and implementation of the new military code raised problematic questions. Finally, the opportunities for European industry to build GPS receivers for the reception of military codes would be significantly enhanced with a European Galileo infrastructure. Europe had industrial weaknesses in areas such as GPS chips and receivers for high-end applications and low-end mass-market applications, whereas its relative position was somewhat better in terms of system integration and service provision. By promoting the domestic European industry, the EC (DG INFSO, DG Research, DG TREN and so on) could boost its own legitimacy and regulatory and institutional 'property rights'.

On 26 March 2002, the EU transport ministers met in Brussels in the EU Transport Council. They endorsed the release of additional public funding (Euro 450 million; US$395 million) for the Galileo program to complement the Euro 100 million already released and thereby formally launched the development phase of Galileo. The European Space Agency (ESA) had already committed Euro 550 million in fall 2001, the release of which, however, had been conditioned on financial commitment by the EU transport ministers for the development phase. European authorities need to launch the first satellites into orbit before mid-February 2006 in order to protect the frequencies allocated in 2000 at the World Radio Communications (WRC) Conference in 2000. In total, Galileo will consist of 30 satellites paced in orbit at an altitude of 24,000 kilometers (and up to 8 spare satellites) that will cover the entire world through a network of ground control stations. In advance of the EU Transport Council meeting, a large number of firms and major trade groups sent letters to the EC and governments in support of the Galileo program, such as the International Air Transport Association (IATA), the European Association of Aerospace Industries (AECMA), the European Information, Communication Technologies Association (EITCA), Eurospace, the European Metalworkers Federation (EMF), the European Association of Remote Sensing Companies (EARSC), as well as the European League for Economic Cooperation (ELEC).

Apart from public funding, the transport ministers reached an agreement on the management structure of the Galileo program along the lines of the proposal for a Council Regulation on the establishment of the Galileo Joint Undertaking to be created for a four-year duration and have its seat in Brussels. The major objectives of the Joint Undertaking, which would likely gradually replace the Galileo Interim Support Structure (GISS), were to manage the Galileo development phase and to prepare the deployment phase, to guide and coordinate the necessary research and development activities, to conclude an agreement with ESA to which the implementation would be entrusted, to mobilize public and private funds, and to establish management structures and 'business plans' for all the phases of the Galileo program. The EU Transport Council requested the EC to set up this Joint Undertaking without delay in cooperation with ESA. The financing of the development phase (subject to the appropriation procedures of the Trans-European Networks budget, TEN) the resources would be allocated to the Joint Undertaking. It was further agreed that EU Member States would not be requested to make direct financial contributions from their national budgets.

The compromise decision among the EU transport ministers also stated that it was important to keep the public sector funding at the lowest possible level throughout the development and implementation of the Galileo program and that the cost-share distribution for the deployment phase should be one-third for the European budget and around two-thirds for the private sector. A much-discussed topic over the years was the structure of the Galileo Joint Undertaking and, in particular, whether private companies would be allowed to participate and have an influence. It was decided that private sector companies would be allowed to become members when the tendering process was completed, expected by the end of 2003 which was aimed at selecting a private consortium in charge of the deployment and exploitation of the Galileo project. The European Investment Bank (EIB) was invited to take part. The EC pointed out that by financially contributing to the Joint Undertaking, the private sector would accept their share of the risks inherent in large-scale industrial and technology infrastructure activities. 'While it is the responsibility of the public sector to carry out forward analyses, to detect future merging markets, and encourage their development (as with Ariane and Airbus) it is also necessary for large firms, which in some cases have until recently been featherbedded by the public authorities, to themselves look beyond the short term if they want to survive in the face of world competition' (EC, 2002c, p.6). The EU Council would review the development of Galileo at the end of 2003 and then subsequently take decisions on the deployment and operational implementation phases.

A common explanatory account for the opposition of certain governments to the Galileo program has been budgetary pressure. An independent study by PricewaterhouseCoopers in late 2001 suggested that EU governments might have to bear the cost of developing and launching the Galileo project and estimated the cost at around Euro 3.6 billion (US$3.1 billion). DG TREN estimated the cost of establishing Galileo at Euro 3.2–3.4 billion. Galileo will have to provide free basic services to match GPS and it remained uncertain how to recoup the public-sector subsidy through royalties on chip sets, which were expected to become cheaper over time. Though tight budgets of course put constraints on national budgets this explanation does not provide a wholly compelling argument for understanding the variance in government preferences in Western Europe. Rather, it is a question about political-economic and strategic orientation and expectations about European distributional effects, although, of course the issue of expected government revenues is central (for example, through tax on the revenues of the private sector). It is arguably more plausible to explain the variation among EU national governments with (1) the relative Atlanticist orientation among EU Member States combined with their 'political-economic conception' of European integration (including views about the role of the public sector in the economy), and (2) the relative weight of domestic economic groups that could benefit from the Galileo program combined with past investment in GPS infrastructure and contracts with the US GPS sector (including the importance of GPS in national defense structures and operations), although it is difficult to collect information to support these alternative arguments.

The UK government (in particular the Treasury) has been critical to the Galileo project, in particular regarding the role of the private sector and the relationship between Galileo and military capabilities and users. Its decision to support the release of additional public funding does not represent a major shift in UK policy toward Galileo. Rather, the UK became increasingly isolated in the EU political negotiations in light of the emerging momentum behind Galileo (the disappearance of a minority voting block under the EU's qualified majority voting rules). It thus decided to exchange its financial support for concessions from other EU Member States in the broader bargaining package, for example with regard to the liberalization of the energy sector. Its domestic industry could also expect a higher share of ESA industrial contracts by committing to the project. 'Because of the importance of Galileo to Europe, we and other Member States such as Germany and the Netherlands have been keen to ensure the viability of the project [...] Our concerns have focused on the management, funding, costs and the benefits to users. I am glad today the Council has accepted our concerns and we have successfully negotiated sound financial and other safeguards,' said John Spellar, the UK transport minister (quoted in Connor, 2002).

Though the UK and the Dutch (and the Danish, German and Swedish) governments demanded an early private-sector funding commitment early on, they were reluctant to grant in particular the satellite industry a potentially influential role in the tendering process. The Galileo project is supported primarily by European aerospace and electronics industries. The major industrial consortium promoting Galileo on the part of the satellite manufacturing industry, Galileo Industries SA, argued that without private-sector influence over decisions and with the prospect of political interference (not least considering the political delays and controversies) there would be little incentive for industry to invest in the Galileo project. An official reason for excluding the satellite manufacturing industry from the Joint Undertaking until the tendering process was put in place was to foster more competition. However, the fact that countries such as France and Italy have strong satellite industries arguably played a role and reduced the interest on the part of the Dutch and the UK governments to promote a more integrated role for the private sector in the Galileo management structure. In order to officially legitimize their position, those governments that opposed and tried to stall the Galileo project argued that Europe did not have to develop its own system since GPS was already operational and its basic services were free of charge.

The EC's DG TREN and other European supporters of Galileo stressed that there was a need to have a long-term perspective and consider all the numerous positive macro-economic, industrial and user benefits and externalities. The French government (supported most actively by the Italian and Spanish governments) stated that Europe needed to ensure that it would not be subjugated and would not become vassals to the US in space matters. The French government has been the staunchest proponent of Galileo. Leading French political figures (President Jacques Chirac, Prime Minister Lionel Jospin, Minister of Research Roger-Gérard Schwartzenberg, Foreign Minister Hubert Védrine and Transport Minister Jean-Claude Gayssot) stressed the need to close the gap in space matters between Europe and the US and to challenge the global US monopoly. 'It [Galileo] will allow the European Union to liberate itself from dependence on the American GPS system,' said Gayssot (quoted in Black, 2002).

The European Galileo project would be the first satellite navigation and positioning systems that from its conception would be designed for primarily civilian purposes. According to DG TREN, it would provide superior accuracy and reliability and ensure genuine continuity of public service. It pointed out that the US GPS infrastructure was characterized by a 'total absence of guarantee and responsibility, this being incompatible with the military objectives of the [GPS] system' (EC, 2002c, p.2). Galileo would help to reduce the problem of intentional or unintentional signal interruptions that had been experienced by aviation authorities and aircraft captains, for example in the mid-US, over the Mediterranean, and for transatlantic flights over Iceland. In addition to becoming an increasingly integrated element of telephone systems and television systems, computer networks, and shipping operations, and so on, Galileo would 'enlarge the airspace' by reducing the need to direct the route of aircraft from one radar control zone to another. In addition, it could, through advanced satellite atomic clocks, also facilitate the interconnection of telecommunications, electricity and banking networks that require high time-synchronization precision levels. DG TREN stressed that Galileo program would promote domestic welfare, jobs, and the overall competitiveness of the European economy and high technology industries:

> It is vital that it [Europe] should not be excluded from what is already looking like being one of the main industrial sectors of the 21st century, which is how the Americans consider it. Without Galileo, the development and even the survival of European new technology sectors would be under very serious threat; with Galileo, the technological advance that the participating European industries will obtain will give them a considerable comparative advantage in this sector and also as regards the numerous ensuing applications (EC, 2002c, p.4).

European authorities and the private sector recognized that it was vital that Galileo needed to enter full operation by 2008 in order to respond in a timely and efficient manner to the expected arrival of the modernized GPS infrastructure (GPS III), planned to enter operation around the same time. It had already been decided in Europe that Galileo, as GPS, would have to provide basic applications and services free of charge, while, according to DG TREN, Galileo would also offer high-quality services that GPS could not provide for which a charge would be made. Europe, she argued (with support from the major Galileo backers), supported the project as a way to enhance European transport infrastructures and policymaking, employment (an estimated 100,000 jobs according) and competitiveness, to reduce the innovation and technological gap with the EU, and to challenge the US global commercial and strategic monopoly as Europe had done with, for example, Airbus aircraft and Ariane satellite launcher programs. 'We now have a "yes" for Galileo, which signifies Europe's wish to be present in the international scene in the areas of research, technology and industrial development,' said De Palacio (quoted in Black, 2002). DG TREN, thus, responded to what it perceived was a US campaign of misinformation and part of broader strategic trade and economic competititon.

> This argument is reminiscent of the early 1960s when the United States offered to launch European satellites free of charge. If it [Europe] had accepted this 'generous offer, Europe would not have won over half the world satellite launching market with Ariane. The United States is showing surprising concern by 'warning' their European friends that, in its view, Galileo is not economic. The credibility of such a message from a threatened competitor is obviously dubious. It simply shows to what extent the United States is afraid that a rival system will be successful (EC, 2002c, p.7).

Apart from debates on the level of public funding, the profitability of the Galileo program, and the specific role of the private sector (and the political control of governments) in the management of Galileo, a major bone of contention was the specific orientation of the Galileo program in terms of civilian versus military usage, which boiled down to strategic-rational interests. Traditionally, US defense authorities call on Europe to develop stronger military and intelligence capabilities and to shoulder a larger burden in operations. Satellite navigation and positioning technology is widely used by the US and NATO and is viewed as a core element in the development of high technology capabilities. Interestingly, US rhetoric regarding Galileo has focused on the sufficiency of one technology infrastructure, GPS, despite the recognition that GPS is subject to potentially severe vulnerabilities. The US military authorities have stressed the right to interfere with what would be the encrypted (Public Regulated Service, PRS) signals of Galileo, which are aimed at ensuring continuity of service for governmental applications, since there is a possibility that such signals could be used for hostile purposes (as could the US GPS signals). DG TREN argued that technical solutions had already been presented to US authorities on how such a problem could be mitigated and that 'Europeans are quite open to joint consideration by equal partners of any problem of concern to them [the US]' (EC, 2002c, p.7).

The potential military services of the Galileo program had become a diplomatic embarrassment for the UK government and Prime Minister Tony Blair due to its desire to maintain the 'special relationship' with NATO and the US. The UK and US governments likely communicated closely about the development of Galileo. 'The United States sees no compelling need for Galileo [...] The jury is still out as to whether a solid basis for cooperation still exists,' said Ralph Braibanti, the US State Department official heading negotiations with the EU on the Galileo project (quoted in Geitner, 2002). Parts of the US government (such as Defense Secretary Donald Rumsfeld, Deputy Defense Secretary Paul Wolfowitz, and Air Force General Richard Meyers, Chairman of the Joint Chiefs of Staff) put pressure on European ministers, not least from December 2001, claiming that the Galileo signals could compromise security ramifications for future NATO operations and the next-generation GPS signals intended for military use (Castle, 2002). The US announced that it would respect the EU's decision and that it would cooperate with the EU to ensure the interoperability and compatibility of the two technology infrastructure projects, while pointing out that various trade-related, technical and security aspects remained to be discussed and solved.

Loyola de Palacio, Commissioner for Transport and Energy and Vice President of the European Commissioned, stressed that the Galileo project would be civilian-oriented and civilian-controlled. At the same time, DG TREN pointed out that it could not be ruled out that military authorities might have an interest in and be allowed to use the Galileo infrastructure.

> Galileo will underpin the common European defense policy that the Member States have decided to establish. There is no question here of coming into conflict with the United States which is and will remain our ally, but simply a question of putting an end to a situation of dependence. If the EU finds it necessary to undertake a security mission that the US does not consider to be in its interest, it will be impotent unless it has the satellite navigation technology that is now indispensable. Although designed primarily for civilian applications, Galileo will also give the EU a military capability (EC, 2002c, p.4).

DG TREN pointed out that a mix of commercial, macroeconomic, technological and political-strategic considerations influenced the support of Galileo. The Spanish government agreed: 'What is really at issue is not whether this is a civil or military project but whether the EU is going to be fully sovereign or whether it will become a subordinated market of 350 million consumers,' said Francisco Alvarez-Cascos, the Spanish Development minister who chaired the EU Transport Council meeting in March 2002 (quoted in Dombey, 2002). In addition, the ESA Executive (and people such as its Director-General General Antonio Rodota and its Director of Applications Claudio Mastracci) stressed that Galileo was a European political imperative and that it signaled to the whole world that Europe was becoming a more coherent and distinct actor in foreign economic and security policymaking. Commissioner Loyola De Palacio and DG TREN promoted the Galileo program as both an indispensable component of the European economy and society and as a globally oriented endeavor. 'More generally, the European do not intend in their turn to adopt a protectionist and monopolistic approach. They do not deny the value of the GPS, even though Galileo is superior; they recognize that an even greater level of security will be obtained as a result of the joint use of the two systems; they are designing their system in such a way as to ensure that the same receiver can be used by a dual user; and they are willing to cooperate actively in all areas with the United States, as well as with Russia or China' (EC, 2002c, p.7). The European high technology policy for satellite navigation is an illustration of a political-economic and globally oriented approach toward the creation of competitive advantage in a context of past comparative disadvantage.

Regarding the wireless sector and the UMTS infrastructure, ETSI provides important political support functions to the EC. The EC was concerned that the established institutional relationship between the EU and ETSI would weaken and thus jeopardize the opportunities for the EC to assert administrative and political influence over standard setting in Europe. The EC has benefited from this relationship in the past and recognized the technical, institutional and regulatory role that ETSI played in Europe and, increasingly, worldwide. Without this institutional arrangement, the EC would have greater difficulties in fostering coherent industrial standards in Europe, which in turn would hamper its opportunities to exert control, to promote the further consolidation of the internal European market, and to support the competitiveness of the European telecom market and all its associated organizations. The EC thus worked hard to maintain this relationship and recognized the need of supporting it. The institutional returns were high in that the EC strived to uphold a beneficial institutional arrangement through which it could retain its regulatory reach. In addition, the success of GSM had proven very beneficial for the reputation of the EC and national governments. The EC recognized that to further nourish this policy success by supporting the system (UMTS) that would eventually replace it would be promising in institutional returns. Proactive EC support could enhance domestic welfare, job creation and the overall competitiveness throughout the value chain in the European high technology industries.

In the radio sector, however, there were no expected institutional returns and the EC expected that EU governments would oppose any attempts to promote a European regulatory authority in the radio sector. The EuroDAB consortium members lobbied the EC and DG EAC in an attempt to receive preferential policy and regulatory treatment. Around 1998, there were indications that staff in DG EAC intended to promote digital radio and DAB. This initiative, however, would 'die' with the EU Santer Commission. Both the political leadership of the DG EAC and staff in DG INFSO were reluctant to specifically support DAB and instead offered it general recognition by including a reference to it in the so-called EU telecom package. There were no indications that the digital radio supporters would receive any substantial and dedicated support from the EC. According to the assumptions guiding this study, there was a common interest at higher EC level as well as among the staff in one of the most powerful DGs in the EC (DG INFSO) to steer clear from specific preferential treatment of the DAB industry since such support was perceived as risky. In short, there were few identified domestic returns on the part of EU Member States. There could have been factors involved that further strengthened this passive posture, such as (1) the investment in resources by DG INFSO in related technology infrastructures (such as DVB and UMTS), (2) the past policy failure in high-definition television, and (3) the interest (DG EAC) to proactively support satellite digital broadcasting services such as the Luxembourg-based Global Radio or WorldSpace that was supported by Alcatel of France. This study cannot validate any of these possibilities.

The EC expected EU governments to perceive domestic political and economic returns in both the case of satellite navigation and mobile communications. In fact, the EC strategically manipulated the development of both these areas by committing itself actively to the projects and by justifying high technology policy based on the support from leading economic actors, the benefits that would accrue EU governments, and the positive consequences it would have for the European economy. In the case of digital radio, however, the EC did not expect that EU governments would support an active high technology policy in the absence of visible potential domestic returns. It thus did not matter that the economic actors who supported digital radio claimed it would have positive effects on the European economy, or result in a more citizen-oriented Europe.

This comparative analysis of EU high technology policy can be concluded and summarized through a qualitative, simplistic categorization of the empirical results informed by the analytical framework (Table 8.5).

Table 8.5: Explaining EU high technology policy.

Policy Sectors	Telecom	Transport	Audiovisual
		European Projects	
Theoretical Modules	GALILEO Satellite Navigation	UMTS Wireless Internet	DAB Digital Radio
Strategic trade theory	High	High	High
Industry characteristics	High	High	Low
Political entrepreneurship	High	High	Low
EU high technology policy	High	High	Low

Despite these findings and arguments, it is important to underline some weaknesses in the analytical framework that focuses on two conditional variables: lobbying strength and political returns. First, the motives for EU high technology policy are not fully captured by a model of political economy and distribution-strategic activities in which the position of economic actors in the structure of international competition (and the association with political support functions), on the one hand, and the expected positive returns from providing high technology policy, on the other, are the driving causal mechanisms. The EC could have other motives for providing or not providing high technology policy. The model in this study is based on certain stringent motivational assumptions, which leaves out other potential motives. Second, the conditional variables could partly be the function of past policies or other structural constraints, which further constrain the validity of the model. Finally, but most importantly, the 'additive' model (presented in Figure 2.2) should not be interpreted as consisting of wholly necessary conditions where the lack of one of them would lead to a no high technology policy outcome. Such a model would unlikely be true, but if it is, however, then only a complete model would predict the true outcome; leaving a single factor unobserved would falsely predict high technology policy where no such policy is correct. Rather, the factors should not be interpreted as strictly necessary. If we can observe their occurrence it will increase the likelihood of high technology policy provision in such an additive model.

A small-n study (that is, an examination of a low number of cases) may not be an inappropriate tool to comparatively analyze the strength of strictly explanatory models. With only three cases, some may argue, there is no fair way for assessing the relative merit of competing models, that is, alternative models may predict the same outcome if there are only three cases. It would make sense to say something about the degree to which the factors differ; for example does 'industry characteristics (or any of its components) increase the likelihood that high technology policy will occur by 20 per cent or by 50 per cent? How important are industry characteristics (or any of its related components) in comparison to political entrepreneurship (or any of its related components)? The empirical examination of the technology infrastructure projects guided by the analytical framework suggests, for example, that the role of import competition in the wireless sector did not constrain leading private sector actors and the EC from pursuing an aggressive and globally oriented strategy. This approach would likely open up the European market to greater penetration of and competition in European markets. This result, though, is still consistent with the assumption that internationally oriented actors would more likely pursue such an aggressive approach.

The problem remains in that it is difficult to weigh the balance between the various components of the conditional variables. This is also true in the examination of the digital radio sector where relatively scarce evidence has impeded a more detailed conclusion of the relative weight of the arguments associated with the analytical framework. In this study, though, it is claimed that this research design may still be useful to evaluate the way in which preferences and structures translate into politically relevant influences and have an impact on variation in the outcome. This model can be further strengthened by adding more cases, by allowing for diachronic or further synchronic variation in the sample of cases, by refining the model and the theoretical arguments, and so forth. The analytical framework has provided a stepping-stone to a more comprehensive political-economic model of high technology policy provision and the distribution-strategic activities of public and private actors. In terms of further research, the historical institutionalist literature, and in particular recent contributions that avoid structural approaches to path dependence, could be incorporated more firmly into comparative studies of the political economy of technology policy and strategy. A more dynamic, process-oriented perspective, which admittedly has been downplayed somewhat in this study, could move the theoretical discussion to a focus on the conditions under which 'lock-in effects' may help to define EU policy, how political institutions respond to technological path dependence and trajectories, and how stability is reproduced. In such a framework, the reproduction and political articulation of path dependence would assume center stage.

This study has offered insights into the political-economic mechanisms and conditions that arguably have a decisive influence on variation in EU high technology policy. The analysis of EU policy for wireless Internet and satellite navigation, in particular, provides an assessment of what may be viewed as a new form of critical industry support in the 'information age', which contrasts with the subsidy-based industry support promulgated by the first wave of strategic trade theorists in the 1980s and 1990s. The definition of the 'new economy' in Europe is not a natural or apolitical process and it is important to gauge the role of political institutions in influencing market adjustment and economic development. This study has demonstrated the distinctive role of EU institutions, and the EC in particular, in structuring global competition. Public actors try to tweak the rules of the world economy in order to structure global competition in ways that enhance job creation, overall competitiveness in cutting-edge high technology sectors and domestic welfare. In particular, in the cases of the evolution of a European strategy for satellite navigation and wireless Internet, this appears to be a plausible conclusion. The development of Galileo, for example, illustrates the dynamic interaction of technology infrastructure policy and political-economic integration in Western Europe and the role of the EU as an actor in foreign economic policymaking, on the one hand, and European security policy integration, transatlantic (and international) security relations and the role of the EU as a more distinct actor in security policymaking, on the other. It has thus shown the differential role of the EU (and the EC) as a more organized supranational political actor by offering a more differentiated account of EU policy than single-case studies would provide.

References

Abbott, Kenneth W. and Duncan Snidal (2001) 'International "Standards" and International Governance', *European Journal of Public Policy* 8, 1, 345–370. Special Issue: Governance and International Standard Setting.

Albury, David and Joseph Schwartz (1982) *Partial Progress: The Politics of Science and Technology* (London: Pluto Press).

Allison, Graham and M. H. Halperin (1972) 'Bureaucratic Politics: A Paradigm and Some Policy Implications', *World Politics* 24, 40–79.

Allison, Graham and P. Zelikow (1999) *Essence of Decision: Explaining the Cuban Missile Crisis*, 2nd ed. (New York: Longman).

Arthur, W. Brian (1989) 'Competing Technologies, Increasing Returns, and Lock-in by Historical Events', *The Economic Journal* 99 (March), 116–131.

Arthur, W. Brian (1994) *Increasing Returns and Path Dependence in the Economy* (Ann Arbor: University of Michigan Press).

Baldwin, Robert E. (1982) 'The Political Economy of Protectionism', in J. N. Bhagwati (ed.), *Import Competition and Response*, 355–363 (Chicago: Chicago University Press).

Baldwin, Robert E. (1985) *The Political Economy of US Import Policy* (Cambridge, MA and London: MIT Press).

Barfield, Claude E. and William A. Schambra (1986) *The Politics of Industrial Policy* (Washington, DC: AEI).

Bator, Francis (1958/1988) 'The Anatomy of Market Failure', reprinted in T. Cowen (ed.), *The Theory of Market Failure*, 35–66 (Fairfax, VA: George Mason University Press).

Bauer, Raymond S., Ithiel de Sola Pool and Lewis A. Dexter (1972) *American Business and Public Policy: The Politics of Foreign Trade* (Chicago: Aldine).

Baumol, William J. (1952/1965) *Welfare Economics and the Theory of the State* (Cambridge: Harvard University Press).

Besen, Stanley M. and Leland Johnson (1986) *Compatibility Standards, Competition, and Innovation in the Broadcasting Industry* (Washington, DC: RAND Corporation).

Blais, Andre (1986) 'Industrial Policy in Advanced Capitalist Democracies', in A. Blais (ed.), *Industrial Policy*, 1–53 (Toronto, Buffalo and London: University of Toronto Press).

Borrus, Michael (1988) *Competing for Control* (Cambridge, MA: Ballinger).

Botelho, Antonio J. (1995) 'The Politics of Resistance to New Technology: Semiconductor Diffusion in France and Japan until 1965', in Martin Bauer (ed.) *Resistance to New Technology: Nuclear Power, Information Technology and Biotechnology*, 227–253 (Cambridge: Cambridge University Press).

Branch, Ann P. and Jakob C. Ohrgaard (1999) 'Trapped in the Supranational-Intergovernmental Dichotomy: A response to Stone Sweet and Sandholtz', *Journal of European Public Policy* 6, 1, 123–143.

Brander, James A. (1986) 'Rationales for Strategic Trade and Industrial Policy', in P. R. Krugman (ed.), *Strategic Trade Policy and the New International Economics* (Cambridge: MIT Press).

Breton, Albert (1996) *Competitive Governments: An Economic Theory of Politics and Public Finance* (Cambridge, NY: Cambridge University Press).

Caporaso, James A. and David P. Levine (1992) *Theories of Political Economy* (Cambridge: Cambridge University Press).

Carliner, Geoffrey (1987) 'Industrial Policies for Emerging Industries', in P. R. Krugman (ed.), *Strategic Trade Policy and the New International Economics*, 147–168 (Cambridge, MA: MIT Press).

Cawson, Alan, et al. (1990) *Hostile Brothers: Competition and Closure in the European Electronics Industry* (Oxford: Clarendon Press).

Cerny, Philip G. (1990) *The Changing Architecture of Politics: Structure, Agency, and the Future of the State* (London: Sage).

Christiansen, Thomas (1998) 'A Maturing Bureaucracy: The Role of the Commission in the Policy Process', in J. Richardson (ed.), *European Union: Power and Policy-making*, 2nd ed. (Routledge: London).

Cobb, Roger W. and Charles D. Elder (1972) *Participation in American Politics: The Dynamics of Agenda-building* (Boston: Allyn and Bacon).

Cobb, Roger W., J. K. Ross and M. H. Ross (1976) 'Agenda Building as a Comparative Political Process', *American Political Science Review* 70, 1, 126–38.

Coen, David (1997) 'The Evolution of the Large Firm as a Political Actor in the European Union', *Journal of European Public Policy* 4, 1, 91–108.

Cohen, Michael D., James G. March, and Johan P. Olsen (1972) 'The Garbage Can Model of Organizational Choice', *Administration Science Quarterly* 17, 1–25.

Collingridge, David (1980) *The Social Control of Technology* (New York: St. Martin's Press).

Commoner, Barry (1979) *The Politics of Energy* (New York: Knopf).

Coombs, Rod, Paolo Saviotti and Vivien Walsh (1987) *Economics and Technological Change* (Totowa, NJ: Rowman & Littlefield).

Cowles, Maria G. (1995) 'Setting the Agenda for a New Europe: The European Round Table and EC 1992', *Journal of Common Market Studies* 33, 4, 501–526.

Cram, Laura (1994) 'The European Commission as a Multi-organization: Social Policy and IT Policy in the EU', *Journal of European Public Policy* 1, 2, 195–217.

Cram, Laura (1997) *Policy-making in the EU: Conceptual Lenses and the Integration Process* (London: Routledge).

Curzon Price, Victoria (1981) *Industrial Policies in the European Community* (London: Macmillan).

Dalum, Bent et al. (1999) 'Europe and the Information and Communication Technologies Revolution', in J. Fagerberg, P. Guerrieri, and B. Verspagen (eds), *The Economic Challenge for Europe*, 106–129 (Cheltenham, UK and Northampton, MA, USA: Edward Elgar).

Dang-Nguyen, Godefroyd, Volker Schneider and Raymund Werle (1993) 'Networks in European Policy-making: Europeification of Telecommunication Policy', in Svein S. Andersen and Kjell A. Eliassen (eds), *Making Policy in Europe: The Europeification of National Policy-making*, 93–114 (London: Sage Publications).

David, Paul A. (1986) 'Understanding the Economics of QWERTY: The Necessity of History', in William N. Parker (ed.), *Economic History and the Modern Economist*, 30–49 (Oxford: Basil Blackwell).

Davies, Andrew (1996) 'Innovation in Large Technical Systems: The Case of Telecommunications', *Industrial and Corporate Change* 5, 4, 1143–1180.

Davies, Andrew (1997) *Complex Product Systems: Europe's Competitive Advantage.* Working Paper prepared for CENTRIM/SPRU Project on Complex Product Systems EPSRC Technology Management Initiative, no. 23.

Dobbin, F. (1994) *Forging Industrial Policy: The United States, Britain and France in the Railway Age* (Cambridge: Cambridge University Press).

Dosi, Giovanni (1981) *Technical Change and Survival: Europe's Semi-Conductor Industry* (European Research Center: University of Sussex).

Dyson, Kenneth and Peter Humphreys (1988) *Broadcasting and New Media Policies in Western Europe* (London: Croom Helm).

Dyson, Kenneth (1991), 'The Nature and Relevance of State Policy for Techno-industrial Innovation', in U. Hilpert (ed.), *State Policies and Techno-industrial Innovation*, 41–63 (London and New York: Routledge).

Dyson, Kenneth (2000) *The Politics of the Euro-Zone: Stability or Breakdown* (Oxford: Oxford University Press).

Eads, George and Richard Nelson (1971) 'Government Support of Advanced Creative Technology', *Public Policy* 19, 113–128.

Eads, George and Richard Nelson (1977) 'Governmental Support of Advanced Civilian Technology: Power Reactor and the Supersonic Transport', *Public Policy* 14, 3, 405–427.

Eden, Lorraine and Maureen A. Molot (1993) 'Insiders and Outsiders: Defining "Who is Us" in the North American Automobile Industry', *Transnational Corporations* 23, 3 (December), 31–61.

Eliasson, Gunnar (1991) 'The International Firm: A Vehicle for Overcoming Barriers to Trade and a Global Intelligence Organization Diffusing the Notion of a Nation', in L. G. Mattsson and B. Stymne (eds), *Corporate and Industry Strategies for Europe*, 139–170 (Amsterdam: Elsevier).

Elster, Jon (1989) *The Cement of Society: A Study of Social Order* (Cambridge: Cambridge University Press).

Fellmeth, Robert (1970) *The Interstate Commerce Omission: Public Interest and the ICC* (New York: Grossman).

Finger, J. Michael, H. Keith Hall and Douglas R. Nelson (1982) 'The Political Economy of Administered Protection', *American Economic Review* 72, 441–466.

Flamm, Kenneth (1988) *Targeting the Computer: Government Support and International Competition* (Washington: Brookings Institution).

Flamm, Kenneth (1996) *Mismanaged Trade? Strategic Policy and the Semiconductor Industry* (Washington, DC: Brookings Institution Press).

Fochs, René (1970) *Europe and Technology: A Political View*, The Atlantic Papers 2 (Paris: Atlantic Institute).

Fries, Sylvia D. (1988) '2001 to 1994: Political Environment and the Design of NASA's Space Station System', *Technology and Culture* 29, 3, 568–593.

Geroski, Paul (1989) 'European Industrial Policy and Industrial Policy in Europe', *Oxford Review of Economic Policy* 5, 20–36.

Gilpin, Robert (1987) *The Political Economy of International Relations* (Princeton, NJ: Princeton University Press).

Gilpin, Robert (2001) *Global Political Economy: Understanding the International Economic Order* (Princeton, NJ and Oxford: Princeton University Press).

Gold, Bonnie (1995) *Politics, Markets, and Security: European Military and Civil Aircraft Collaboration, 1954–1994* (Lanham, MD: University Press of America).

Gourevitch, Peter A. (1982) 'Making Choices in France: Industrial Structure and the Politics of Economic Policy', in S. Cohen and P. Gourevitch (eds), *France in the Troubled World Economy* (London: Butterworth).

Gourevitch, Peter A. (1986) *Politics in Hard Times* (Ithaca, NY: Cornell University Press).

Grande, Edgar (1996) 'The State and Interest Groups in a Framework of Multi-level Decision-making: The Case of the European Union', *Journal of European Public Policy* 3, 3, 318–338.

Grant, Wyn (1989) *Government and Industry: A Comparative Analysis of the US, Canada and the UK* (Aldershot: Edward Elgar).

Grant, Wyn (1997) *The Common Agricultural Policy* (New York: St. Martin's Press).

Grant, Wyn (2000) *Pressure Groups and British Politics* (Basingstoke: Palgrave).

Grossman, Gene M. and Elhanan Helpman (1994) 'Protection for Sale', *The American Economic Review* 84, 833–850.

Guerrieri, Paolo and Carlo Milana (1998) 'High-technology Industries and International Competition', in D. Archibugi and J. Michie (eds), *Trade, Growth and Technical Change*, 188–207 (Cambridge: Cambridge University Press).

Hall, Peter A. (1986) *Governing the Economy* (Cambridge: Cambridge University Press).

Hall, Peter A. and R.C.R. Taylor (1996) 'Political Science and Three New Institutionalisms', *Political Studies* 44, 952–973.

Hamlett, Patrick W. (1992) *Understanding Technological Politics: A Decision-Making Approach* (Englewood Cliffs, NJ: Prentice Hall).

Harcourt, A. J. and Claudio M. Radaelli (1999) 'Limits to EU Technocratic Regulation?', *European Journal of Political Research* 35, 107–122.

Hart, Jeffrey (1992) *Rival Capitalists: International Competitiveness in the United States, Japan and Western Europe* (Ithaca, NJ: Cornell University Press).

Hart, Jeffrey and John C. Thomas (1995) 'European Policies toward HDTV', *Communications & Strategy* 20, 23–55.

Hart, Jeffrey A. and Assem Prakash (1999) 'Globalization, Governance, and Strategic Trade and Investment Policies', in A. Prakash and J. A. Hart (eds), *Globalization and Governance* (London and New York: Routledge).

Hart, Michael (2000) *The American Internet Advantage: Global Themes and Implications of the Modern World* (Lanham, MD: University Press of America).

Harte, Lawrence, Richard Levine and Steve Prokup (1997) *Cellular and PCS: The Big Picture* (London: McGraw-Hill).

Hawkins, Richard W. (1994) *Regional Technical Infrastructures: The European Position in the Evolving Geo-politics of Telecommunication*. Working Paper for the European Network for Communication and Information Perspectives.

Hawkins, Richard W. (1997) 'Prospects for a Global Communication Infrastructure in the 21st Century: Institutional Restructuring and Network Development', in A.

Sreberny-Mohammadi et al. (eds), *Media in Global Context: A Reader* (London: Arnold).
Hayward, Keith (1994) 'European Union Policy and the European Aerospace Industry', *Journal of European Public Policy* 1, 3, 347–365.
Head, John G. (1974) *Public Goods and Public Welfare* (Durham: Duke University Press).
Héritier, Adrienne (1996) 'The Accommodation of Diversity in European Policy-making and its Outcomes: Regulatory Policy as Patchwork', *Journal of European Public Policy* 3, 2 (June), 149–167.
Hills, Jill (1981) 'Government Relations with Industry: Japan and Britain–A Review of Two Political Arguments', *Polity* 14, 2, 15–24.
Hills, Jill and Michalis, M. (2000) 'Restructuring Regulation: Technological Convergence and European Telecommunications and Broadcasting Markets', *Review of International Political Economy* 7, 3, 434–464.
Hilpert, Ulrich (1991) 'The State, Science and Techno-industrial Innovation: A New Model of State Policy and a Changing Role of the State', in U. Hilpert (ed.), *State Policies and Techno-industrial Innovation*, 3–40 (London and New York: Routledge).
Hindess, Barry (1989) *Political Choice and Social Structure: An Analysis of Actors, Interests and Rationality* (Aldershot: Edward Elgar).
Hine, Robert C. (1992) 'Regionalism and the Integration of the World Economy', *Journal of Common Market Studies* 30, 2 (June), 115–123.
Hirschman, Albert O. (1970) *Exit, Voice and Loyalty: Responses to Decline in Firms, Organizations and States* (Cambridge, MA: Harvard University Press).
Hood, Christopher C. (1986) *The Tools of Government* (Chatham, NJ: Chatham House Publishers).
Howells, Jeremy and Michelle Wood (1993) *The Globalisation of Production and Technology* (London: Belhaven Press).
Howlett, Michael and M. Ramesh (1995) *Studying Public Policy: Policy Cycles and Policy Subsystems* (New York; Oxford; Toronto: Oxford University Press).
Hughes, Thomas P. (1983) *Networks of Power: Electrification in Western Society, 1880–1930* (Baltimore: Johns Hopkins University Press).
Jenkins, William I. (1978) *Policy Analysis: A Political and Organizational Perspective* (London: Martin Robertson).
Johnson, Chalmers (1982) *MITI and the Japanese Miracle* (Stanford: Stanford University Press).
Justman, Moshe and Morris Teubal (1996) 'Technological Infrastructure Policy (TIP): Creating Capabilities and Building Markets', in M. Teubal et al. (eds), *Technological Infrastructure Policy: An International Perspective*, 21–58 (Dordrecht: Kluwer Academic Publishers).
Kaitatzi-Whitlock, Sophia (1998) 'HDTV and Standardization Policymaking in Europe', in M. Dupagne and P. B. Seel (eds), *High-definition Television: A Global Perspective*, 95–133 (Ames, IA: Iowa State University Press).
Kenis, Patrick and Volker Schneider (1987) 'The EC as an International Corporate Actor: Two Case Studies in Economic Diplomacy', *European Journal of Political Research* 15, 4, 437–457.
Kenis, Patrick and Volker Schneider (1991) 'Policy Networks and Policy Analysis: Scrutinizing a New Analytical Toolbox', in Bernd Marin and Renate Mayntz

(eds), *Policy Networks: Empirical Evidence and Theoretical Considerations*, 25–36 (Boulder, CO: Westview Press).

Knight, Jack (1992) *Institutions and Social Conflict* (Cambridge; New York: Cambridge University Press).

Kozmetzky, George and Piyu Yue (1997) *Global Economic Competition: Today's Warfare in Global Electronics Industries and Companies* (Boston, Dordrecht and London: Kluwer).

Krasner, Stephen D. (1978) *Defending the National Interest: Raw Materials, Investments and US Foreign Policy* (Princeton: Princeton University Press).

Krauss, Melvyn B. (1978) *The New Protectionism: The Welfare State and International Trade* (New York: New York University Press).

Krugman, Paul R. (ed.) (1986) *Strategic Trade Policy and the New International Economics* (Cambridge, MA: MIT Press).

Krugman, Paul R. (1992) 'Technology and International Competition: A Historical Perspective', in M. Harris and G. E. Moore (eds), *Linking Trade and Technology Policies* (Washington, DC: National Academy of Engineering).

Laffran, Brigid (1997) 'From Policy Entrepreneur to Policy Manager: The Challenge Facing the European Commission', *Journal of European Public Policy* 4, 3, 422–438.

Lavergne, Real P. (1983) *The Political Economy of U.S. Tariffs: An Empirical Analysis* (Toronto: Academic Press).

Lawton, Thomas C. (1997) *Technology and the New Diplomacy: The Creation and Control of EC Industrial Policy for Semiconductors* (Aldershot, England: Avebury).

Lawton, Thomas C. (1999) 'Introduction: Concepts Defined and Scenes Set', in Thomas C. Lawton (ed.), *European Industrial Policy and Competitiveness: Concepts and Instruments*, 1–22 (Houndmills, UK: Macmillan Press).

Lawton, Thomas C. (2000) 'The Evolving Global Production Structure: Implications for International Political Economy', in T. C. Lawton, J. N. Rosenau and A. C. Verdun (eds), *Strange Power: Shaping the Parameters of International Relations and International Political Economy* (Aldershot: Ashgate).

Lembke, Johan (2001) 'Harmonization and Globalization: UMTS and the Single Market', *INFO: The Journal of Policy, Regulation and Strategy for Telecommunications, Information and Media* 3, 1, 15–26.

Lembke, Johan (2002a) 'The EU Regulatory Strategy for Mobile Internet', *Journal of European Public Policy* 9, 2, 273–291 (March).

Lembke, Johan (2002b) 'Global Competition and Strategies in the Information and Communications Technology Industry: A Liberal Strategic Perspective', *Business and Politics* 4, 1, 41–69 (April).

Lembke, Johan (2002c) 'EU Critical Infrastructure and Security Policy: Capabilities, Strategies and Vulnerabilities', *Current Politics and Economics in Europe* 11, 2, 99–129 (Summer).

Lembke, Johan (2002d) *Defining the New Economy: EU Technology Infrastructure Policy* (Stockholm: University of Stockholm).

Lewis, Eugene (1980) *Public Entrepreneurship: Toward a Theory of Political Power* (Bloomington and London: Indiana University Press).

Liebowitz, Stan J. and Stephen E. Margolis (1990) 'The Fable of the Keys', *Journal of Law and Economics*, 33, 1–25.

McKay, D. and Wyn Grant (1983) 'Industrial Policies in OECD Countries: An Overview', *Journal of Public Policy* 3 (February), 1–12.

MacKenzie, Donald and Judy Wajcman (eds) (1985) *The Social Shaping of Technology* (Milton Keynes: Open University Press).

MacKenzie, Donald (1990) *Inventing Accuracy: A Historical Sociology of Nuclear Missile Guidance* (Cambridge, MA: MIT Press).

Magaziner, Ira and Mark Patinkin (1989) *The Silent War: Inside the Global Business Battles Shaping America's Future* (New York: Random House).

Majone, Giandomenico (1992) 'Regulatory Federalism in the European Union', *Government and Policy* 10, 299–316.

Majone, Giandomenico (1994) 'The Rise of the Regulatory State in Europe', *West European Politics* 17, 3, 77–101.

Malerba, Franco (1983) *Technical Change, Market Structure and Government Policy: The Evolution of the European Semi-Conductor Industry*, unpublished Ph.D. dissertation, Yale University).

Marcus, Alfred A. (1991) 'Airline Deregulation, Business Strategy and Regulatory Theory', in M. J. Dubnick and A. R. Gitelson (eds), *Public Policy and Economic Institutions*, 325–350 (Greenwich, CO and London: JAI Press).

Mazey, Sonia and Jeremy Richardson (1996) 'The Logic of Interest Groups', in J. Richardson (ed.), *European Union: Power and Policy-making* (London: Routledge).

Menon, Anand and Jack Hayward (1996) 'States, Industrial Policies and the European Union', in H. Kassim and A. Menon (eds), *The European Union and National Industrial Policy*, 267–290 (London: Routledge).

Milner, Helen V. (1988) *Resisting Protectionism: Global Industries and the Politics of International Trade* (Princeton, NJ: Princeton University Press).

Mokyr, Joel (1990) *The Lever of Riches* (Oxford: Oxford University Press).

Mokyr, Joel (1998) 'The Political Economy of Technological Change: Resistance and Innovation in Economic History', in M. Berg and K. Bruland (eds), *Technological Revolutions in Europe: Historical Perspectives*, 39–64 (Cheltenham, UK: Edward Elgar).

Mosher, Frederick C. (1978) 'Professionals and the Public Service', *Public Administration Review* 38, 144–50.

Muller, Joachim W. (1989) 'Policy Options for Government Funding of Advanced Technology: The Case of International Collaboration in the European Telecommunications Satellite Programme', *Research Policy* 28, 33–50.

Muller, Joachim W. (1990) *European Collaboration in Advanced Technology* (Amsterdam: Elsevier).

Nelson, Richard R. (1984) *High-technology Policies: A Five-nation Comparison* (Washington, DC: American Enterprise Institute for Public Policy Research).

Nelson, Richard R. (1986) 'The State and Private Enterprise in High Technology Industries', in Roy M. MacLeod (ed.), *Technology and the Human Prospect*, 71–97 (London and Wolfeboro, NH: Frances Pinters).

Nelson, Richard. R. and G. Wright (1992) 'The Rise and Fall of American Technological Leadership: The Postwar Era in Historical Perspective', *Journal of Economic Literature 30*, 1931–1964.

Noll, Roger G. and Bruce M. Owen (eds) (1983) *The Political Economy of Deregulation: Interest Groups in the Regulatory Process* (Washington and London: American Enterprise for Public Policy Research).

Nordlinger, Eric (1981) *On the Autonomy of the Democratic State* (Cambridge, MA: Harvard University Press).

Olsen, Johan P. (1983) *Organizational Democracy*, 148–187 (Bergen: Universitetsforlaget).

Olson, Mancur (1965/1971) *The Logic of Collective Action: Public Goods and the Theory of Goods* (Cambridge, MA: Harvard University Press).

Olson, Mancur (1982) *The Rise and Decline of Nations: Economic Growth, Stagflation, and Social Rigidities* (New Haven: Yale University Press).

Ostry, Sylvia (1990) *Governments & Corporations in a Shrinking World: Trade and Innovation Policies in the United States, Europe and Japan* (New York, NY and London: Council on Foreign Relations Press).

Patel, Pari and Keith Pavitt (1987) 'Is Western Europe Losing the Technological Race?', *Research Policy* 16, 59–99.

Patterson, L. Ann (1998) *Regulating Competitiveness: The Development and Elaboration of Biotechnology Regulatory Policy in the European Union*, Ph.D. dissertation, University of Pittsburgh.

Pavitt, Keith and Margaret Sharp (1993) 'Technology Policy in the 1990s: Old Trends and New Realities', *Journal of Common Market Studies*, June.

Peltzman, Sam (1974) 'Constituent Interests and Congressional Voting', *Journal of Law and Economics* 5, 211–241.

Peters, B. Guy (1989) *The Politics of Bureaucracy*, 3rd Ed (Longman: New York and London).

Peters, B. Guy (1992) 'Bureaucratic Politics and the Institutions of the European Community', in A. Sbragia (ed.), *Euro-politics: Institutions and Policymaking in the 'New' European Community* (Washington, DC: Brookings Institute).

Peters, B. Guy (2000) 'Governance and Comparative Politics', in J. Pierre (ed.), *Debating Governance*, 36–53 (Oxford; New York: Oxford University Press).

Peterson, John (1992) 'The European Technology Community: Policy Networks in a Supranational Setting', in D. Marsh and R. A. W. Rhodes (eds), *Policy Networks in British Government* (Oxford: Oxford University Press).

Peterson, John (1993) *High Technology and the Competition State* (London: Routledge).

Peterson, John (1996) 'Research and Development Policy', in H. Kassim and A. Menon (eds), *The European Union and National Industrial Policy* (London: Routledge).

Pierre, Jon (2000) *Giving Structures to Values: The Institutional Politics of European Aviation Safety*, Paper, Southern Political Science Association, annual conference, Atlanta, 9–12 November.

Pierson, Paul (1996) 'The Path to European Integration: A Historical Institutionalist Perspective', *Comparative Political Studies* 29, 123–163.

Pierson, Paul (1997) *Increasing Returns, Path Dependence and the Study of Politics*, Jean Monnet Chair Papers 44, The Robert Schuman Centre at the European University Institute.

Porter, Michael (1990) *The Competitive Edge of Nations* (New York: Free Press).

Posner, Richard A. (1974) 'Theories of Economic Regulation', *Bell Journal of Economics and Management* 5, 335–358.

Potters, Jan and Randolph Sloof (1996) 'Interest Groups: A Survey of Empirical Models that Try to Assess Their Influence', *European Journal of Political Economy* 12, 403–422.

Price, Lee (1992) 'A Constructive Role for Government in the International Economy', in V. A. Canto and J. K. Dietrich (eds), *Industrial Policy and International Trade*, 11–36 (Greenwich, CO; London: JAI Press).
Reagan, Michael D. (1987) *Regulation: The Politics of Policy* (Boston: Little Brown).
Reece, Ray (1979) *The Sun Betrayed: A Report on the Corporate Seizure of U.S. Solar Energy Development* (Boston: South End Press).
Rhodes, Martin and Bastiaan van Apeldoorn (1997) *The Transformation of West European Capitalism?* EUI Working Papers, Robert Schuman Centre, RSC 97/60, Badia Fiesolana, San Domenico.
Richardson, J. D. (1990) 'Strategic Trade Policy', *International Organization* 44, 107–135.
Roobeek, Annemieke J. M. (1990) *Beyond the Technology Race: An Analysis of Technology Policy in Seven Industrial Countries* (Amsterdam: Elsevier).
Rosenberg, Nathan (1992) 'Science and Technology in the Twentieth Century', in G. Dosi, R. Giannetti and P. A. Toninelli (eds), *Technology and Enterprise in a Historical Perspective*, 63–96 (Oxford: Clarendon Press).
Ross, George (1993) 'Sliding into Industrial Policy: Inside the European Commission', *French Politics and Society* 11, 1, 20–44.
Sally, Razeen (1997) 'Public Policy and the Janus Face of the Multinational Enterprise: National Embeddedness and International Production', in Philip G. Cerny et al. (eds), *Globalization and Public Policy* (Cheltenham: Edward Elgar).
Samuels, Richard (1987) *The Business of the Japanese State* (Ithaca: Cornell University Press).
Sandholtz, Wayne (1992a) 'ESPRIT and the Politics of International Collective Action', *Journal of Common Market Studies* 30, 1, 1–21.
Sandholtz, Wayne (1992b) *High-tech Europe: The Politics of International Co-operation* (Berkeley, USA: University of California Press).
Sandholtz, Wayne (1993) 'Institutions and Collective Action: The New Telecommunications in Western Europe', *World Politics* 45, 242–270.
Sandholtz, Wayne and John Zysman (1989) '1992: Recasting the European Bargain', *World Politics* 42, 95–118.
Sandholtz, Wayne and Alec Stone Sweet (1997) 'European Integration and Supranational Governance', *Journal of European Public Policy* 4, 3, 297–317.
Sbragia, Alberta (2000) 'The European Union as Coxswain: Governance by Steering', in J. Pierre (ed.), *Debating Governance* (Oxford: Oxford University Press).
Scherer, F. M. (1992) *International High-technology Competition* (Cambridge, MA: Harvard University Press).
Schnee, J. (1978) 'Government Programmes and Growth of High-technology Industries', *Research Policy* 8, 204–43.
Servan-Schreiber, Jean Jacques (1969) *The American Challenge* (Harmondsworth: Pelican Books).
Sharp, Margaret and Chris Shearman (1987) *European Technological Collaboration* (London: Routledge & Kegan Paul).
Soete, Luc (1997) 'Technology Policy and the International Trading System: Where Do We Stand?', in Horst Siebert (ed.), *Towards a New Global Framework for High-technology Competition* (Tübingen: Institut für Weltwirtschaft an der Universität Kiel).
Spencer, B. J. and J. A. Brander (1983) 'International R&D Rivalry and Industrial Strategy', *Review of Economic Studies* 50, 707–722.

Spiller, Pablo T. (1983) 'The Differential Impact of Airline Regulation on Individual Firms and Markets: An Empirical Analysis', *The Journal of Law and Economics* 26, 655–689.

Stegeman, K. (1989) 'Policy Rivalry among Industrial States: What Can We Learn from Models of Strategic Trade Policy', *International Organization* 43, 73–100.

Stigler, George (1975) 'The Theory of Economic Regulation', in G. J. Stigler, *The Citizen and the State: Essays on Regulation*, 114–45 (Chicago: University of Chicago Press).

Strange, Susan (1988/1994) *States and Markets* (London: Pinter).

Street, John (1992) *Politics and Technology* (New York: The Guilford Press).

Sung, Liching (1997) 'Standards Competition in Wireless: Regionalism vs. Globalism', in E. M. Noam and A. J. Wolfson (eds), *Globalism and Localism in Telecommunications*, 323–341 (Amsterdam: Elsevier).

Thompson, Chris (1989) 'High-technology Theories and Public Policy', *Environment and Planning C: Government and Policy* 7, 121–152.

Thompson, Victor (1961) *Modern Organizations* (New York: Knopf).

Thornton, David W. (1995) *Airbus Industrie: The Politics of an International Industrial Collaboration* (New York: St Martin's Press).

Thurow, Lester C. (1996) *The Future of Capitalism: How Today's Economic Forces Shape Tomorrow's World* (New York: W. Morrow & Co).

Tyson, Laura D. (1992) *Who's Bashing Whom? Trade Conflict in High-technology Industries* (Washington, DC: Institute for International Economics).

Tyson, Laura D. and Pei-Hsiung Chin (1992) 'Industrial Policy and Trade Management in the Commercial Aircraft Industry', in Laura D. Tyson (ed.), *Who's Bashing Whom? Trade Conflict in High-technology Industries*, 155–216 (Washington, DC: Institute for International Economics).

Van Tulder, Rob and Gerd Junne (1988) *European Multinationals in Core Technologies* (Chichester and New York: John Wiley & Sons).

Volti, Rudi (1988) *Society and Technological Change* (New York: St. Martin's Press).

Warnecke, Steven J. (1978) 'Government Intervention and an Open Global Trading System', in S. J. Warnecke (ed.), *Industrial Policies in Western Europe* (New York: Holmes and Meier).

Weber, Steve and John Zysman (1992) 'The Risk that Mercantilism Will Define the Next Security System', in W. Sandholtz et al. (eds), *The Highest Stakes*, 167–196 (New York: Oxford University Press).

Werle, Raymund (2001) 'Institutional Aspects of Standardization: Jurisdictional Conflicts and the Choice of Standardization Organizations', *European Journal of Public Policy* 8, 1, 392–410. Reprinted [1998]. Special Issue: Governance and International Standard Setting.

Westrum, Ron (1991) *Technologies and Society: The Shaping of People and Things* (Belmont, CA: Wadsworth Publishing Company).

Williams, Roger (1973) *European Technology: The Politics of Collaboration* (London: Croom Helm).

Williams, Roger (1984) 'The International Political Economy of Technology', in S. Strange (ed.), *Paths to International Political Economy* (London: George Allen & Unwin).

Wilson, James Q. (1980) *The Politics of Regulation* (New York: Basic Books).

Wyatt-Walter, Andrew (1995) 'Globalization, Corporate Identity and European Technology Policy', *Journal of European Public Policy* 2, 3, 427–46.
Yin, Robert K. (1994) *Case Study Research: Design and Methods*, 2nd ed. (Thousand Oaks, CA: Sage).
Yoffie, David B. (1993) 'Introduction: From Comparative Advantage to Regulated Competition', in D. B. Yoffie (ed.), *Beyond Free Trade: Firms, Governments, and Global Competition*, 1–25 (Boston, Mass.: Harvard Business School Press).
Zysman, John (1977a) *Political Strategies for Industrial Order: State Power and Industry in France* (Berkeley: University of California Press).
Zysman, John (1977b) *Governments, Markets, and Growth: Financial Systems and the Politics of Industrial Change* (Ithaca, NY: Cornell University Press).
Zysman, John and Laura D. Tyson (eds) (1983) *American Industry in International Competition* (Ithaca: Cornell University Press).

EU Documents

Allgeier, Herbert J., SAG Chairman and DG Research/JRC Director (2000) 'Contribution of Space Applications to Improve Everyday Life', *Keynote speech, Conference on Satellites in Improving Everyday Life*, Istanbul, 1 March.
Argyris, Nicholas (1998) *Mobile Multimedia in the Information Society*, UMTS 1998 Conference, Director for Telecommunications and Postal Services (DG INFSO/DGXIIIA), Rome 23 September.
Arroyo-Fernandez, Bartolome (1998) 'European Collaborative Research on Wireless ATM: From ACTS to the 5th Framework Program Europe', *First International Workshop on Wireless Mobile ATM Implementations*, DG INFSO, Hangzhou, China, 6–10 April.
Brittan, Leon, EC Vice-President (1997) *Ensuring European Competitiveness in the Global Information Economy*, Speech in London, 17 January.
Busquin, Philippe, Commissioner for Research (2000) *Renforcer la Coherence de l'Approche Européenne de l'Espace*, Introductory speech, Munich, 22 February.
CDAB, Conference on Digital Audio Broadcasting (2000) 'Conclusions', Portuguese Presidency, Lisbon, 6–7 April.
Council (1992) *Guidelines for the Development of Transeuropean Transport Networks*, Council Decision EC/92/1692 of 23 July 1996 (OJ L228), Brussels, 9 September.
Council (1993a) *Council Decision of 25 February 1992 on Radio-navigation Systems for Europe*, 92/143/EEC, Brussels.
Council (1993b) *Council Conclusions of the Research Council on 29 April 1993 on the European Community and Space*, Brussels.
Council (1993c) *Council Resolution of 22 July on Mobile and Personal Communications*, 93/C/213/01, Brussels.
Council (1994) *Council Resolution of 19 December 1994 on the European Contribution to the Development of a Global Satellite Navigation Satellite System (GNSS)*, 94/C379/2, Brussels.
Council (1995) *Council Resolution of 29 June on the Further Development of the Mobile and Personal Communications Sector in the European Union*, 95/C/188/02, Brussels.

Council (1997) *Council Conclusions on The EU and Space: Fostering Applications, Markets and Industrial Competitiveness*, Brussels.

Council (1998a) *Council Resolution of 22 June 1998 on Cooperation between the EU and ESA*, OJ C224, Luxembourg, 17 July.

Council (1998b) *Council Resolution of 2 December on Developing a Coherent European Space Strategy*, Brussels.

Council (1998c) *Decision No. 52/98 of the European Parliament and of the Council on the Coordinated Introduction of a Third-generation Mobile and Wireless Communications System (UMTS) in the Community*. Common Position, Adopted by the Council on 24 September. Inter-institutional File No. 98/0051 (COD), Brussels, 25 September.

Council (1999) *Council Decision on the Coordinated Introduction of Third Generation Mobile and Wireless Communications System (UMTS) in the Community*, 128/1999/EEC, Brussels.

Council, Council of the European Union (2001) *Council Decision of 20 December 2000 on the Implementation of a Program to Encourage the Development, Distribution and Promotion of European Audiovisual Works (MEDIA PLUS: Development, Distribution and Promotion, 2001–2005)* (2000/821/EC), Official Journal of the European Communities, 17 January.

DaSilva, Joao and Bosco E. Fernandes (1995) 'The European Research Program for Advanced Mobile Systems: Addressing the Needs of the European Community', *IEEE Personal Communications* 2, 1, 14–19 (February).

DaSilva, Joao et al. (1995) 'Mobile and Personal Communications: From RACE to ACTS', *RACE Mobile Telecommunications Summit 1995*, 9–22, EC report, Cascais, Portugal, 22–24 November.

DaSilva, Joao et al. (1996) 'European Third Generation Mobile Systems', *ACTS Mobile Telecommunications Summit Report*, EC report.

DaSilva, Joao et al. (1997a) 'European Third Generation Mobile Systems', *IEEE Communications Magazine* 34, 10 (October), 68–83.

DaSilva, Joao et al. (1997b) 'Evolution Towards UMTS', HF Electronics/Communications 1, *Special Issue on Mobile Communications, Systems and Services*, 37–54.

DaSilva, Joao et al. (1997c) *Mobile and Personal Communications: ACTS and Beyond*, Paper.

DaSilva, Joao et al. (1997d) 'European R&D Programs on Third Generation Mobile Communications Systems', *IEEE Personal Communications* 4, 1 (February).

DaSilva, Joao et al. (1998a) 'ACTS Mobile Programme in Europe', *IEEE Communications Magazine*, editorial, 36, 2 (February).

DaSilva, Joao et al. (1998b) 'Mobile and Personal Communications in Europe: The Road Ahead', *3rd Business Forum: Mobile Systems 1998*, Moscow, Russia, 23–27 March.

DaSilva, Joao et al. (1998c) 'Towards the Wireless Information Society: A Contribution from the ACTS Programme', *ACTS Mobile Telecommunication Summit*, 8–11 June, EC report.

DaSilva, Joao et al. (1998d) 'Towards a Wireless Communications Society', *Paper presented at the FITCE Forum*, Russia.

EC, Commission of the European Communities (1988) *The European Community and Space: A Coherent Approach*, COM (88) 417, Brussels.

EC, Commission of the European Communities (1991a) *The European Community: Crossroads in Space*, EUR-14010, Brussels.
EC, Commission of the European Communities (1991b) *Proposal for a Council Decision on the Loran-C Radio-Navigation System, Communication from the Commission*, COM (91) 1, Brussels.
EC, Commission of the European Communities (1991c) *Amendment to the Proposal for a Council Decision on the Loran-C Radio-navigation System, Communication from the Commission*, COM (91) 445, Brussels.
EC, Commission of the European Communities (1992) *The European Community and Space: Challenges, Opportunities and New Actions*, COM (92) 360, Brussels, 23 September.
EC, Commission of the European Communities (1993a) *Industrial Policy Elements in a European Strategy for Space: A First Reflection*, DG III, Brussels, February.
EC, Commission of the European Communities (1993b) *The European Community and Space: Challenges, Opportunities and New Actions*. Reprinted version of COM (92) 360, Brussels, 23 September, 1992. EUR 14965, Luxembourg: Office for Official Publications of the European Communities.
EC, Commission of the European Communities (1993c) *Report from the Commission to the Council on Radio-navigation Systems for Europe*, COM (93) 62, Brussels.
EC, Commission of the European Communities (1993d) *Growth, Competitiveness and Employment: The Challenges Forward to the 21st Century*, COM (93) 700, Brussels.
EC, Commission of the European Communities (1993e) 'Mobile Communications', Pieter Weltevreden, Director of telecommunications policy, DG Information Society, *XIII Magazine* 3 (October), 4–7.
EC, Commission of the European Communities (1993f) *Development of Mobile Communications: A Regulatory Perspective*, DG XIIIA, Presentation, Famous meeting, Washington, 24–25 May.
EC, Commission of the European Communities (1994a) *Satellite Navigation Services: A European Approach*, COM (94) 248. Including a Draft Council Resolution on the European Contribution to the Development of a Global Satellite Navigation System, Brussels.
EC, Commission of the European Communities (1994b) *Proposal for a decision by the EP and Council concerning Community Guidelines for the Development of the Trans-European Transport Network*, COM (94) 106, Brussels.
EC, Commission of the European Communities (1994c) *Proposal for a Council Regulation on Financial Assistance by the Community in the Field of Trans-European Networks*, COM (94) 62, Brussels.
EC, Commission of the European Communities (1994d) *Proposal for Specific Programs for the Fourth EU R&D Program: Telematics and Transport*, COM (94) 68, Brussels.
EC, Commission of the European Communities (1994e) *Research and Technological Development: Achieving Coordination through Cooperation*, COM (94) 438, Brussels.
EC, Commission of the European Communities (1994f) *Industrial Policy Aspects in a European Strategy for Space*, Reflection paper, DG Industry, Brussels.
EC, Commission of the European Communities (1994g) *EC/ESA/Space Industry Round Table*, 24 October 1994.

EC, Commission of the European Communities (1994h) *An Industrial Competitiveness Policy for the European Union*, COM (94) 319, Brussels, 14 September.

EC, Commission of the European Communities (1994i) *UMTS: A 21st Century Vision*, RACE Mobile Project Line.

EC, Commission of the European Communities (1994j) 'RACE Mobile Project Line Unveils its Vision of the 3rd Generation Mobile System', *I&T Magazine* (Winter issue), DG INFSO (DG XIII).

EC, Commission of the European Communities (1994k) *Towards the Personal Communications Environment: Green Paper on a Common Approach in the Field of Mobile and Personal Communications in the European Union*, COM (94) 145.

EC, Commission of the European Communities (1994l) *Communication to the European Parliament and Council on the Consultation on the Green Paper on Mobile and Personal Communication*, COM (94) 492, Brussels, 23 November.

EC, Commission of the European Communities (1995a) *Invitation Letter to Industrialists for First High Level Group Meeting*, Brussels.

EC, Commission of the European Communities (1995b) *Proposal for a Decision on Measures at the Union Level in the Field of Satellite Services for Personal Communication in the EU*, COM (95) 529, Brussels, 8 November.

EC, Commission of the European Communities (1995c) *RACE: Research and Technology Development in Advanced Communications Technologies in Europe*, February, Brussels.

EC, Commission of the European Communities (1995d) 'UMTS & MBS System Requirements', *RACE Mobile Telecommunications Summit 1995*, 1–8, EC report, Cascais, Portugal, 22–24 November.

EC, Commission of the European Communities (1995e) *RACE Vision of UMTS, Workshop on Third Generation Mobile Systems*, DGXIIIB, Brussels, January.

EC, Commission of the European Communities (1995f) 'Mobile and Personal Communications: Commission Wants to Open Market', *I&T Magazine*, July.

EC, Commission of the European Communities (1996a) *Space-related Activities in the Commission Services: Progress Report 1992–1995*, DG XII: Science, Research and Development, Joint Research Center, Brussels, 14 March.

EC, Commission of the European Communities (1996b) *Draft Recommendation for a Council Decision Authorizing a European Organization Responsible for Civil Aviation Safety*, SEC 2152, Brussels, 16 December.

EC, Commission of the European Communities (1996c) *Standardization and the Global Information Society: The European Approach*, COM (96) 359, Brussels, 24 July.

EC, Commission of the European Communities (1996d) *Satellite Communications: Toward a European Strategy*, IP/96/914, Brussels, 11 October.

EC, Commission of the European Communities (1996e) *Intervention of Edith Cresson at the Space Reunion*, Paris, 10 October.

EC, Commission of the European Communities (1996f) *Amendments to the Decision on a Common Framework for a Harmonized Development in the Field of Satellite Services for Personal Communications in the EU*, COM (96) 467, Brussels, 14 October.

EC, Commission of the European Communities (1996g) *Background Document on the Achievements in the RACE Programme*. Final Report on Phase II of RACE (1992–1995), Brussels, March.

EC, Commission of the European Communities (1996h) *ACTS Work Plan*, Brussels.

EC, Commission of the European Communities (1996i) *UMTS Task Force Report: The Road to UMTS – 'In Contact Anytime, Anywhere, With Any One'*, Brussels, 1 March.

EC, Commission of the European Communities (1996j) *Europe at the Forefront of the Global Information Society: 'Rolling Action Plan'*, COM (96) 607, Brussels.

EC, Commission of the European Communities (1996k) *Standardization and the Global Information Society*, COM (96) 359, Brussels.

EC, Commission of the European Communities (1997a) *The European Union and Space: Fostering Applications, Markets and Industrial Competitiveness*. Published by Directorate-General XII: Science, Research and Development. (Luxembourg: Office for Official Publications of the European Communities).

EC, Commission of the European Communities (1997b) *Proposal for a Council Decision on the Agreement between the European Community, the European Space Agency and the European Organization for the Safety of Air Navigation on a European Contribution to the Development of a Global Navigation Satellite System*, COM (97) 442, Brussels, 23 September.

EC, Commission of the European Communities (1997c) 'Should Europe Develop an Independent Satellite Navigation System?', *Press Release*, 21 October.

EC, Commission of the European Communities (1997d) *The Competitiveness of the European Information and Communication Technologies (ICT) Industries*, ICTCM19, Brussels, 14 April.

EC, Commission of the European Communities (1997e) *The Satellite Action Plan: Satellite Communications in the Information Society*, COM (97) 91, Brussels, 5 March.

EC, Commission of the European Communities (1997f) *Terms of Reference for the Space Advisory Group*, Brussels, 17 February.

EC, Commission of the European Communities (1997g) *EU Action Plan: Satellite Communications in the Information Society*, COM (97) 91, Brussels, 3 March.

EC, Commission of the European Communities (1997h) *Further Development of Mobile and Wireless Communications: Challenges and Choices for the European Union*, COM (97) 217, Brussels, 29 May.

EC, Commission of the European Communities (1997i) 'Commission Adopts Paper on Mobile Communications', *Press Release*, Brussels, 29 May.

EC, Commission of the European Communities (1997j) *Strategy and Policy Orientations with Regard to the Further Development of Mobile and Wireless Communications (UMTS)*, COM (97) 513, Brussels, 15 October.

EC, Commission of the European Communities (1997k) 'Commission Adopts Paper on Mobile Multi-media Telephones', *Press Release*, Brussels, 15 October.

EC, Commission of the European Communities (1997l) *Second European Report on S&T Indicators*, Brussels.

EC, Commission of the European Communities (1998a) *Towards a Trans-European Positioning and Navigation Network: A European Strategy for Global Navigation Satellite Systems*, COM (1998) 29, Brussels, 21 January.

EC, Commission of the European Communities (1998b) *Global Navigation Satellite System* (Brussels: European GNSS Office).

EC, Commission of the European Communities (1998c) 'Commission Urges Europe not to Fall Behind in the Race to Develop and Exploit State-of-the-art Global Navigation Satellite Systems', *Press Release*, 21 January.

EC, Commission of the European Communities (1998d) *Implementation of a Civil-military Interface for Satellite-based Navigation and Positioning Services*, Unpublished paper, Brussels.

EC, Commission of the European Communities (1998e) *GNSS-2 Forum Working Group 2 Final Report: Technical and Financial Issues*, DG VII and DG XIII, Brussels, 15 December.

EC, Commission of the European Communities (1998f) *Green Paper on Radio Spectrum Policy in the Context of European Community Policies such as Telecommunications, Broadcasting, Transport and R&D*, COM (98) 596, Brussels, 9 December.

EC, Commission of the European Communities (1998g) *Draft Report on Research and Development, Satellite Working Group*, DG Information Society, Brussels, 19 January.

EC, Commission of the European Communities (1998h) 'Evolving the UMTS Vision', Mobile Communications, 82–102, *ACTS Mobility, Personal & Wireless Communications Domain*, Intracom, Athens.

EC, Commission of the European Communities (1998i) *Draft Decision on the Coordinated Introduction of Mobile and Wireless Communications (UMTS) in the Community*, COM (98) 58, Brussels, February.

EC, Commission of the European Communities (1998j) 'EP Committee Approves Proposal on Mobile Phones', *Press Release*, Brussels, 8 June.

EC, Commission of the European Communities (1998k) *Amended Proposal for a European Parliament and Council Decision on the Coordinated Introduction of Mobile and Wireless Communications in the Community*, COM (98) 496, Brussels, 27 July.

EC, Commission of the European Communities (1998l) *Radio Frequency Requirements for Community Policies in the Context of the World Radiocommunications Conference 1999 (WRC–99)*, COM (1998) 298, Brussels, 13 May.

EC, Commission of the European Communities (1998m) *Radio in the Digital Era.* Report on Meeting Organized by DG EAC, Centre A. Borschette, Brussels, 5 March.

EC, Commission of the European Communities (1998n) *Audiovisual Policy: The Next Steps*, COM (98) 446, Brussels, 14 July.

EC, Commission of the European Communities (1998o) *Report from the High Level Group on Audiovisual Policy*, October.

EC, Commission of the European Communities (1999a) *Theme: Navigation Satellites*, DG TREN presentation to an inter-group meeting of members of the EU Parliament, Strasbourg, Brussels, 13 January.

EC, Commission of the European Communities (1999b) *Galileo: Involving Europe in a New Generation of Satellite Navigation Services*, COM (99) 54, Brussels, 10 February.

EC, Commission of the European Communities (1999c) *Galileo: Global Satellite Navigation Services for Europe.* Meeting of the Council of EU Ministers of Transport, Luxembourg, 17 June.

EC, Commission of the European Communities (1999d) *Towards a Coherent European Approach for Space*, SEC (99) 789, Brussels, 7 June.

EC, Commission of the European Communities (1999e) 'Galileo: Organizational Framework', *Air & Space Europe*, 1, 237–241.

EC, Commission of the European Communities (1999f) 'Galileo, The Way Forward: Implementing the Strategy', *Air & Space Europe* 1.

EC, Commission of the European Communities (1999g) 'Ambitious Project Would Involve Europe in New Generation of Satellite Navigation Services', *ICAO Journal* 54, 9, 4–5 and 29. Article by N. Warinsko, DG TREN.

EC, Commission of the European Communities (1999h) 'Dossier Galileo: A Satellite Navigation System for Europe', *CNES Magazine* 6 (August), 12. Article by Matthias Ruete, Director General DG TREN.

EC, Commission of the European Communities (1999i) *Note by the European Commission to the United States Government on the Introduction of a Third Generation Mobile and Wireless Communications System (UMTS) in the Community*, Brussels, 12 January.

EC, Commission of the European Communities (1999j) *Letter to William Daley, Secretary of Commerce, and Charlene Barshefsky, USTR Ambassador*, 22 July.

EC, Commission of the European Communities (1999k) *Moving Forward: The Achievements of the European Common Transport Policy*, Brussels and Luxembourg.

EC, Commission of the European Communities (1999l) 'How Will Galileo Serve Civil Aviation?', *CNES Magazine* 6 (August), N. Warinsko, DG TREN.

EC, Commission of the European Communities (1999m) *Draft Communication on the European Radio Industry and Its Transition to Digital*, DG EAC, 25 February.

EC, Commission of the European Communities (1999n) *Workshop Principles for the Implementation of Digital Radio, Proceedings*, Brussels, March.

EC, Commission of the European Communities (2000a) 'Galileo Status Presentation', *ICAO GNSSP WGB*, Ad Hoc Meeting, DG TREN Presentation, 25 July.

EC, Commission of the European Communities (2000b) *Commission Communication to the European Parliament and the Council on Galileo*, COM (2000) 750, Brussels, 22 November.

EC, Commission of the European Communities (2000c) 'Continuation of the Galileo Project: The Commission Underlines the Need for Rapid Decisions', *Press Release*, Brussels, 22 November.

EC, Commission of the European Communities (2000d) *Galileo Definition Phase: Initial Results*, Brussels, 7 June.

EC, Commission of the European Communities (2000e) 'Projet Galileo: Loyola de Palacio Favorable à Une Décision Claire Dans le Prochaines Semaines', *Press Release*, Brussels, 22 December.

EC, Commission of the European Communities (2000f) 'Commission Call for Adequate Radio Spectrum to Ensure Growth of 3G Mobile Phones and Satellite Navigation', *Press Release*, Brussels, 8 March.

EC, Commission of the European Communities (2000g) *Europe and Space: Turning to a New Chapter*, COM (2000) 597, Brussels, 27 September.

EC (2001) *Summary of Market Analysis: Results for Galileo*, Information material, January.

EC, Commission of the European Communities (2002a) *Galileo: An Imperative for Europe*. Information Note, DG Energy and Transport, 18 January.

EC, Commission of the European Communities (2002b) *The European Dependence on US-GPS and the Galileo Initiative*. Technical Note, DG Energy and Transport, 14 February.

EC, Commission of the European Communities (2002c) *Galileo: The European Project on Radio Navigation by Satellite*, Information Note, DG Energy and Transport, 26 March.

EC/ESA (1998) 'Draft Program Proposal for a European Global Navigation Satellite System: EUROSATNAV', Discussion paper, Brussels and Paris, November.

EC/ESA (2000) *Mandate of the Galileo Program Office*, Brussels, 1 September.

EC/ESA/Eurocontrol (1998a) 'A European View on the Legal and Institutional Issues Relating to GNSS', *ICAO Council*, Rio de Janeiro, 11 May.

EC/ESA/Eurocontrol (1998b) 'Legal and Institutional Issues Relating to GNSS Implementation', *ICAO Assembly*, 22 September–2 October.

EP (2000) EP Resolution of 18 May on the Communication of the Commission 'Towards a Coherent Approach for Space', SEC–C5–0336/1999–1999/2213.

EP (2001) *Galileo: Time to Choose: European Parliamentarians' Appeal to the Laeken European Council*, November–December.

EP CCYEMS, European Parliament Committee on Culture, Youth, Education, the Media and Sport (2000a) *Radio in the Digital Age: Spheres of Activity of the European Union*, Public Hearing, 24 May.

EP CERT, European Parliament Committee on Research, Technological Development and Energy (1997) *The European Union and Space: Fostering Applications, Markets and Industrial Competitiveness*, Report on the Communication from the Commission to the Council and the EP, December.

EP CIETRE, Committee on Industry, External Trade, Research and Energy (2000a) *Report on the Commission Communication on The development of the Market for Digital Television in the European Union. Report in the context of Directive 95/47/EC of the European Parliament and of the Council of 24th October 1995 on the Use of Standards for the Transmission of Television Signals*, COM (99) 540–C5–0114/2000–2000/2074 (COS). Rapporteur: Astrid Thors.

EP CIETRE, Committee on Industry, External Trade, Research and Energy (2000b) *Report on the Commission communication entitled: 'Towards a New Framework for Electronic Communications Infrastructure and Associated Services: The 1999 Communications Review'*, COM (1999) 539–C5–0141/2000–2000/2085 (COS). Rapporteur: W. G. van Velzen.

Huber, Roland, DG INFSO (DG XIIIB) Director (1995) 'Foreword', RACE Mobile Telecommunications Summit 1995 Report, Cascais, 22–24 November.

Kinnock, Neil, EU Commissioner for Transport (1995) *Three EU Transport Challenges for the Nineties and Beyond: TENs, Environment and Safety*, Speech at EU Transport Seminar, Stockholm, 29 June.

Kinnock, Neil, EU Commissioner for Transport (1997) *Mobility for All, Speech, Conference on Intelligent Transport Systems*, Berlin, 21 October.

Kinnock, Neil, EU Commissioner for Transport (1998) *European Strategy for GNSS (Global Navigation Satellite System)*, Speech at the GNSS 98 Symposium organized by CNES and the European Group of Institutes of Navigation, 20 October.

Niepold, Ruprecht (1999) *The European Vision for UMTS*, Head of DG INFSO mobile and satellite communications unit, UMTS 1999 Conference, Brussels, 28 May.

Pompidou, Alain (1998) 'Aerospace Technology: Wings for Employment', *Article*, published by the EP.

Tytgat, Luc and Pascal Campagne (2000) 'Galileo: A New GNSS Designed with and for the Benefit for all Kind of Users', *ION GPS 2000*, 19–22 September.

Verrue, Robert (1999) *UMTS and the Information Society*. Paper, ISIMM-OMNITEL Conference, Rome, 17 February.

Other Empirical Documents

AEA, Association of European Airlines (1999) *Global Navigation Satellite Systems: A European User Point of View*. AEA Position Paper.

AER, Association of European Radios (2000) 'What Community Measures to Support Radio in the Services of the Citizen?', AER Secretary General Sergio Nantucci, Public Hearing, in EP CCYEMS, *Radio in the Digital Age: Spheres of Activity of the European Union*, 24 May.

AER et al. (2000) *Key Points of the Presentation*, Presentation before DG EAC and DG INFSO, October.

Alcatel Espace (1995) *Conditions for Consensus*, Presentation at the ESA/EUROSPACE/Industry Workshop on European Space Policy and Program after 1995, ESA/ESRIN, Frascati, Italy, 5–6 July.

Alenia Spazio (2000) 'Le Industrie Spaziali Leader in Europa Costituiscono Una Nuova Societa Per La Navigazione Satellitare', *Press Release*, 25 May.

Allemand, Christophe, CNES Galileo Project Manager (1999) 'Galileo Joins GPS and Glonass', *CNES Magazine* 6 (August), 13–16.

Anderson, Per-Olof, Ericsson Radio Systems AB (1998) 'FRAMES Contributions to the UMTS Standardization Process', *Research and Definition of IMT-2000/UMTS Mobile Radio*, Report from FRAMES Workshop held in Conjunction with International Conference on Communication Technology (ICCT 1998), Beijing, 21 October.

Anderson, Per-Olof et al. (1999) *Investigations in the FRAMES Project on the UMTS Radio Interface*, Industry paper.

Astrium (2000) 'Galileo Industries is Established in Brussels by Europe's Leading Space Companies to Support the European Satellite Navigation System', *Press Release*, 25 May.

Astrium (2001) *Galileo Update: The Program, Financing Aspects and Political Status*, Presentation material, November.

Baier, A. and H. Panzer, Philips Kommunikations Industrie (1993) 'Multi-Rate DS-CDMA Radio Interface for 3G Cellular Systems', *Seventh IEE European Conference on Mobile and Personal Communications*, 255–259, 13–15 December.

Barbarance, K. et al. (ESA) (1996) 'Satellite Navigation Activities: The International Context', *Space Communications* 14, 155–161.

Barshefsky, Charlene (1998) *Letter to U.S. House Representative Robert T. Matsui*, 14 September.

BBC (2000) 'Radio in the Digital Era: The Broadcasting Perspective', Radio Director Jenny Abramsky, Public Hearing, European Parliament Committee on Culture, Youth, Education, the Media and Sport (2000a), *Radio in the Digital Age: Spheres of Activity of the European Union*, 24 May.

Berg, Greger, Ericsson Radio Systems (1996) 'The GSM Evolution Towards UMTS', *Telecommunications Online*, October.

Berruto, E. (CSELT) et al. (1998) 'Research Activities on UMTS Radio Interface, Network Architectures, and Planning', *IEEE Communications Magazine*, 82–102 (February).

Campagne, Pascal (2000) 'The Road Ahead: Galileo Economics', *Galileo's World*, Special issue (Autumn): Building and Using Europe's Global Navigation Satellite System, 18–21.

CCIA, Computer and Communications Industry Association (1998) *Letter to Ambassador McCann*, Department of State, 18 June.

CDG, CDMA Development Group (1998) *Third Generation Systems*, White Paper.

CDG, CDMA Development Group (1999a) 'Global CDMA Operators offer Comprehensive Proposal to Harmonize 3G Wireless Standards: CDG Calls for Global 3G Summit to Work Collectively on Unified Resolution', *Press Release*, 14 January.

CDG, CDMA Development Group (1999b) 'The CDMA Development Group Comments on Recent TABD Trade Meeting: Universal Support for Convergence Could Lead to Common Worldwide 3G CDMA Approach', *Press Release*, 22 February.

CEA, Council of Economic Advisers (2000) *The Economic Impact of Third Generation Wireless Technology*, October.

Chaudhury, Prodip et al. (1999) 'The 3GPP Proposal for IMT-2000', *IEEE Communications Magazine* (December), 72–81.

DA, Deutsche Aerospace AG et al. (1994a) 'Assessment of the Potential of Satellite Navigation Systems for Europe: Results of a Study for the European Commission', *Journal of Navigation* 47, 3, 295–304.

DA, Deutsche Aerospace AG (1994b) 'Work Done at Eurospace Regarding European Space Policy: A Status Presentation', W. P. Denker, *ESA/Industry Workshop*, 5–6 July, Frascati, Italy.

Daschle, Tom (1999) *Letter to USTR Ambassador Barshefsky*, 25 February.

DisplaySearch (2001) 'New DisplaySearch Quarterly Report Tracks Worldwide Mobile Phone Demand, Penetration, Display Procurement, Revenues and Prices', *Press Release*, Austin, Texas, 20 December.

Dupuis, Philippe, ETSI SMG Chairman (1995) 'A European View on the Transition Path Towards Advanced Mobile Systems', *IEEE Personal Communications* 2, 1 (February).

EBU, European Broadcasting Union (2000) 'European Broadcasters and Manufacturers Issue Joint Appeal on DAB', *Press Release*, 22 May.

EBU/WorldDAB (2000) 'Digital Radio: A Pan-European Problem', Head of New Technology David Wood, in EP CCYEMS, *Radio in the Digital Age: Spheres of Activity of the European Union*, 24 May.

ECTEL (1997) *UMTS and GSM*, January.

ENS, Ericsson, Nokia and Siemens (1996) *Requirements for the UMTS Regulatory Process*, Brussels, 3 December.

ERC (1996) *UMTS*, Copenhagen, September.

ERC (1997) *ERC Decision of 30 June 1997 on the Frequency Bands for the Introduction of the Universal Mobile Telecommunications System (UMTS)*, ERC/DEC/(97) 07.

ITU, International Telecommunication Union (1999b) *IMT-2000: New Dimensions, New Horizons*, Presentation material, Geneva.
Levin, Sander M. (1999) *Letter to USTR Ambassador Barshefsky*, 24 February.
Matsui, Robert (1998) *Letter to USTR Ambassador Charlene Barshefsky* (Summer).
Mehrota, Asha (1994) *Cellular Radio: Analogue and Digital Systems* (London: Artech House).
Mohr, Werner, Siemens AG (1998) 'Research and Definition of IMT-2000/UMTS Mobile radio Systems', *Research and Definition of IMT-2000/UMTS Mobile Radio*, Report from FRAMES Workshop held in Conjunction with International Conference on Communication Technology (ICCT'98), Beijing, 21 October.
Mohr, Werner et al. (1999) *ACTS FRAMES Project towards IMT-2000/UMTS*, Presentation material.
Morella, Constance (1998) 'Opening Statement', US House Technology Subcommittee, *Hearing on International Standards Part II: The Impact of Standards on the Digital Economy*, 4 June.
MPT, Japanese Ministry of Post and Telecommunications (1997) 'Report on Next-generation Mobile Communications Systems Compiled', *Press Release*, 28 July.
NAPA/NRC, National Academy of Public Administration/National Research Council (1995) *The Global Positioning System: Charting the Future*, Report for the US Congress and Department of Defense (Washington, DC: National Academy Press).
Nokia (1997) 'Nokia, Alcatel, Ericsson and Siemens Agree on Core Network for Third-generation Wideband Wireless Multimedia', *Press Release*, 14 August.
Paetsch, Michael (1993) *The Evolution of Mobile Communications in the U.S. and Europe: Regulation, Technology, and Markets* (London: Artech House).
Parkinson, Bradford W. et al. (1995) 'A History Of Satellite Navigation', *Navigation* 42, 1, Special Issue, Journal of the Instititute of Navigation (March).
PEC, President's Export Council (1999) *Letter to President William J. Clinton*, Susan Corralez-Diaz (PEC), 11 January.
Qualcomm (1998a) 'Testimony of John Major', Executive Vice President, US House Technology Subcommittee, *Hearing on International Standards Part II: The Impact of Standards on the Digital Economy*, 4 June.
Qualcomm (1998b) *Comments of Qualcomm Incorporated Pursuant to 63 FR 31546: United States/European Union Transatlantic Economic Partnership*. Before the Executive Office of the President, Office of the United States Trade Representative, July.
Rapeli, Juha (1995a) 'Standardization for Global Mobile Communications in the 21st Century', *European Telecommunications Standardization and the Information Society: The State of the Art 1995*, Atalink/ ETSI, Sophia Antipolis.
Rapeli, Juha (1995b) 'UMTS Targets, System Concept and Standardization in a Global Framework', *IEEE Personal Communications* 2, 1 (February).
Rapeli, Juha (1995c) *A Path Toward 3rd Generation UMTS/FPLMTS Global Mobile Communications for the 21st Century*, ICT'95 Conference, Bali, 3–7 April.
Reay, Lord (1991) 'UK Space Policy', *Space Policy* 7, 4, 307–315.
Redl, Siegmund M. et al. (1995) *An Introduction to GSM* (London: Artech House).
Salomon, Arnaud (1999) 'A Question of Independence and Sovereignty', *CNES Magazine* 6 (August), 20–21. Director of Space and Aeronautics, French Ministry of Education, Research and Technology.

ERC (1999a) *Adjacent Band Compatibility Between UMTS and Other Services in the 2GHz band*, ERC Report 65, Menton, May.
ERC (1999b) *Draft ERC Decision on the Licensing and Global Circulation and Use of IMT-2000 Terrestrial and Satellite Mobile Terminals*, 28 December.
ESA, European Space Agency (1993) *Europe at the Crossroads: The Future of its Satellite Communications Industry* (Noordwijk: ESA Publications Division).
ESA, European Space Agency (1994a) 'Status of Long-term Space Policy Committee Activities', *ESA/Industry Workshop*, Frascati, Italy, 5–6 July. Presentation made by ESA officers J. J. Dordain and G. Naja.
ESA, European Space Agency (1994b) 'An Integrated Approach for the Preparation of the 95 Ministerial Council', J. J. Dordain, *ESA/Industry Workshop*, Frascati, Italy, 5–6 July.
ESA, European Space Agency (1995) *Report on ESA/EC Joint Working Groups*, ESA/IRC (95) 8, Paris, 7 June.
ESA, European Space Agency (1996) 'Europe Provides Satellite Navigation Services', *Press Release*, Paris, 28 June.
ESA (1997a) *Status of the European Space Sector*, ESA Council, 3 October, Paris.
ESA, European Space Agency (1997b) *Diversité des Stratégies Spatiales: Les Actions Menées par les Principaux Acteurs du Secteur*, Speech, Jean-Jacques Dordain, ESA Industrial Asffairs Department, Paris, 18 December.
ESA, European Space Agency (1999a) *Egnos: The European Geostationary Navigation Overlay Service: Europe's First-generation Regional Contribution to GNSS* (ESTEC, Noordwijk: ESA Publications Division).
ESA, European Space Agency (1999b) *Resolution on the European Space Agency's Contribution to Galileo*, Paris, 9 April .
ESA, European Space Agency (1999c) 'The European Space Agency and the European Commission Sign Contracts for the Galileo Program', *Press Release*, Paris, 7 December.
ESA, European Space Agency (2000a) *Declaration Covering the GalileoSat Program*, Revised Version, Paris, March.
ESA, European Space Agency (2000b) *Galileo & Egnos: Les Initiatives Européennes en Matière de Navigation par Satellite*, ESA presentation material, November.
ESA, European Space Agency (2000c) 'ESA and the EU Adopt a Common Strategy for Space', *Press Release*, Paris, 16 November.
ESA, European Space Agency (2000d) *Towards a Space Agency for the EU*, Report by Carl Bildt, Jean Peyrelevade, and Lothar Späth to the ESA Director General, Paris, 9 November.
ESA/Eurocontrol (1998) 'Go-ahead For a European Global Navigation Satellite System', *Joint Press Release*, Paris, 18 June.
ETSI, European Telecommunication Standards Institute (1996) *Global Multimedia Mobility Report*, Sophia Antipolis.
ETSI, European Telecommunication Standards Institute (1998a) *Communication on Standardization Efficiency to EC*, Sophia Antipolis.
ETSI, European Telecommunication Standards Institute (1998b) 'Results from the ETSI Vote on the Choice of Radio Interface for the Third Generation System UMTS', *Press Release*, Sophia Antipolis, 28 January.
ETSI, European Telecommunication Standards Institute (1998c) *Third Generation Partnership Project (3GPP)*, Sophia Antipolis.

ETSI, European Telecommunication Standards Institute (1998d) *Framework Agreement between ARIB/TTC and ETSI on 3GPP*, Sophia Antipolis.

ETSI, European Telecommunication Standards Institute (1998e) *Further Development of the Internal Structure and Scope of 3GPP*, Sophia Antipolis.

ETSI, European Telecommunication Standards Institute (1998f) *Final Vote Results*, Information Note, Sophia Antipolis.

ETSI, European Telecommunication Standards Institute (1998g) *Letter to Chairwoman Morella*, US House Subcommittee on Technology, 19 June.

ETSI, European Telecommunication Standards Institute (1998h) *Letter to Chairman Crane*, US House Subcommittee on Trade, Committee on Ways and Means, 18 August.

ETSI, European Telecommunication Standards Institute (2000a) *3rd Generation Mobile: Universal Mobile Telecommunications System (UMTS)*, Sophia Antipolis.

ETSI, European Telecommunication Standards Institute (2000b) 'Transposition of 3GPP Deliverables into ETSI Deliverables', *Mobile News*, ETSI, Special Edition, 2000 GSM World Congress (Spring).

ETSI, European Telecommunication Standards Institute (2000c) 'GSM Evolution Now Part of 3rd Generation Partnership Project (3GPP)', *Press Release*, 25 July.

Eurospace (1999) *The European Space Industry*, Information Note, Paris.

EURP, Eureka Project (1987) *Description of the Eureka Project*, Information Note.

FAMOUS (1995) *Draft 'FAMOUS' Resolution on Cooperation in Future Advanced Mobile Universal Systems*, Lisbon, 11–12 May.

FCC, Federal Communications Commission (1993) *Third Trilateral FAMOUS Meeting*, Washington, DC, 24–25 May.

FCC, Federal Communications Commission (1996) *US Delegation's Summary Notes on FAMOUS Talks*, Dallas, 28–29 March.

FCC, Federal Communications Commission (1999a) 'United States Welcomes EC Statement of Support for ITU Process on Setting New Mobile Telecommunications Standards', *Press Release*, 20 January.

FCC, Federal Communications Commission (1999b) *US and EU Telecommunications Policy*, FCC Commissioner Ness, Speech, European Institute, Washington DC, 12 January

Ferrazzani, Marco (1998) *The European Initiative and Programs for Satellite Navigation*, Paper submitted to the 41st Colloquium on the Law of Outer Space, Melbourne, Australia, 1 October.

Gaubert, Alain M., Secretary General (1999) 'Europe Needs a Space Policy', *Air & Space Europe* 1, 1, 24–25.

GO, Genesis Office (2001) *Galileo Newsletter* 13, May.

GSMA, GSM Association (1995a) *Input on FPLMTS/UMTS*, Cape Town, South Africa, 22–24 February.

GSMA, GSM Association (1995b) *The GSM Success*, Presentation material, Brussels, February.

GSMA, GSM Association (1996a) 'GSM', *The 1996 GSM World Congress Conference Report*, Cannes, 20–22 February.

GSMA, GSM Association (1996b) *The GSM/UMTS/FPLMTS Relationship*, Presentation material, Edinburgh, January.

GSMA, GSM Association (1998) 'Historical Background: The GSM Story', *1998 GSM World Congress Conference Report*, Cannes, 17–19 February.

Helms, Jesse, Trent Lott, Richard Selby, Mike Dewine et al. (1999) *Letter to President Clinton*, 4 March.

Hillebrand, Friedhelm (2000) 'Globalized GSM and UMTS Specification Work Makes Progress', *Mobile News*, ETSI, Special Edition – 2000 GSM World Congress, spring 2000.

HLG, High Level Group (1996) *Development and Competitiveness of Space Industries in Europe*, Report of the Industry's High Level Group to the European Commission, January.

Hollings, Ernest (1998) *Letter to USTR Ambassador Charlene Barshefsky*, 13 August.

Hutchinson, Kay B., Dick Armey and Sam Johnson (1998) *Letter to Ambassador Vonya McCann: Department of State on US Proposed Wireless Standards to ITU*, 29 June.

ICAO, International Civil Aviation Organization (1997) 'Growth in Air Traffic to Continue: ICAO Releases Long-Term Forecasts', *Press Release*, Montreal, March.

ICAO, International Civil Aviation Organization (1998a) *Final Report, Panel of Legal and Technical Experts on the Establishment of a Legal Framework with Regard to GNSS (LTEP/3)*, ICAO Council, Montreal, 9 March.

ICAO, International Civil Aviation Organization (1998b) *Declaration on Global Air Navigation Systems for the 21st Century*, ICAO Council, Rio de Janeiro, 15 May.

IMO, International Maritime Organization (1997) *Resolution A.860 (20) on Maritime Policy for a Future Global Navigation Satellite System (GNSS)*, 27 November.

IMO, International Maritime Organization (1998) *Resolution A.815 (19) on Worldwide Radio-navigation System*, 23 November.

IMO, International Maritime Organization (1999) *Maritime Policy for a Future Global Navigation Satellite System (GNSS)*, Resolution A860, November.

IPP, Industry Position Paper (1997a) *Positions and Responses to the Executive Summary of COM (97) 217*, European wireless manufacturers.

IPP, Industry Position Paper (1997b) *Letter on UMTS to DG INFSO (DG XIII)*, European wireless manufacturers, August.

IPP, Industry Position Paper (1997c) *Response to COM (97) 513 on UMTS*, European wireless manufacturers, October.

IPP, Industry Position Paper (1998a) *Position on Proposal by the European Commission for a Decision on the Coordinated Introduction of Mobile and Wireless Communications*, UMTS Alliance industry position, April.

IPP, Industry Position Paper (1998b) *Background Briefing Memorandum: Position on the Draft EU UMTS Decision and the ETSI–CDMA Standards Process*, November, European wireless manufacturers.

ITIC, Information Technology Industry Council (1998) 'Statement of Oliver Smoot', US House Technology Subcommittee, *Hearing on International Standards Part II: The Impact of Standards on the Digital Economy*, 4 June.

ITU, International Telecommunication Union (1992) *Future Public Land Mobile Telecommunication Systems*, CCIR Recommendation 687, Geneva, Switzerland.

ITU, International Telecommunication Union (1998) *IMT-2000: The Global Standard for Personal Communications*, Information brochure, Geneva, Switzerland.

ITU, International Telecommunication Union (1999a) *World Telecommunication Development Report: Mobile Cellular*, Geneva.

Samukic, Antun (1998) 'UMTS Universal Mobile Telecommunications System: Development of Standards for 3G', *IEEE Transactions on Vehicular Technology* 47, 4 (November).

Samukic, Antun and P. Polese, ETSI Secretariat (1995) 'Migration/Evolution towards UMTS: A Market-driven Approach in Research and Standardization', *RACE Mobile Telecommunications Summit Report*, EC publication, Cascais, Portugal, 22–24 November.

Sidenbladh, Thomas, Ericsson AB (1997) 'Economic and Policy Aspects of Spectrum Management: A Manufacturer's View', *Spectrum Management Seminar*, EC publication, Brussels, 23 September.

TABD (1999a) *TABD Statement of Conclusions from the 1998 Conference in Charlotte*, November.

TABD (1999b) *Statement of Conclusions and Recommendations, TABD EETIS group*, Washington D.C., 17 February.

TABD (1999c) *TABD Mid-Annual Report*, Washington DC, June.

TIA, Telecommunications Industry Association (1998) 'Testimony of Jesse E. Russell' [Chairman of Wireless Communications Division and 3G Ad Hoc Group], US House Technology Subcommittee, *Hearing on International Standards Part II: The Impact of Standards on the Digital Economy*, Jesse E. Russell, 4 June.

Tyssen Krupp Multimedia (2000) 'DAB: On Its Way to Mass Market?', Director H. Amor, in EP CCYEMS, *Radio in the Digital Age: Spheres of Activity of the European Union*, 24 May.

UMTSF, UMTS Forum (1996) *The UMTS Forum Articles of Association: Towards the Personal Communications Environment*, Zurich, Switzerland, 9 December.

UMTSF, UMTS Forum (1997a) *A Regulatory Framework for UMTS*, Report 1, London, 1 March.

UMTSF, UMTS Forum (1997b) 'UMTS: Technology Vision", IBC UMTS Conference, *UMTS: The Next Generation of Mobile*, Motorola, UMTSF Technology Aspects Group, Dorchester, UK, 27–29 October.

UMTSF, UMTS Forum (2000) *The UMTS Forum Articles of Association: Towards the Personal Communications Environment*, Revised by General Assembly, Helsinki, Finland, 13–15 September.

Urie, Alistair, Alcatel (1995) 'Advanced GSM: A Long Term Future Scenario for GSM', *7th World Telecommunication Forum: Convergence of Technologies, Services and Applications* 2, Geneva, 3–11 October.

Urie, Alistair, M. Streeton, and C. Mourot [Alcatel] (1995) 'An Advanced TDMA Mobile Access System for UMTS', *IEEE Personal Communications Magazine* 2, 1 (February).

USTR, US Trade Representative (1998a) *Letter to Commissioner Martin Bangemann*, United States Trade Representative, Madeleine K. Albright, Charlene Barshefsky, William M. Daley and William E. Kennard, 19 December.

USTR, US Trade Representative (1998b) 'United States Presses Europe to Adopt Fair Standards For Wireless Communications', *Press Release*, 22 December.

USTR, US Trade Representative (1999a) *Letter to Commissioner Karel Van Miert*, William Daley and Charlene Barshefsky, 2 July.

USTR, US Trade Representative (1999b) *Letter to Erkki Liikanen*, William Daley, Charlene Barshefsky and William E. Kennard, 13 October.

UWCC, Universal Wireless Communications Consortium (1998) 'Universal Wireless Communications Consortium (UWCC) Announces UWC-136, the TDMA IS-136 Solution for Third Generation', *UWCC Press Release*, 23 February.
UWCC, Universal Wireless Communications Consortium (1999) *Letter to Secretary of Commerce William M. Daley*, 22 April.
WDABF, WorldDAB Forum (1998a) *Radio in Europe*, Information material.
WDABF, WorldDab Forum (1998b) *Green Paper on the Convergence of the Telecommunications, Media and Information Technology Sectors, and the Implications for Regulation towards an Information Society Approach: A Submission to DG INFSO and DG EAC of the European Commission*, 12 May.
WDABF, WorldDAB Forum (1999a) *DAB: A Digital Revolution for Europe and the World*, Progress report, April.
WDABF, WorldDAB Forum (1999b) *Green Paper on Radio Spectrum Policy in the Context of European Community policies such as Telecommunications, Broadcasting, Transport, and R&D: A Submission to the European Commission by WorldDAB*, 15 April.
WDABF, WorldDAB Forum (2000a) *Towards a New Framework for Electronic Communications Infrastructure and Associated Services: The 1999 Communications Review: A Response by WorldDAB*, 15 February.
WDABF, WorldDAB Forum (2000b) *DG Information Society Working Documents: The 1999 Communications Review. Comments by WorldDAB*, 19 May.
WDABF, WorldDAB Forum (2000c) 'Digital Gurus Outline Plans for Further World Terrestrial DAB Penetration', *Press Release*, 16 January.
West, Mary B. (1999) 'US Committed to Maintaining Free GPS Service', *CNES Magazine* 6 (August), 28–29. Deputy Assistant Secretary for Oceans, Fisheries, and Space (Bureau of Oceans, International Environment and Scientific Affairs), US Department of State.
Wireless Strategy Analytics (2001) *2.5G and 3G Wireless Applications*, Presentation material. Boston, MA.

Newspaper Articles and Professional Press

AFX Europe (2001) 'US warns EU of possible conflicts between Galileo project, NATO systems', 18 December.
AIW, Asia Intelligence Wire (1997a) 'Commissioner Paves the Way to Third-Generation Mobile Communications', *European Report*, 1 March.
AIW, Asia Intelligence Wire (1997b) 'European Manufacturers Unite To Push Future Mobile Standard', *Exchange Telecommunications Newsletter*, 22 August.
AIW, Asia Intelligence Wire (1997c) 'EU Council of Ministers', *European Report*, 1 March.
AP, Associated Press (2001) 'EU Asks Europe to Ease Up on Telecoms', 28 March.
AWST, Aviation Week & Space Technology (1999) 'Europe Launches Satnav Project', 5 July 5.
BC, Broadcasting and Cable (1999a) 'Radio Stations Make Digital Push', 11 January.
BC, Broadcasting and Cable (1999b) 'First Step for Digital Radio', 8 November.
BC, Broadcasting and Cable (1999c) 'Digital Radio Industry Grows Smaller', 20 December.
BC, Broadcasting and Cable (2000) 'Radio Dreams of Digital', 6 March.

Black, Ian (2002) 'European Satellite Plan Ruffles US Feathers', *The Guardian*, 27 March.

BW, Business Week Online (2001) 'Weaker Signals for Mobile Phone Firms: Sales Growth in the Worldwide Market is Slackening', 6 April.

Castle, Stephen (2002) 'Galileo satellite project goes into orbit: Blair changes tune, supporting a £2bn European space system opposed by US military', *The Independent*, 10 March.

CI, Communications International (1998) 'Huffing and Puffing', 8–16 (October).

Connor, Steve (2002) 'Europe Agrees to Fund Rival to US Navigation System', *The Independent*, 27 March.

CWI, Communications Week International (2000) 'GSM Could be Music to the Ears of Digital Audio Industry', 3 April.

Dombey, Daniel (2002) 'EU to Go Ahead with Satellite System: Galileo Project Plan Has Divided European Countries and Aroused US Concern Over NATO', *Financial Times*, 27 March.

EB, Electronic Business (2000) 'Digital Radio Makes Waves', March.

EDN (1999) 'Hello...Is Anybody Out There', 23 December.

EE, EETimes (1998) 'Alliance's Stand Sends Noisy Signal to Qualcomm', 25 September.

FT, Financial Times (1997) 'Top Four Agree Mobile Phone Standard', 15 August.

FT, Financial Times (1998) 'Compromise Over Mobile Phone Standards', 30 January.

FT, Financial Times (2000) 'Headlong Scramble for Spectrum as Demand for Services Soar', FT Telecoms section, Regional Focus: North America, 21 June.

FT, Financial Times (2001a) 'UK and France in Talks on Defense Links', 9 February.

FT, Financial Times (2001b) 'Prodi Hits at EU Slowness To Liberalize', 8 February.

FT, Financial Times (2001c) 'Small Players at Risk in Mobile Market', 6 April.

Geitner, Paul (2002) 'Galileo May Put Europe Back In Orbit: The European Union Is Pushing Ahead With a $3-billion Rival to GPS Despite U.S. Opposition With Military Overtones', *Los Angeles Times* [Associated Press], 18 March.

GPSW, GPS World, (1999a) 'ESA Backs GNSS', June.

GPSW, GPS World, (1999b) 'Defining Galileo', June 28.

GPSW, GPS World (1999c) 'EU Ministers Expected to Endorse Galileo', July.

GPSW, GPS World (1999d) 'EU Ministers Give GNSS and Galileo the Green', August.

GPSW, GPS World (1999e) 'Europe Advances GPS Counterpart, Galileo', November.

GPSW, GPS World (1999f) 'More on Galileo: EU Takes Its Plan to NATO, Western European Union', May 1999.

GPSW, GPS World (1999g) 'Galileo Enthusiasm and Money Propel Europe's GNSS', November.

GPSW, GPS World (2000a) 'Galileo Contracts, Talks, and Opinions', February.

GPSW, GPS World (2000b) 'U.S. and Europe Continue GNSS Talks', May.

GPSW, GPS World (2000c) 'WRC Signals Success: GNSS Wins Allocations', July.

GPSW, GPS World (2000d) 'Galileo: Negotiations Moving Forward', November.

GPSW, GPS World (2000e) 'European Transport Ministers to Decide Fate of Proposed Galileo System', December.

MW, Marketing Week (1999) 'Quiet Countdown to UK Digital Radio', 16 September.

MW, Marketing Week (2000a) 'Digital Radio Gets Its Wires Crossed', 26 October.

MW, Marketing Week (2000b) 'Digital Explosion', 9 November.
RCR News (1998a) 'Lawmaker Presses for 3G Answers', 4 May.
RCR News (1998b) 'USTR Prepared to War Over 3G', 21 September.
RT, Reuter Textline (1997) 'EU, Mobile Industry Unite on New Technology', 26 February.
RT, Reuter Textline (1998a) 'EU Unveils High-Tech Mobile Phone Strategy', 12 February.
RT, Reuter Textline (1998b) 'Proposal: ESC Opinion on Proposal for Introducing UMTS in the Community', *Spicers Centre for Europe*, 15 June.
RT, Reuter Textline (1998c) 'EU Backs Plans for New Mobile Phone Standard', 19 May.
RT, Reuter Textline (1998d) 'Focus: 21st Century Standard Set for Mobile Phones', 29 January.
RT, Reuter Textline (1998e) 'EU's Bangemann Hails Deal on Mobile Phone Standard', 4 February.
SD, Screen Digest (2000) 'European Digital Radio Market Expected to Reach 4.7 Billion in Five Years', *Press Release*, 23 March.
SN, Space News (1998a) 'Europeans See Profit Potential In Rival to GPS', 9, 23, 8 June.
SN, Space News (1998b) 'European Commitment To Separate GPS Grows', 9, 26, 22 June.
SN, Space News (1998c) 'European GPS Demand Expected to Quadruple'', 9, 25, 6 July.
SN, Space News (1999a) 'Upgraded GPS Constellation Will Not Carry Civilian Signals', 10, 4, 1 February.
SN, Space News (1999b) 'US Air Force Weighs GPS Upgrade Options', 7 February.
SN, Space News (1999c) 'EC Urges Independent Satellite Navigation System: Non-government Funding Key to Project', 10, 7, 22 February.
SN, Space News (1999d) 'Widespread Reliance on GPS Raises Liability Concerns', 10, 12, 29 March.
SN, Space News (1999e) 'Firms Want To Invest in EU Navigation System', 10, 18, 10 May.
SN, Space News (1999f) 'Britain To Bolster Funding Of Pan-European Programs', 10, 19, 17 May.
SN, Space News (1999g) 'ESA Gets Increase for Satellite Navigation; Other Programs Cut', 10, 20, 24 May.
SN, Space News (2000a) 'Europe's GPS Enhancement Plan Faces Delay', 11, 10, 13 March.
SN, Space News (2000b) 'Pentagon Seeks Additional Funds to Modernize GPS', 11, 14, 10 April.
SN, Space News (2000c) 'Galileo Appraisal Due by Summer', 11, 16, 24 April.
SN, Space News (2000d) 'GPS Regionals Miss Deadlines As Costs Rise', 11, 19, 15 May.
SN, Space News (2000e) 'Europe Cheers While Questioning End of GPS Selective Availability', 11, 19, 15 May.
SN, Space News (2000f) 'Europe Debates US Galileo Participation', 11, 19, 15 May.
SN, Space News (2000g) 'Pilot Group Against Satellite Navigation', 11, 19, 15 May.
SN, Space News (2000h) 'Russia Plans To Modernize Glonass', 11, 19, 15 May.
TD, Television Digest (1999),'Digital Audio Growing', 23 August.

UWCC, Universal Wireless Communications Consortium (1998) 'Universal Wireless Communications Consortium (UWCC) Announces UWC-136, the TDMA IS-136 Solution for Third Generation', *UWCC Press Release*, 23 February.

UWCC, Universal Wireless Communications Consortium (1999) *Letter to Secretary of Commerce William M. Daley*, 22 April.

WDABF, WorldDAB Forum (1998a) *Radio in Europe*, Information material.

WDABF, WorldDab Forum (1998b) *Green Paper on the Convergence of the Telecommunications, Media and Information Technology Sectors, and the Implications for Regulation towards an Information Society Approach: A Submission to DG INFSO and DG EAC of the European Commission*, 12 May.

WDABF, WorldDAB Forum (1999a) *DAB: A Digital Revolution for Europe and the World*, Progress report, April.

WDABF, WorldDAB Forum (1999b) *Green Paper on Radio Spectrum Policy in the Context of European Community policies such as Telecommunications, Broadcasting, Transport, and R&D: A Submission to the European Commission by WorldDAB*, 15 April.

WDABF, WorldDAB Forum (2000a) *Towards a New Framework for Electronic Communications Infrastructure and Associated Services: The 1999 Communications Review: A Response by WorldDAB*, 15 February.

WDABF, WorldDAB Forum (2000b) *DG Information Society Working Documents: The 1999 Communications Review. Comments by WorldDAB*, 19 May.

WDABF, WorldDAB Forum (2000c) 'Digital Gurus Outline Plans for Further World Terrestrial DAB Penetration', *Press Release*, 16 January.

West, Mary B. (1999) 'US Committed to Maintaining Free GPS Service', *CNES Magazine* 6 (August), 28–29. Deputy Assistant Secretary for Oceans, Fisheries, and Space (Bureau of Oceans, International Environment and Scientific Affairs), US Department of State.

Wireless Strategy Analytics (2001) *2.5G and 3G Wireless Applications*, Presentation material. Boston, MA.

Newspaper Articles and Professional Press

AFX Europe (2001) 'US warns EU of possible conflicts between Galileo project, NATO systems', 18 December.

AIW, Asia Intelligence Wire (1997a) 'Commissioner Paves the Way to Third-Generation Mobile Communications', *European Report*, 1 March.

AIW, Asia Intelligence Wire (1997b) 'European Manufacturers Unite To Push Future Mobile Standard', *Exchange Telecommunications Newsletter*, 22 August.

AIW, Asia Intelligence Wire (1997c) 'EU Council of Ministers', *European Report*, 1 March.

AP, Associated Press (2001) 'EU Asks Europe to Ease Up on Telecoms', 28 March.

AWST, Aviation Week & Space Technology (1999) 'Europe Launches Satnav Project', 5 July 5.

BC, Broadcasting and Cable (1999a) 'Radio Stations Make Digital Push', 11 January.

BC, Broadcasting and Cable (1999b) 'First Step for Digital Radio', 8 November.

BC, Broadcasting and Cable (1999c) 'Digital Radio Industry Grows Smaller', 20 December.

BC, Broadcasting and Cable (2000) 'Radio Dreams of Digital', 6 March.

Samukic, Antun (1998) 'UMTS Universal Mobile Telecommunications System: Development of Standards for 3G', *IEEE Transactions on Vehicular Technology* 47, 4 (November).

Samukic, Antun and P. Polese, ETSI Secretariat (1995) 'Migration/Evolution towards UMTS: A Market-driven Approach in Research and Standardization', *RACE Mobile Telecommunications Summit Report*, EC publication, Cascais, Portugal, 22–24 November.

Sidenbladh, Thomas, Ericsson AB (1997) 'Economic and Policy Aspects of Spectrum Management: A Manufacturer's View', *Spectrum Management Seminar*, EC publication, Brussels, 23 September.

TABD (1999a) *TABD Statement of Conclusions from the 1998 Conference in Charlotte*, November.

TABD (1999b) *Statement of Conclusions and Recommendations, TABD EETIS group*, Washington D.C., 17 February.

TABD (1999c) *TABD Mid-Annual Report*, Washington DC, June.

TIA, Telecommunications Industry Association (1998) 'Testimony of Jesse E. Russell' [Chairman of Wireless Communications Division and 3G Ad Hoc Group], US House Technology Subcommittee, *Hearing on International Standards Part II: The Impact of Standards on the Digital Economy*, Jesse E. Russell, 4 June.

Tyssen Krupp Multimedia (2000) 'DAB: On Its Way to Mass Market?', Director H. Amor, in EP CCYEMS, *Radio in the Digital Age: Spheres of Activity of the European Union*, 24 May.

UMTSF, UMTS Forum (1996) *The UMTS Forum Articles of Association: Towards the Personal Communications Environment*, Zurich, Switzerland, 9 December.

UMTSF, UMTS Forum (1997a) *A Regulatory Framework for UMTS*, Report 1, London, 1 March.

UMTSF, UMTS Forum (1997b) 'UMTS: Technology Vision", IBC UMTS Conference, *UMTS: The Next Generation of Mobile*, Motorola, UMTSF Technology Aspects Group, Dorchester, UK, 27–29 October.

UMTSF, UMTS Forum (2000) *The UMTS Forum Articles of Association: Towards the Personal Communications Environment*, Revised by General Assembly, Helsinki, Finland, 13–15 September.

Urie, Alistair, Alcatel (1995) 'Advanced GSM: A Long Term Future Scenario for GSM', *7th World Telecommunication Forum: Convergence of Technologies, Services and Applications* 2, Geneva, 3–11 October.

Urie, Alistair, M. Streeton, and C. Mourot [Alcatel] (1995) 'An Advanced TDMA Mobile Access System for UMTS', *IEEE Personal Communications Magazine* 2, 1 (February).

USTR, US Trade Representative (1998a) *Letter to Commissioner Martin Bangemann*, United States Trade Representative, Madeleine K. Albright, Charlene Barshefsky, William M. Daley and William E. Kennard, 19 December.

USTR, US Trade Representative (1998b) 'United States Presses Europe to Adopt Fair Standards For Wireless Communications', *Press Release*, 22 December.

USTR, US Trade Representative (1999a) *Letter to Commissioner Karel Van Miert*, William Daley and Charlene Barshefsky, 2 July.

USTR, US Trade Representative (1999b) *Letter to Erkki Liikanen*, William Daley, Charlene Barshefsky and William E. Kennard, 13 October.

ERC (1999a) *Adjacent Band Compatibility Between UMTS and Other Services in the 2GHz band*, ERC Report 65, Menton, May.

ERC (1999b) *Draft ERC Decision on the Licensing and Global Circulation and Use of IMT-2000 Terrestrial and Satellite Mobile Terminals*, 28 December.

ESA, European Space Agency (1993) *Europe at the Crossroads: The Future of its Satellite Communications Industry* (Noordwijk: ESA Publications Division).

ESA, European Space Agency (1994a) 'Status of Long-term Space Policy Committee Activities', *ESA/Industry Workshop*, Frascati, Italy, 5–6 July. Presentation made by ESA officers J. J. Dordain and G. Naja.

ESA, European Space Agency (1994b) 'An Integrated Approach for the Preparation of the 95 Ministerial Council', J. J. Dordain, *ESA/Industry Workshop*, Frascati, Italy, 5–6 July.

ESA, European Space Agency (1995) *Report on ESA/EC Joint Working Groups*, ESA/IRC (95) 8, Paris, 7 June.

ESA, European Space Agency (1996) 'Europe Provides Satellite Navigation Services', *Press Release*, Paris, 28 June.

ESA (1997a) *Status of the European Space Sector*, ESA Council, 3 October, Paris.

ESA, European Space Agency (1997b) *Diversité des Stratégies Spatiales: Les Actions Menées par les Principaux Acteurs du Secteur*, Speech, Jean-Jacques Dordain, ESA Industrial Asffairs Department, Paris, 18 December.

ESA, European Space Agency (1999a) *Egnos: The European Geostationary Navigation Overlay Service: Europe's First-generation Regional Contribution to GNSS* (ESTEC, Noordwijk: ESA Publications Division).

ESA, European Space Agency (1999b) *Resolution on the European Space Agency's Contribution to Galileo*, Paris, 9 April .

ESA, European Space Agency (1999c) 'The European Space Agency and the European Commission Sign Contracts for the Galileo Program', *Press Release*, Paris, 7 December.

ESA, European Space Agency (2000a) *Declaration Covering the GalileoSat Program*, Revised Version, Paris, March.

ESA, European Space Agency (2000b) *Galileo & Egnos: Les Initiatives Européennes en Matière de Navigation par Satellite*, ESA presentation material, November.

ESA, European Space Agency (2000c) 'ESA and the EU Adopt a Common Strategy for Space', *Press Release*, Paris, 16 November.

ESA, European Space Agency (2000d) *Towards a Space Agency for the EU*, Report by Carl Bildt, Jean Peyrelevade, and Lothar Späth to the ESA Director General, Paris, 9 November.

ESA/Eurocontrol (1998) 'Go-ahead For a European Global Navigation Satellite System', *Joint Press Release*, Paris, 18 June.

ETSI, European Telecommunication Standards Institute (1996) *Global Multimedia Mobility Report*, Sophia Antipolis.

ETSI, European Telecommunication Standards Institute (1998a) *Communication on Standardization Efficiency to EC*, Sophia Antipolis.

ETSI, European Telecommunication Standards Institute (1998b) 'Results from the ETSI Vote on the Choice of Radio Interface for the Third Generation System UMTS', *Press Release*, Sophia Antipolis, 28 January.

ETSI, European Telecommunication Standards Institute (1998c) *Third Generation Partnership Project (3GPP)*, Sophia Antipolis.

ETSI, European Telecommunication Standards Institute (1998d) *Framework Agreement between ARIB/TTC and ETSI on 3GPP*, Sophia Antipolis.

ETSI, European Telecommunication Standards Institute (1998e) *Further Development of the Internal Structure and Scope of 3GPP*, Sophia Antipolis.

ETSI, European Telecommunication Standards Institute (1998f) *Final Vote Results*, Information Note, Sophia Antipolis.

ETSI, European Telecommunication Standards Institute (1998g) *Letter to Chairwoman Morella*, US House Subcommittee on Technology, 19 June.

ETSI, European Telecommunication Standards Institute (1998h) *Letter to Chairman Crane*, US House Subcommittee on Trade, Committee on Ways and Means, 18 August.

ETSI, European Telecommunication Standards Institute (2000a) *3rd Generation Mobile: Universal Mobile Telecommunications System (UMTS)*, Sophia Antipolis.

ETSI, European Telecommunication Standards Institute (2000b) 'Transposition of 3GPP Deliverables into ETSI Deliverables', *Mobile News*, ETSI, Special Edition, 2000 GSM World Congress (Spring).

ETSI, European Telecommunication Standards Institute (2000c) 'GSM Evolution Now Part of 3rd Generation Partnership Project (3GPP)', *Press Release*, 25 July.

Eurospace (1999) *The European Space Industry*, Information Note, Paris.

EURP, Eureka Project (1987) *Description of the Eureka Project*, Information Note.

FAMOUS (1995) *Draft 'FAMOUS' Resolution on Cooperation in Future Advanced Mobile Universal Systems*, Lisbon, 11–12 May.

FCC, Federal Communications Commission (1993) *Third Trilateral FAMOUS Meeting*, Washington, DC, 24–25 May.

FCC, Federal Communications Commission (1996) *US Delegation's Summary Notes on FAMOUS Talks*, Dallas, 28–29 March.

FCC, Federal Communications Commission (1999a) 'United States Welcomes EC Statement of Support for ITU Process on Setting New Mobile Telecommunications Standards', *Press Release*, 20 January.

FCC, Federal Communications Commission (1999b) *US and EU Telecommunications Policy*, FCC Commissioner Ness, Speech, European Institute, Washington DC, 12 January

Ferrazzani, Marco (1998) *The European Initiative and Programs for Satellite Navigation*, Paper submitted to the 41st Colloquium on the Law of Outer Space, Melbourne, Australia, 1 October.

Gaubert, Alain M., Secretary General (1999) 'Europe Needs a Space Policy', *Air & Space Europe* 1, 1, 24–25.

GO, Genesis Office (2001) *Galileo Newsletter* 13, May.

GSMA, GSM Association (1995a) *Input on FPLMTS/UMTS*, Cape Town, South Africa, 22–24 February.

GSMA, GSM Association (1995b) *The GSM Success*, Presentation material, Brussels, February.

GSMA, GSM Association (1996a) 'GSM', *The 1996 GSM World Congress Conference Report*, Cannes, 20–22 February.

GSMA, GSM Association (1996b) *The GSM/UMTS/FPLMTS Relationship*, Presentation material, Edinburgh, January.

GSMA, GSM Association (1998) 'Historical Background: The GSM Story', *1998 GSM World Congress Conference Report*, Cannes, 17–19 February.

Helms, Jesse, Trent Lott, Richard Selby, Mike Dewine et al. (1999) *Letter to President Clinton*, 4 March.

Hillebrand, Friedhelm (2000) 'Globalized GSM and UMTS Specification Work Makes Progress', *Mobile News*, ETSI, Special Edition – 2000 GSM World Congress, spring 2000.

HLG, High Level Group (1996) *Development and Competitiveness of Space Industries in Europe*, Report of the Industry's High Level Group to the European Commission, January.

Hollings, Ernest (1998) *Letter to USTR Ambassador Charlene Barshefsky*, 13 August.

Hutchinson, Kay B., Dick Armey and Sam Johnson (1998) *Letter to Ambassador Vonya McCann: Department of State on US Proposed Wireless Standards to ITU*, 29 June.

ICAO, International Civil Aviation Organization (1997) 'Growth in Air Traffic to Continue: ICAO Releases Long-Term Forecasts', *Press Release*, Montreal, March.

ICAO, International Civil Aviation Organization (1998a) *Final Report, Panel of Legal and Technical Experts on the Establishment of a Legal Framework with Regard to GNSS (LTEP/3)*, ICAO Council, Montreal, 9 March.

ICAO, International Civil Aviation Organization (1998b) *Declaration on Global Air Navigation Systems for the 21st Century*, ICAO Council, Rio de Janeiro, 15 May.

IMO, International Maritime Organization (1997) *Resolution A.860 (20) on Maritime Policy for a Future Global Navigation Satellite System (GNSS)*, 27 November.

IMO, International Maritime Organization (1998) *Resolution A.815 (19) on Worldwide Radio-navigation System*, 23 November.

IMO, International Maritime Organization (1999) *Maritime Policy for a Future Global Navigation Satellite System (GNSS)*, Resolution A860, November.

IPP, Industry Position Paper (1997a) *Positions and Responses to the Executive Summary of COM (97) 217*, European wireless manufacturers.

IPP, Industry Position Paper (1997b) *Letter on UMTS to DG INFSO (DG XIII)*, European wireless manufacturers, August.

IPP, Industry Position Paper (1997c) *Response to COM (97) 513 on UMTS*, European wireless manufacturers, October.

IPP, Industry Position Paper (1998a) *Position on Proposal by the European Commission for a Decision on the Coordinated Introduction of Mobile and Wireless Communications*, UMTS Alliance industry position, April.

IPP, Industry Position Paper (1998b) *Background Briefing Memorandum: Position on the Draft EU UMTS Decision and the ETSI–CDMA Standards Process*, November, European wireless manufacturers.

ITIC, Information Technology Industry Council (1998) 'Statement of Oliver Smoot', US House Technology Subcommittee, *Hearing on International Standards Part II: The Impact of Standards on the Digital Economy*, 4 June.

ITU, International Telecommunication Union (1992) *Future Public Land Mobile Telecommunication Systems*, CCIR Recommendation 687, Geneva, Switzerland.

ITU, International Telecommunication Union (1998) *IMT-2000: The Global Standard for Personal Communications*, Information brochure, Geneva, Switzerland.

ITU, International Telecommunication Union (1999a) *World Telecommunication Development Report: Mobile Cellular*, Geneva.

ITU, International Telecommunication Union (1999b) *IMT-2000: New Dimensions, New Horizons*, Presentation material, Geneva.
Levin, Sander M. (1999) *Letter to USTR Ambassador Barshefsky*, 24 February.
Matsui, Robert (1998) *Letter to USTR Ambassador Charlene Barshefsky* (Summer).
Mehrota, Asha (1994) *Cellular Radio: Analogue and Digital Systems* (London: Artech House).
Mohr, Werner, Siemens AG (1998) 'Research and Definition of IMT-2000/UMTS Mobile radio Systems', *Research and Definition of IMT-2000/UMTS Mobile Radio*, Report from FRAMES Workshop held in Conjunction with International Conference on Communication Technology (ICCT'98), Beijing, 21 October.
Mohr, Werner et al. (1999) *ACTS FRAMES Project towards IMT-2000/UMTS*, Presentation material.
Morella, Constance (1998) 'Opening Statement', US House Technology Subcommittee, *Hearing on International Standards Part II: The Impact of Standards on the Digital Economy*, 4 June.
MPT, Japanese Ministry of Post and Telecommunications (1997) 'Report on Next-generation Mobile Communications Systems Compiled', *Press Release*, 28 July.
NAPA/NRC, National Academy of Public Administration/National Research Council (1995) *The Global Positioning System: Charting the Future*, Report for the US Congress and Department of Defense (Washington, DC: National Academy Press).
Nokia (1997) 'Nokia, Alcatel, Ericsson and Siemens Agree on Core Network for Third-generation Wideband Wireless Multimedia', *Press Release*, 14 August.
Paetsch, Michael (1993) *The Evolution of Mobile Communications in the U.S. and Europe: Regulation, Technology, and Markets* (London: Artech House).
Parkinson, Bradford W. et al. (1995) 'A History Of Satellite Navigation', *Navigation* 42, 1, Special Issue, Journal of the Instititute of Navigation (March).
PEC, President's Export Council (1999) *Letter to President William J. Clinton*, Susan Corralez-Diaz (PEC), 11 January.
Qualcomm (1998a) 'Testimony of John Major', Executive Vice President, US House Technology Subcommittee, *Hearing on International Standards Part II: The Impact of Standards on the Digital Economy*, 4 June.
Qualcomm (1998b) *Comments of Qualcomm Incorporated Pursuant to 63 FR 31546: United States/European Union Transatlantic Economic Partnership*. Before the Executive Office of the President, Office of the United States Trade Representative, July.
Rapeli, Juha (1995a) 'Standardization for Global Mobile Communications in the 21st Century', *European Telecommunications Standardization and the Information Society: The State of the Art 1995*, Atalink/ ETSI, Sophia Antipolis.
Rapeli, Juha (1995b) 'UMTS Targets, System Concept and Standardization in a Global Framework', *IEEE Personal Communications* 2, 1 (February).
Rapeli, Juha (1995c) *A Path Toward 3rd Generation UMTS/FPLMTS Global Mobile Communications for the 21st Century*, ICT'95 Conference, Bali, 3–7 April.
Reay, Lord (1991) 'UK Space Policy', *Space Policy* 7, 4, 307–315.
Redl, Siegmund M. et al. (1995) *An Introduction to GSM* (London: Artech House).
Salomon, Arnaud (1999) 'A Question of Independence and Sovereignty', *CNES Magazine* 6 (August), 20–21. Director of Space and Aeronautics, French Ministry of Education, Research and Technology.

TT, *Total Telecom* (2001a) 'EU Members Call for Study into Effects of 3G Auction Costs', 26 March.

TT, *Total Telecom* (2001b) 'Europe's EIB Set to Bail Out 3G license-holders', 9 April.

TW, Traffic World (2000), 'Galileo Spending Starts', 3 January.

TWB, TechWeb (1998a) 'Euro Vote Likely To Shape Mobile Communications Future', 27 January.

TWB, TechWeb (1998b) 'European Divided on Cellular Standards', 29 January.

TWB, TechWeb (1998c) 'ETSI Pushes for World Mobile Telecom Standard', 2 February.

TWB, TechWeb (2001a) 'Nokia Says It Holds Close To 40 Percent Of Market', 21 March.

TWB, TechWeb (2001b) 'U.S. Mobile Profits Won't Amount To Much', 2 April.

WP, The Washington Post Online (2001) 'A Push For More Frequencies: Wireless Firms Say They Can't Advance Until Government Opens Up the Airwaves', 28 February.

Index

Donated by
Ian Thomson

Not for loan

Competition for Technological Leadership